ADVANCES IN PARALLEL ALGORITHMS

ADVANCES IN
PARALLEL ALGORITHMS

Edited by
LYDIA KRONSJÖ
MSc, PhD

and

DEAN SHUMSHERUDDIN
BSc, MB ChB, MSc, PhD

School of Computer Science,
University of Birmingham, UK

Halsted Press: an Imprint of JOHN WILEY & SONS, INC.
NEW YORK   TORONTO

Published in the Americas by Halsted Press,
an Imprint of John Wiley & Sons, Inc., New York

Copyright © Blackwell Scientific Publications 1992

Library of Congress Cataloging in Publication Data available.
ISBN: 0-470-21907-6

First published in Great Britain 1992 by
Blackwell Scientific Publications, Osney Mead, Oxford OX2 0EL

Printed in Great Britain

Contents

Foreword

Over the past ten years very high performance computing has become an essential tool in many areas of scientific research, resource management and product design. It is also now widely accepted that given the scale of the problems which need to be addressed, only parallel computing systems will be able to provide the necessary computing power in a cost-effective and timely way. However, although considerable progress had been made over the last five years in developing novel and powerful hardware architectures, software remains a major limiting factor in exploiting these high performance systems. In particular no dominant programming paradigm has yet emerged for parallel computing and there is a considerable shortage of standard and scalable software tools and application packages. There is also much work to be done in improving parallel programming languages, compiler technology, program development environments and parallel algorithms.

It is in this latter area that this book focuses with the purpose of providing a clear description and state of the art overview of current concepts and work on parallel algorithms. As such it will be of interest to both students and professionals who wish to understand this new technology and its significance.

I very much welcome the book as an important and timely contribution to the literature.

Peter Jarratt
Professor of Computer Science
March 1992

Editors' Preface

Parallel computers are rapidly coming into widespread use. They are now a cost-effective means of obtaining the processing power needed to solve complex real-world problems within a reasonable length of time. However, in order to exploit the full potential of parallel computers we need new parallel software based on new parallel algorithms.

Advances in Parallel Algorithms surveys current developments across the broad and rapidly expanding field of parallel algorithms and parallel software. It begins with an overview of parallel software paradigms. This is followed by discussions of the divide and conquer paradigm as a basis for parallel language design and the neural network paradigm. It goes on to review general methods for solving problems with parallel computers and applications of parallel algorithms. The book concludes with a discussion of the analysis of complexity and performance of parallel algorithms and software. It is designed to provide a wide audience of computing professionals and students with an in-depth survey of the latest developments and advances in exploration of parallel processing.

The material of the book can be roughly divided into three sections: software paradigms (Chapters 1 to 3), general methods (Chapters 4 to 10), applications (Chapters 11 to 14) and a chapter on computational complexity (Chapter 15).

We feel that our enthusiasm for parallel processing methods and applications is justified. Extremely interesting developments have taken place in neural networks, parallel simulation, genetic algorithms, branch-and-bound algorithms and dynamic programming.

We would like to sincerely thank all the people who helped with proposals for interesting authors, or critically read various chapters and portions of the manuscript. In particular, we would like to thank Professor Vic Rayward-Smith and all at Blackwell Scientific Publications, especially Robin Arnfield and Julia Burden. We also wish to thank Professor Aaron Sloman, Professor Peter Jarratt, Rachid Anane, Neville Thomas and Ian Fitchet. Our sincere thanks also go to the hard working authors who have all written excellent contributions under tight deadlines.

The chapters of this book were transferred to us by electronic mail. The book was typeset by Dean Shumsheruddin using Leslie Lamport's LaTeXdocument preparation system.

Lydia Kronsjö
Dean Shumsheruddin
Birmingham, March 1992

The Authors

Tom W. Archibald

Mr Tom W. Archibald is with the University of Edinburgh, Scotland. He graduated in 1988 from the University of Edinburgh with an Honours degree in Mathematics. For the following three years he was employed as a Research Assistant working on parallel implementation of Markov decision processes in association with Professor L. C. Thomas and Professor K. I. M. McKinnon. He is currently developing course material for management science courses at Edinburgh. He has recently been made a member of the Edinburgh Parallel Computing Centre.

Tom H. Axford

Dr Tom H. Axford has an MSc in Theoretical Physics (University of Melbourne, 1962) and a PhD in Mathematical Physics (University of Birmingham, 1968). He has lectured in Computer Science at the University of Birmingham since 1967, and was appointed senior lecturer in 1972. He is the author of Concurrent Programming: Fundamental Techniques for Real-Time and Parallel Software Design (John Wiley, 1989). His current research interests are in functional languages and their interpretation, parallel programming models and concurrent programming.

David W. Bustard

Professor David W. Bustard is a professor of Computing Science at the University of Ulster, Coleraine. His broad research interest is in the engineering of concurrent software-based systems. For ten years (1973–83) his main concern was the representation of concurrent systems, which involved the development of the general purpose programming language Pascal Plus and related software. Since then his focus has shifted to system specification and design issues, and to techniques for presenting concurrent behaviour visually.

Murray I. Cole

Dr Murray I. Cole gained his BSc (Hon.) in Computer Science in 1984 and a PhD in Computer Science in 1988, both at the University of Edinburgh. From 1987 to 1989 he lectured in the Computing Science Department at the University of Glasgow, and subsequently held a post-doctoral research fellowship in the same Department. Since 1990 he has been a lecturer in Computer Science at the University of Edinburgh, where he is also a member of the Edinburgh Parallel Computing Centre. His current research interests focus on the design and implementation of high level programming abstractions for parallel computers, with a particular emphasis on the use of functional programming languages.

Michel Dubois

Michel Dubois is an assistant professor in the Department of Electrical Engineering of the University of Southern California. Before joining U.S.C. in 1984, he was a research engineer at the Central Research Laboratory of Thomson-CSF in Orsay, France. His main interests are Computer Architecture and Parallel Processing. He has published more than 50 technical papers on multiprocessor architecture, performance, and algorithms. Dubois holds a PhD from Purdue University, an MS from the University of Minnesota, and an engineering degree from the Faculte Polytechnique de Mons in Belgium, all in Electrical Engineering. He is a member of the ACM and the Computer Society of the IEEE.

Jürgen-Friedrich Hake

Dr Jürgen-Friedrich Hake is Assistant Director of the Forschungszentrum Jülich (KFA) where he is involved in the formulation of KFA's high-performance computing strategy. He served previously as the Mathematical Software Coordinator at the Zentralinstitut für Angewandte Mathematik (ZAM) of KFA. He has an MS in mathematics from Bielefeld University.

Yijie Han

Dr Yijie Han has a BS from the University of Science and Technology of China, an MA and a PhD from Duke University. He has been an assistant professor at the Department of Computer Science of the University of Kentucky. He has research interest in the design of parallel algorithms and algorithms for robot motion planning. He has obtained fast and efficient parallel algorithms for linked list problems and several graph problems.

Phil Husbands

Dr Phil Husbands is a lecturer in AI in the School of Cognitive and Computing Sciences, University of Sussex. He has held research positions in industry and spent three years as a research fellow at the University of Edinburgh. His research interests centre around the theory, implementation and application of evolutionary methods in problem solving and in the development of autonomous 'intelligent' systems.

E. V. Krishnamurthy

Professor E. V. Krishnamurthy is presently with the Computer Science Laboratory, Research School of Physical Sciences and Engineering, Australian

National University, Canberra. He was previously (1984–91) professor and head of the Computer Science Department, University of Waikato, Hamilton, New Zealand. He has held visiting positions at University of Chicago, Weizmann Institute of Science, Technion-Israel Institute of Technology, University of Tennessee, University of Maryland, University of Alaska, Johannes Kepler University, Frederich Schiller University, University of Lagos and University of Sheffield. He has published several books and papers in numerical and combinatorial algorithms, digital picture processing and computer science. His current interests are in parallel and distributed computing and information systems.

Lydia I. Kronsjö

Dr Lydia I. Kronsjö is a lecturer at the School of Computer Science, University of Birmingham, England. She is the author of Algorithms: Their Complexity and Efficiency (John Wiley, 2nd Ed., 1987) and Computational Complexity of Sequential and Parallel Algorithms (John Wiley, 1985). Her current interests lie in the computational complexity of numerical and combinatorial algorithms and the development of efficient tools for monitoring the performance of parallel and distributed algorithms.

Geoff P. McKeown

Dr Geoff P. McKeown is a senior lecturer in Computing at the University of East Anglia with active research interests in numerical methods, functional programming and combinatorial optimisation. As with all members of the Mathematical Algorithms Group, his prime interests lie in the application of modern computing techniques to the solution of mathematical problems.

Sanguthevar Rajasekaran

Dr Sanguthevar Rajasekaran is currently an assistant professor of Computer and Information Science at the University of Pennsylvania. He received his ME degree in Automation from the Indian Institute of Science in 1983, and his PhD degree in Computer Science from Harvard University in 1988. His research interests include parallel computing, randomised algorithms, optimisation, learning theory, computer simulation, motion planning, and computational biology.

Vic J. Rayward-Smith

Professor Vic J. Rayward-Smith is a professor of Computer Science at the University of East Anglia and has published widely in computing and mathematics. He has authored a number of undergraduate texts on mathematics, formal languages, logic and complexity as well as numerous research papers. Most of his research has been concerned with combinatorial optimisation, especially scheduling and complexity theory.

Steven A. Rush

Mr Steven A. Rush is a Research Assistant at the University of East Anglia working on a transputer based implementation of branch-and-bound. He did his first degree at Exeter University and is expecting to submit his PhD thesis during the next year.

Dean Shumsheruddin

Dr Dean Shumsheruddin gained his medical degree in 1981. After working as a doctor, he obtained an MSc and PhD in Computer Science. He is currently a lecturer in Computer Science at the University of Birmingham. His main research interests are neural networks for motor control and medical expert systems.

Aydın Üresin

Dr Aydın Üresin received the BS degree in electrical engineering from Technical University of Istanbul, Turkey, and the MS and PhD degrees in computer engineering in 1984 and 1990, respectively, from the University of Southern California. He is currently an assistant professor in the Department of Computer Science at York University, Canada. His research interests include parallel processing, and computer architecture.

James S. Weston

Dr James S. Weston obtained a B.A. in Applied Mathematics in 1962, and a PhD in Computer Science in 1988, both at the Queen's University of Belfast. From 1979 to 1989 he was a senior lecturer in Mathematics and Computer Studies at St. Mary's College of Education, Belfast. Since January 1990 he has been a senior lecturer in the Department of Computing Science at the University of Ulster at Coleraine. His current interests include the design of parallel algorithms and their evaluation on a variety of parallel architectures.

Andrew Wing

Mr Andrew Wing studied for his first degree in Computer Systems Engineering at the University College of North Wales. After graduating in 1987, he worked as a Software Engineer for GEC Avionics. Andrew started a PhD at the University of East Anglia in 1988. His research, which examines the application of parallel simulation to the printed circuit board industry, is currently nearing completion.

CHAPTER 1

Parallel Software Paradigms

Murray Cole
University of Edinburgh, UK

1.1 Introduction

The diversity of research into what might be broadly termed 'parallel computing' is large and ever growing. A brief foray into the library would reveal topics ranging from the highly abstract and theoretical to the intensely specific and practical, apparently united only by an interest in what happens and what can be made to happen when many computational devices interact. In this chapter we aim to examine topics somewhere towards the middle of this spectrum, where language design and implementation attempt to bridge the gap between the ways in which we might like to think about parallelism and the manner in which it is exhibited by real computers. Rather than becoming embroiled in a catalogue of syntactic detail, we will attempt to concentrate on the broad principles which divide approaches to the problem. Thus, we attempt to classify and contrast a range of parallel software *paradigms*, each of which tackles the issues involved in a fundamentally different way.

Parallel software is of interest for two reasons. Firstly, and more traditionally, it is of importance when modelling or interacting with environments which themselves exhibit real parallelism. This theme has been most visible in the development of theories and tools used in the construction of operating systems, whether to handle the actual parallelism exhibited by a collection of devices (and perhaps even several processors) or the imagined, but conceptually convenient parallelism implied by the time-slicing of a single processor between many processes. Secondly, parallel software is the means by which we can harness the computational power of parallel computers, hopefully allowing faster solution of fixed size problems and making solution of increasingly large problem instances feasible in realistic time (while bearing in mind the implications of Snyder 1986). We will consider the various paradigms from the latter angle.

Our experiences with sequential computers have made us aware of a conflict between the twin goals of portability and efficiency of code. While it might be argued that advanced compiler technology is beginning to make these achievable simultaneously, it remains broadly true that one can sacrifice portability to gain speed by exploiting detailed machine characteristics. For parallel computers, the variety of fundamentally distinct machine architectures is much wider and the trade-off between portability and machine specific efficiency correspondingly more pronounced. The range of paradigms considered will exploit various points of this trade-off.

Finally, we note that as in all software design there are two important issues to be addressed, namely correctness and efficiency. It is important to know that one's program is performing the computation which was intended, as well as performing that computation quickly. The introduction of parallelism brings with it new perils in the area of correctness which any useful parallel programming paradigm must address. While the ensuing survey will be primarily concerned with the trade-off between portability and efficiency, we will also consider the ways in which correctness oriented issues arise.

1.2   Producing a Parallel Computation

In this section we consider the work which has to be carried out in order to solve a problem (usually specified without explicit reference to parallelism) on a parallel computer. We will differentiate between the various paradigms on the basis of the way each approaches these issues, and in particular on the extent to which the associated tasks are automated (as opposed to being under the explicit control of the programmer). We also note that the eventual target architecture will have an impact on the practical relevance of some of the tasks and will discuss such situations as they arise in the context of specific implementations. For example, the problem of synchronisation disappears (at least from the programmer's world) in SIMD architectures, but is crucial in a MIMD machine with physically shared memory. Similarly, the act of sharing some piece of data is trivial in a shared memory context but not so in a sparsely connected message passing setting.

The fundamental task in programming any parallel machine is the identification of potential parallel activity within a solution of the given problem. We will refer to this as *problem partitioning*. It may involve producing a completely static description of the parallel tasks available, a completely dynamic scheme for identifying new parallel activities at run-time, or some combination of these. As we shall see, particular paradigms require varying degrees of programmer intervention at even this most basic level.

Given a problem partition, the next task is to devise a *scheduling* of the implied processes to the available processors, so as to reduce (and ideally minimise) overall execution time. Such schemes may be static or dynamic (possibly, but not necessarily, in correspondence with the partitioning method (Sarkar 1989)) and may have to pay varying degrees of attention to machine topology, depending upon the extent to which the locality of one process with respect to another, or with respect to some item of data, is a meaningful concept.

The final task concerns the implementation of *coordination* between the various processes and has two aspects. Since it is unlikely that a partition will produce a set of completely independent processes, there must be some means within a paradigm of describing or inferring the ways in which *data-sharing* occurs. Depending upon the divergence between the way such sharing is described and the corresponding facilities provided by a real machine, satisfying these requirements may be a simple or complex task. Similarly, it is likely that some degree of *synchronisation* between processes will be required, which may or may not fit closely with the facilities for synchronisation presented by the underlying machine.

In the remainder of the chapter we will categorise a range of parallel software paradigms according to the extent to which they apportion the tasks of problem partitioning, process distribution, data-sharing and synchronisation between the programmer and the supporting software (whether at compile-time or run-time). Inevitably, any attempt to classify real systems on such a basis will be at best an approximation, as the peculiarities of this or that implementation on this or that machine make definitive categorisation impossible. However, we hope that the broad principles outlined will act as a useful starting point for more detailed understanding.

1.3 Highly Abstract Paradigms

In this category we consider paradigms in which the bulk of the effort in producing parallel software is made by the supporting compiler and run-time system. In fact, at face value the approaches considered involve no explicit reference to parallelism whatsoever on the part of the programmer! However, we shall see that in each case the best results will be obtained by programming in a certain style (dependent upon the paradigm). Thus, although not confronting the range of problems associated with parallel programming directly, the programmer is still well advised to formulate programs with certain principles in mind. We will encounter instances of this phenomenon in both declarative and sequential imperative contexts.

1.3.1 Functional Programming

Functional programming languages and logic programming languages (of which more later) are the primary representatives of the so-called *declarative* approach to programming. The intention behind the name is to convey the feeling that a program need only declare the relationships between its input and its output, rather than specifying some precise sequence (or sequences in the parallel case) of actions by which the relationship can be realised, (as occurs in *imperative* languages). The extent to which such a stark contrast really exists is open to a debate which is beyond the scope of this chapter. It is certainly the case that the use of a declarative language does not obviate the need for a careful choice of algorithm, nor is it difficult to specify many quantitatively different algorithms which implement the same input-output relationship in such a language. Equally, and more importantly from our perspective, it is certainly true that declarative languages facilitate a programming style which discourages both over-specification of algorithms and the introduction of sequential dependencies where none need exist.

```
squarelengths xs = map square (map length xs)

square x = x*x

length []      = 0
length (x:xs) = 1 + length xs

map f []       = []
map f (x:xs) = (f x):(map f xs)
```

Figure 1.1

An excellent introduction to pure functional programming is presented in Bird and Wadler (1988). The essential characteristic of the paradigm is the absence of any notion of assignment to variables. Instead, we find only values (whether simple data items, data structures or even functions) and their use in the construction (or definition) of new values. Thus, instead of regarding the execution of a program as a means of changing the state of a collection of variables, we should think of execution (or more appropriately *evaluation*) of a functional program as defining and constructing some *new* collection of data items in terms of existing ones.

A functional program is a collection of function definitions (one of which may usually be regarded as the definition of 'the main program'), each defined in terms of the others and of some pre-defined set of simple functions. Each function describes (either directly using primitive data constructors, or indirectly by reference to other functions) the means by which a new value representing the result of a call of the function should be constructed from the values of the arguments presented when the call is made.

One of the commonest data structures in functional programming is the list, representing a sequence (of arbitrary length) of items of some type. The simple functional program in Figure 1.1 produces a list of integers, corresponding to the squares of the lengths of the lists in its argument list (thus its argument is a list of lists). The symbol :, pronounced 'cons' is used to denote construction of a list from an element (the left operand) and a list (the right operand). The symbol [], pronounced 'nil' denotes the empty list. Note that the type of the items in the lists is left unstated since it is of no consequence to this program. This 'polymorphism' is a useful feature of modern functional languages. The program is represented by the function squarelengths. square is a function which returns the square of its argument, length returns the length of its argument (which must be a list), and map is a function which returns the results of applying its first argument (a function) to every item in its second argument (an appropriately typed list). For example applying squarelengths to the list of integer lists [[1,2,3],[],[8,1]] would produce the integer list

```
silly x y  = square x

timebomb x = (timebomb x) + (timebomb x)
```

Figure 1.2

value [9,0,4], (which is just a syntactic variation of 9:(0:(4:[]))).

A detailed discussion of sequential implementation techniques for functional languages (upon which parallel implementations are based) can be found in Peyton-Jones (1987). It is sufficient here to note the essential characteristics of these methods. In functional programming the equivalent of running a program on some data is to demand evaluation of a function given appropriate arguments. This function application may be thought of (indeed is) an expression which must be given some internal representation. Evaluation then amounts to repeatedly rewriting this representation, according to the definitions of the functions involved, until it represents a more convenient, but semantically equivalent, version of the expression (in the sense that 12 is a more convenient representation of (square 2) * (length [32,4,4]).

From the parallel point of view, the essential feature of this process of expression rewriting (or 'reduction') is that the rewriting options available at any point in evaluation may be performed in any order, including concurrently, without affecting the final result, provided only that the order chosen eventually terminates. This is known as the Church-Rosser property and is a direct result of the absence of variables and assignment. In the previous example, we may safely evaluate the square and length sub-expressions concurrently or in either order.

If all we ever do is replace values with equivalent ones, how can we possibly go wrong? The answer lies in the cautionary remark about eventual termination. Consider, the (rather pointless) functional program in Figure 1.2 and the call silly 2 (timebomb 3). If we repeatedly choose to rewrite the timebomb 3 sub-expression first then the evaluation will never terminate (or more pragmatically, will fall over when the system runs out of memory). On the other hand, rewriting the silly first, leads quickly to the result 4. In real functional programs such dangers may be buried rather more deeply , and trouble sequential and parallel implementations equally.

We can now consider the functional paradigm in terms of the manner in which the tasks involved in the production of a piece of parallel software are concerned. Superficially, it might appear that the programmer has no rôle to play beyond providing the functional program itself. The flaw in this view is that one can quite easily (and not necessarily deliberately) write functional programs which exhibit little or no scope for concurrent evaluation as a simple result of their structure, rather than any inherent sequentiality in the problem itself. For example, contrast the scope for concurrency in these two versions of

```
facA 0 = 1
facA n = n * facA (n-1)

facB 0   = 1
facB n   = prod 1 n
prod m n = if m=n then m
     else (prod m halfway) * (prod halfway+1 n)
     where halfway = m + ((n-m) div 2)
```

Figure 1.3

factorial calculation in Figure 1.3. The obvious definition of facA is almost entirely sequential, whereas the more contorted facB is explosively parallel. In the context of more realistically sized problems and programs, it seems that the programmer must be aware of the fact that certain styles of program are more likely to result in parallel evaluations than others. On the other hand these styles lie entirely within the functional paradigm, and are entirely devoid of explicit reference to parallelism and the associated difficulties. Furthermore, the amenability of declarative programs in general to semantics preserving transformation makes the task of trying to improve the potential parallelism of programs at least a little more tractable than for procedural languages.

The remaining problem in the *identification* of parallelism concerns the possibility of non-termination. This means that it is not in general safe for the system to start a parallel task for every sub-expression which becomes available for reduction. Closer inspection reveals that the source of the problem concerns the evaluation of sub-expressions, and particularly of arguments, which do not actually contribute to the result. A function is said to be *strict* in those arguments which must be evaluated in order to produce a result. It is always therefore safe to evaluate a strict argument in parallel with the rest of the function body. There are two solutions to this problem which place responsibility in the hands of the programmer and system respectively. The first option is to ask the programmer for strictness information (by means of some annotations) or to make the semantics of the language such that all arguments are evaluated before the function body (in which case non-termination of our example is to be expected, but is the programmer's fault!). Alternatively, the compilation process may include a 'strictness analysis' phase in which an attempt is made to infer such information automatically.

We now turn to the problem of scheduling the potential parallelism. Since opportunities for concurrent activity only emerge as the expression evolves, this is an entirely dynamic activity. The most common approach to parallel implementation of functional languages extends the sequential technique of *graph reduction*. Essentially, the expression is represented in memory as a graph, with function calls and data items at nodes and edges indicating arguments. At any point in time, a number of nodes may be ready for rewriting. These nodes effectively form a task pool, from which processors can draw work. The issues to be addressed in scheduling a reduction concern the granularity of such tasks and the physical distribution of ready tasks through the system. In both cases, the decisions made are significantly affected by the characteristics of the eventual target architecture. For granularity, the important question is whether the amount of work packaged up within a node (which will emerge as the node and its descendants are repeatedly rewritten) is large enough to justify the cost of marking it as a member of the task pool and (in a non-shared memory architecture) of transferring its description and copies of any required arguments to another processor. Existing systems have largely relied on rather crude complexity analyses in an attempt to solve this problem, and the relative success of shared memory implementations as against non-shared memory suggests that there is still much work to be done. The distribution problem is really only a serious problem for non-shared memory machines. For these, the key is to try and ensure that the nodes representing available tasks are placed on processors which are physically close to those on which the arguments to the function reside, and to those which will eventually access the result of the call. Failure to achieve a reasonable level of locality results in the evolution of highly scattered distributions in which repeated communication of values across the machine dominates evaluation time. Thus real systems have tended to adopt a 'local diffusion' model for scheduling, in which evaluable nodes are only shifted to immediate or near neighbours. Unfortunately, for significantly sized machines this tends to ignore the need for more widespread spawning of tasks in the initial phases at least, in order to utilise the whole machine as early in the computation as possible. The tension between these requirements is obviously not restricted to the functional paradigm.

The task coordination issues of data-sharing and synchronisation are easy to handle. Sharing of data occurs when a process rewriting some function node requires access to the value of some argument node which may have been produced by another processor. Since nodes are never updated in the sense of conventional assignment, it is never possible to read a 'stale' piece of data. The worst that could happen would be that a process received an unevaluated copy of some node and would proceed to duplicate the work of evaluation being carried out by some other process. This would create no semantic errors but would be a (possibly significant) waste of time. Consequently, it makes more sense to suspend the demanding process until the required value is ready. This introduces the need for a simple synchronisation mechanism. Since we are often likely to have more processes available than processors, the suspended processor can usually proceed with some other piece of work in the meantime.

Parallel aspects of graph reduction have been surveyed by Peyton-Jones (1989) and Szymanski (1991), while Augustsson and Johnsson (1989), Darlington *et al.* (1989) and Goldberg (1988) have discussed specific projects. It should not be surprising that the languages in this 'implicitly parallel' category are the most attractive in terms of ensuring correctness. Amongst these, the

```
sort([],[]).
sort([X|Xs], Ys) :- sort(Xs, Zs),
    insert(X,Zs,Ys).

insert(X,[],[X]).
insert(X,[Y|Ys],[X,Y|Ys]) :- X<=Y.
insert(X,[Y|Ys],[Y|Zs] :- X>Y, insert(X,Ys,Zs)
```

Figure 1.4

functional style is the most amenable to powerful transformation and verification techniques (Darlington *et al.* 1990).

1.3.2 Logic Programming

Languages based on logic programming form the other major branch of the declarative programming family. In place of functions are first order predicates used to define *clauses* which describe relationships between data items. A clause has a 'head' which is a single predicate and a body which is a conjunction of predicates. The same head may be reused with different bodies. Such a clause represents the information that the head is satisfiable when all the conjuncts of any of its defining bodies are satisfiable. Since heads and bodies can involve references to unbound variables, the process of evaluation corresponds to finding a consistent assignment of values such that the 'goal' predicate is satisfied. Clauses with a head but no body are used to present the initial collection of facts from which such inferences may be drawn. The program in Figure 1.4 describes the standard insertion sort algorithm. The head and body of each clause are separated by the symbol ':-', each clause terminates with a '.', and conjuncts within the body are separated by commas. The list constructed from elements E, F and a further list L is represented by '[E,F|L]'.

For example, the second clause for sort may be read 'The list Ys represents a sorted copy of the list constructed from X and Xs if there is a list Zs which is a sorted copy of Xs and inserting X into Zs produces Ys'. Note that the predicates '<=' and '>' are built-in. The analogy for running a program is to ask for evaluation of a 'query', which is a head, possibly with some arguments bound to real values. The result of the run will be a binding of values to the other variables in the head such that the predicate is satisfied or an indication that no such binding can be found. For example, asking for 'sort([21,4,3,5],S).' would produce the binding 'S=[3,4,5,21]'. Since in general an identical head may have many bodies there may be more than one possible collection of satisfying bindings. In its purest form a logic language would produce any one of these non-deterministically. However, with efficient sequential implementation in mind, the commonest logic language, Prolog, imposes a depth-first, left to right search on clauses, considered in textual order, with backtracking to investigate alternatives on failure.

Considered as candidates for parallel evaluation, we see that logic programs without artificially sequentialising semantics contain two obvious sources of potential concurrency. When a goal has a structure which matches the heads of several clauses then there is no reason not to spawn processes which attempt to find success via each of the options simultaneously. Since our eventual requirement is for only one successful binding this is known as 'OR-parallelism' (Crammond 1985). It has the convenient feature that the various processes are completely independent (their bindings are never used simultaneously) but also implies a possible degree of redundant computation against a sequential, clause by clause search (i.e. if the sequential search succeeds with the first clause tried, then the other parallel processes have just been wasting computational resources). The other source of parallelism is in the examination of the various conjuncts which make up the body of some clause. Since a successful computation requires all of these to succeed this is known as 'AND-parallelism'. The contrast with 'OR-parallelism' is clear. In a successful attempt, all the work done in parallel now contributes to the solution (though we may still waste work against the sequential version in a failed attempt). On the other hand, the processes may no longer be completely independent, since variables shared by more than one conjunct must be bound consistently.

In terms of the issues concerning construction of a parallel computation, the position is similar to that of the functional approach. Thus, although pure logic programs have no explicit parallelism, there is nevertheless a style of programming which produces good potential for concurrent activity. Again, it is quite possible to write inherently sequential logic programs (for example, consider coding up the two factorial algorithms in this paradigm). As before, we must confront the issue of process granularity in the context of the underlying architecture. For scheduling, OR-parallelism is rather like the functional case, in that the processes are independent. For AND-parallelism the situation is complicated by the simultaneous binding problem, which provides the primary source of data-sharing and synchronisation in the model. In practice, the complications involved are often held to be so serious (especially for non-shared memory machines) that the issue is avoided by the introduction of new language features (see Section 1.4.4), thereby handing the problem back to the programmer (and effectively making the languages explicitly parallel). The bulk of research in the field takes this approach. Clearly, the abstract style encouraged by logic programming is beneficial in terms of program correctness—the conceptual distance between the specification of some problem and its realisation as a logic program is often small.

1.3.3 Automatic Parallelisation

The automatic parallelisation of existing sequential code represents both the hardest and (at least traditionally) the most heavily worn path to the production

```
a := b+c;   (1)
b := a-e;   (2)
c := b*d;   (3)
b := a+e    (4)
```

Figure 1.5

of parallel software from non-parallel specifications. The difficulty is a direct product of the need to separate out those data and control dependencies which are inherent in the algorithm from those which are introduced artificially as a result of the language. The significant amount of effort which has been expended on the approach is a result of the so called 'dusty deck' problem. Huge quantities of 'tried and tested' sequential code (primarily Fortran) already exist. To the users of this code, who may often be scientists and engineers in fields other than computer science, the cost of rewriting everything in some more suitable language is perceived to be unacceptable whether in terms of the time required to learn and use such a language, the money required to pay someone else to do so, or both. Hence, the temptation is to encourage the once and for all effort of producing an acceptably efficient 'parallelising compiler'.

The task of a parallelising compiler is to determine the dependencies which exist between statements in a sequential program and to re-schedule the statements for parallel execution in a manner which does not disrupt these dependencies. Since a statement in a sequential language can range from a simple assignment to a heavily parameterised procedure call, the task is highly complex in its most general form. As a result, most attention has been paid to the task of parallelising simple (procedure call free) sequences of statements and loops. This is a sensibly pragmatic approach given that the bulk of the work carried out by the 'scientific' programs which characterise the area is largely described by fairly straightforward nested loops operating across regular data structures (in contrast say, to the procedurised and recursive style of language processing programs). In this context there are two areas of interest, involving the dependencies between different statements in 'basic blocks' (statement sequences with no internal jumps) and between instances of the same statements in different loop iterations.

An ideal application of the 'basic-block' method would show that all statements within the block could be scheduled in parallel. There are three types of dependency which can prevent this. Consider the basic block in Figure 1.5. Statement (2) must follow statement (1) in order to access the new value of a (i.e. a read following a write). Statement (3) must follow statement (1) in order not to overwrite the value of c too soon (a write following a read). Statement (4) must follow statement (2) in order to leave the correct value in b (a write following a write). Determination of such dependencies is complicated by the introduction of function calls and subscripted data structures.

```
a:=b+c;
d:=e*f;
if e > f then begin
   g:= c*10;
   h:= b/f
end else begin
   g:=c+1;
   h:=b+f
end
```

Figure 1.6

```
if e > f then begin
   a:=b+c;
   d:=e*f;
   g:= c*10;
   h:= b/f
end else begin
   a:=b+c;
   d:=e*f;
   g:=c+1;
   h:=b+f
end
```

Figure 1.7

This method is limited by the fact that a typical basic block tends to contain only a few statements. Basic block size can be extended artificially by moving conditional statements and introducing new code to compensate for any inconsistencies which arise. For example, consider the fragment in Figure 1.6 which contains no basic block of more than two statements. The code can be re-arranged to produce basic blocks of size four, as illustrated in Figure 1.7.

The goal in the analysis of loops is to determine the dependencies between the variable references from one iteration to the next. Since most loops act across subscripted data structures this is usually a question of analysing the subscripts and their relationship to the loop variable or variables. The exploitation of the analysis is determined by the target architecture. A pipelined processor begins computation of one iteration before that of the preceding iteration has completed. In contrast, a simple SIMD approach executes all iterations simultaneously. For example, consider the loops in Figure 1.8. In (1), the subscripting implies that a pipelined approach is unsuitable—items generated by one iteration are needed in the next. The simplistic SIMD ap-

```
for i:=1 to n do A[i] := A[i-1] + B[i]      (1)

for i:=1 to n do A[i] := A[i] + B[i]        (2)

for i:=1 to n do A[i] := A[i+1] + B[i]      (3)
```

Figure 1.8

```
set A to all zeros;
for i:=1 to n do
  for j:=1 to n do
    for k:=1 to n do A[i,j] :=
        A[i,j] + (B[i,k] * C[k,j])
```

Figure 1.9

proach is inappropriate for the same reason (although a more radical analysis might recognize the possibility of execution on a SIMD tree). In (2), we can imagine either form being appropriate—one processor per element of A or a pipelined addition unit, or even several pipelined units acting independently on sub-sequences of A . Statement (3) is again ideal for a single pipelined unit. A parallel implementation is also possible, but note that we would have to take care to ensure correct synchronisation between the reads of A[i+1] and the writes of A[i]—no write should precede the read of the same element in its other guise.

The task is complicated by the introduction of nested loops and more complicated subscript expressions. In the former case, it is sometimes possible to devise source code transformations which improve the dependency properties with respect to some architecture. For example, the obvious matrix multiplication loops in Figure 1.9 are already in a suitable form for replication style parallelism (one processor for each element of A), but not for a pipelined approach (since an updated value is passed from each iteration of the innermost loop to the next). Restructuring the code as illustrated in Figure 1.10 allows the innermost loop to be pipelined since there is now no dependency between the elements of A written and read on successive iterations.

In terms of the relative responsibilities of programmer and system, the onus in this approach lies heavily with the system, but again we note that the programmer is well advised to be aware of the type of program (now defined at the simple statement and loop level) which is likely to prove successful. The fact that this style is defined within a context which imposes artificial sequentiality only adds to the difficulty. Issues of scheduling, synchronising and sharing data in the resulting partition are entirely in the hands of the system.

```
set A to all zeros;
for k:=1 to n do
   for  i:=1 to n do
     for j:=1 to n do A[i,j] :=
          A[i,j] + (B[i,k] * C[k,j])
```

Figure 1.10

As we have seen, the question of correctly enforcing synchronisation can be especially difficult. Good starting points for further reading in this area are provided by Nicolau *et al.* (1991) and Wolfe (1989).

1.3.4   Data Parallelism

The data parallel paradigm recognises and attempts to exploit the rich source of parallelism inherent in many simple operations on bulk data structures such as arrays, matrices, lists and sets. In contrast to the automated approaches outlined in the previous section, which effectively try to re-parallelise programs upon which an artificial sequentiality was forced by the nature of the language, data parallel languages (Blelloch 1990; Hillis 1985; Iverson 1962; Metcalf and Reid 1990; Parkinson and Litt 1990; Perrot 1979; Skillicorn 1990) provide the ability to operate upon whole data sets in a single logical step. Since the control structure of the language remains sequential, the programmer is not forced to handle parallelism explicitly, but is once again advised to adopt a certain style in order to produce performance gains. For example, the doubly nested loop required to add corresponding elements of a pair of two dimensional matrices becomes a single statement exploiting an overloaded '+' operator. Similarly it will typically be possible to apply some function to all elements of a structure simultaneously to create a new structure. The analogy with the corresponding sequential loop requires that the function has no side effects which would carry from one iteration to the next. Further new operations allow reduction of structures to single values (e.g. sum all elements, find the maximum element) and the ability to create new structures as re-organisations of existing ones (e.g. copy a matrix column to a vector, extract a sub-matrix, make an $m \times n$ matrix from m copies of an n element vector).

As an example, consider the problem of multiplying together two $n \times n$ element matrices A and B into C. It would be quite possible for a data parallel language to provide such an operation as a primitive. Alternatively, we can build our own procedure as follows, using a sequence of simpler array manipulations:

1. Construct an $n \times n$ matrix of n element vectors A', where $A'[i, j]$ is a copy of the ith row of A.

2. Similarly construct B' so that $B'[i, j]$ is a copy of the jth column of B.

3. Pairwise multiply the elements of $A'[i, j]$ by $B'[i, j]$, into $A'[i, j]$.

4. Set $C[i, j]$ to be the sum ('+ reduction') of $A'[i, j]$.

The relative responsibilities placed upon programmer and system in producing a good partition and scheduling depend to a large extent upon the sophistication of the primitives provided. It will take more effort on the programmer's part to produce the multiplication scheme above rather than the obvious triply nested sequential loop, but the individual manipulations will be easier to map onto most architectures (particularly the SIMD machines with which such languages are primarily associated) than a single step matrix multiplication operation. The more general and powerful the primitives become, the harder they may be to implement, particularly if the required data transfers are not obviously supported by the underlying architecture. For example, consider the difficulties introduced by provision of an operator allowing array elements to be permuted against a template of destinations. It is likely that such an array may be distributed element by element across the processors of the machine as the result of some previous step. Implementation of the permutation operation then requires provision of a general purpose inter-processor routing mechanism.

In this approach, data sharing (as occurs for example in the 'sum' step) is entirely up to the system, as is the local synchronisation potentially required within the parallel implementation of each operation (e.g. the reductions), and the global synchronisation between operations.

In terms of program correctness, the data parallel approach is as good or bad (depending upon one's point of view) as the sequential language within which it is embedded.

1.4 Moderately Abstract Paradigms

This category includes software paradigms in which the programmer is responsible for defining the partition into parallel processes (whether statically or dynamically), but in which the system provides a more friendly environment for these processes than will exist on a typical machine. We investigate three variations on this approach. The first is characterised by the provision of some form of idealised shared memory, equally accessible to all processes. The second discards shared memory in favour of message passing abstractions in which processes may communicate with each other in arbitrary patterns, without consideration of the physical topology of the underlying communications medium. The final technique introduces a novel mechanism which may be considered as a shared, structured, associative memory.

1.4.1 Shared Memory Models

The various models which have emerged in the shared memory paradigm may be conveniently partitioned into two groups according to their origin. Those in the first group were conceived as models in their own right with subsequent consideration given to implementation, while those in the second have evolved from, or in tandem with, their realisation.

The most pervasive idealised model in this area is the PRAM (Parallel Random Access Machine) which is discussed in detail in Chapter 15 and by Gibbons and Rytter (1988), Karp and Ramachandran (1990) and Valiant (1990). Although a range of minor variations have been proposed the essential features are as follows. A PRAM is a collection of 'standard' sequential processors (each with its own program counter and register set) and a global memory to which all processors have unit time access. Processors are synchronised between every instruction (as in SIMD except that, by virtue of the individual program counters, each processor may be executing a different instruction). Clearly a shared memory model must define the behaviour of conflicting accesses to memory. In the PRAM world there are three interesting flavours of the model which address this issue. In the so called EREW (Exclusive Read Exclusive Write) PRAM, the problem is avoided by requiring programs to access a set of distinct memory addresses at each program step. The CREW variant (Concurrent Read Exclusive Write) relaxes the situation to allow multiple reads of the same location during a program step, while a CRCW PRAM allows any collection of accesses to be made, with conflicting writes being resolved by one of a variety of schemes, including 'lowest numbered processor succeeds', 'randomly chosen processor succeeds' and 'largest/smallest value succeeds'.

The PRAM programmer's task is to produce a program for each processor, or more likely design a single program to be executed by every processor in what often amounts to a SIMD style with a rather coarse grain of instruction (i.e. what might normally constitute a entire procedure). It is often possible to exploit the clashing write resolution rules of the CRCW models to produce fast solutions (at least at this abstract level). For example, consider the problem of finding the largest value in an array of n distinct elements. With an EREW PRAM we must evaluate a tree of comparisons in $O(\log n)$ time using $n/\log n$ processors. Given a CRCW PRAM with conflicting writes resolved in favour of the smallest value written, we can solve the problem in constant time using n^2 processors. Since no widely recognised high level PRAM programming language exists, we follow the normal convention and express the algorithm in a (hopefully clear) pseudo-code in Figure 1.11. Note that the two statements inside the last conditional will be executed by n processors, where one would suffice.

It is the task of the underlying system to support the PRAM model on a more realistic architecture. The usual assumption is that this will allow only local memory (one chunk per processor) and a restricted (low or constant degree) point-to-point communications network between processors. A substantial theoretical research effort has gone into the design of the techniques which can support such an emulation, with key results involving the mapping from global addresses to local memory modules and the routing algorithms which can support the implied message exchanges. Note that the most obvious approach

```
Global Data
  A : array [0..n-1] of integer;
      /the input data
  B : array [0..n-1] of 0..1;
      /internal storage
  largestvalue, largestindex : integer;
      /the results

Local Data
      /Each processor has a distinct
      /instance in local registers
  i, j : integer;
      /the indices of elements to be compared

for p:=0 to n^2 -1 do in parallel
      /p is a processor's unique identifier
begin
  i := p mod n;
  j := p div n;
  if A[i]>A[j] then begin
    B[i] := 1; B[j] := 0
  end else begin
    B[i] := 0; B[j] := 1
  end;
  if B[i] = 1 then begin
    largestindex := i;
    largestvalue := B[i]
  end
end
```

Figure 1.11

to the first problem, involving an 'address mod (number of modules)' mapping, would have a worst case time penalty of $\Omega(p)$ for a p processor machine for unfortunate steps in which all accesses map to the same module. This has been attacked in two ways—by randomisation of the mapping, and by the introduction of multiple physical copies of each logical location. In the former case, the essential idea is to choose the mapping randomly from a class of hash functions selected to produce low expected worst case contention (Melhorn and Vishkin 1984). This does not change the absolute worst case behaviour, it simply makes it 'highly unlikely'. The alternative approach (Alt *et al.* 1987) introduces several copies of each logical location distributed to distinct modules. Each copy is timestamped with the step at which it was last updated.

The protocol then dictates that to implement a write it is only necessary to update and timestamp a majority of the copies. To implement a read requires a majority of the copies to be read, with the latest time-stamped value accepted as the 'correct' value. The freedom implied by the requirement for access to any simple majority allows the worst-case time penalty caused by contention at modules to improve to $O(\log p)$, when about $\log p$ copies of each location are in use.

In the area of routing it has been shown that the worst case hot-spots which can occur at intermediate nodes using deterministic routing schemes can be similarly avoided by the introduction of randomisation (see Chapter 9 and Valiant 1983)—instead of routing directly from source to destination, we route (deterministically) to a randomly chosen intermediate node, then on (deterministically) to the actual target. This approach brings expected worst case latency down to a level at which it can effectively be hidden altogether by time-slicing each processor between a number of PRAM processes. This exploitation of 'parallel slackness' is not unrealistic given that the emerging PRAM programming style encourages the use of large numbers of processes (usually a function of the size of a large data set). Successful practical realisation of these techniques is likely to have a substantial impact on the way parallel machines are used.

Of course, in parallel with the largely pencil and paper work of the PRAM community, many real machines with globally shared memory have been constructed and programmed. Working from the machine up, the most immediate issue concerns introducing some form of controllable synchronisation between what will in practice be a collection of asynchronous processors. Many useful concepts (indivisible instructions, locks, monitors, semaphores) already exist from work on multi-tasking operating systems for sequential machines and have been introduced in a straightforward way to concurrent versions of existing languages, such as Pascal, and to newer languages like Ada. More interestingly, new primitives such as the 'fetch and add' operation of the NYU Ultracomputer (Gottlieb *et al.* 1983) allow many useful synchronisations to be performed in a distributed way in a suitably designed communications network, thereby avoiding the time penalties implied by the sequentialization of traditional methods.

1.4.2 Message Passing Models

The message passing approach discards globally accessible memory in favour of the ability to exchange messages between any pair of processes. The main issue in this context is the choice between synchronous and asynchronous protocols. The operations of sending and receiving messages may be individually defined to be either 'blocking' (meaning that the active process must wait if a corresponding receive or send has not been executed) or 'non-blocking' (in which case the process may proceed whatever). A choice of blocking both send and receive defines a synchronous model as provided by the language occam

```
-- First the two process type declarations

PROC stage1 (CHAN OF INT in, out)
  INT item:
  WHILE TRUE
    SEQ
      in ? item
      process item ......
      out ! item
  :

PROC stage2 (CHAN OF INT in, out)
  INT item:
  WHILE TRUE
    SEQ
      in ? item
      process item ......
      out ! item
  :

-- Now the main program

-- the channels
CHAN OF INT from.world, to.world,
           connect.stages:
PAR
  stage1(from.world, connect.stages)
  stage2(connect.stages, to.world)
```

Figure 1.12

(Inmos 1984). This is certainly advantageous from the system's viewpoint, in that it is never necessary to buffer more than one message between a pair of processes. It may also held to be a more tractable protocol from the point of view of program verification (as is emphasised by occam's origins in CSP (Hoare 1978)). On the other hand, a non-blocking send has the advantage that the busy sender can proceed with more work immediately, rather than repeatedly waiting for a slower receiver. Similarly, a non-blocking receive allows a waiting receiver to take another course of action in the absence of a message.

As a simple example, consider the two process occam pipeline in Figure 1.12, in which 'c ? v' denotes input of a value to variable 'v' over channel 'c' (declared as a CHAN), and 'c ! e' denotes output of the value of expression 'e' over channel 'c'. The synchronous model of communication means that stage1 will never accept a new item from the outside world until the previous one has been passed on to stage2. A non-blocking version of '!' could allow a queue of items to build up on the connect.stages channel. These would have to be buffered by the system. On the other hand, in a system with many such processes, the non-blocking version might make it easier for an implementation with a good dynamic scheduling policy to proceed with available work immediately, before redeploying resources to speed up those areas of the process network which have become bottlenecks. The analyses of randomised routing algorithms discussed in the preceding section are clearly of importance when consideration is given to the task of actually implementing the arbitrary communication patterns allowed by the model.

1.4.3 Linda

Rather than introducing a completely new language, the Linda approach (Carriero and Gelernter 1989; Carriero and Gelernter 1990) is to define a parallel programming model which can be derived from any base sequential language by the addition of a small set of simple operations. At the heart of the Linda model is the notion of 'tuple-space', a form of shared memory which provides the means whereby an otherwise uncoordinated set of independent processes may share information and synchronise. The tuple space contains an unordered collection of 'tuples', where each tuple contains an ordered collection of data fields. For example, ('here',1,true) and ('there', 2.5, 7, 'one') are tuples with three and four fields respectively.

Processes interact with the tuple space (and thereby indirectly with each other) by means of six operations which are embedded in the host language as procedure calls. The operation out(t) inserts a copy of the tuple t into tuple space. The fields of t are described by expressions in the host language and are evaluated before insertion. For example, in Pascal-Linda, out('this',2*i,A[i]+7) makes reference to variables A and i (which are local to the outputting process).

Conversely, the in operation is used to remove tuples from tuple space, assigning some or all of the field values to local variables. Tuples are selected by pattern matching against a template provided as a parameter to the call. The template can include actual values, expressed as constants or expressions, and formal parameters, each a local variable. A match occurs between a template and a tuple when all corresponding fields have the same type and actual values in the template match corresponding values in the tuple. An in will suspend its process until a matching tuple can be found in the tuple space, at which time the formal parameters in the template are assigned the corresponding values from the tuple, which is then deleted from tuple space. The system may make an arbitrary choice in the presence of multiple matches. Since both actual values and formal parameters in a template may be simple local variables, a ? is introduced to indicate a formal parameter. For example if i is a local integer variable and j is a local real variable with current value 3.5, then in('yes',

`?   i,  j`) could match against the tuple (`'yes'`, `4,  3.5`), resulting in the assignment of the value 4 to `i`. The `rd` (for read) operation functions in a similar way to `in` except that the matching tuple is not removed from tuple space. This simple set of operations is extended by the variants *eval* (which is like `out` except that new processes are created to evaluate the fields of the tuple) and `inp` and `rdp` (which are 'predicated' versions of `in` and `rd`, returning 1 if a match is made immediately, otherwise returning 0 and terminating).

Problem partition is performed explicitly by the Linda programmer who specifies the initial collection of processes. A typical program will involve some processes whose role is to fill tuple space with the data describing the problem, some processes which perform the actual work and one process which collects the results for down-loading to the file system (or perhaps several such processes if results may be distributed across several files). As an alternative to the tuple set-up processes, it is also possible to envisage the maintenance of persistent tuple spaces whose lifetime extends beyond a single program execution.

The flexibility of tuple matching means that Linda can easily be used to write programs in either message passing or shared memory styles. For message passing, the programmer could use one or more fields to indicate the identities of the interacting processes. Point to point message passing would be achieved by matching calls of 'in' and 'out' while a broadcast could be achieved with many 'rd's and one 'out'. The predicated versions could be used to choose between blocking and non-blocking receive. In contrast, a shared memory style is achieved by using one field to act as the global address of its tuple. Locking can then be implemented by using 'in' to remove tuples for exclusive update, followed by an 'out' of the new version. Of course, in either scenario it is the responsibility of the programmer to ensure that the tuples are used in a manner which is consistent with the intended interpretation. Careless programming could allow a message to be stolen by a process for which it was not intended or for some shared address to be either destroyed, or perhaps worse, replicated.

The Linda system is faced with the problem of implementing the tuple space in as efficient a manner as possible on each given architecture, a task complicated by the associative nature of tuple access. Clearly a single centralised implementation will rapidly become a bottleneck as machine size grows. Many distributed schemes have been considered and implemented. One extreme option is to replicate the tuple space at each processor. This makes 'rd' very easy and means that 'out' is supported by a simple broadcast (which may or may not be expensive). On the other hand implementation of 'in' requires a careful protocol to ensure that a tuple is only removed once (in other words a synchronisation problem is introduced). At the other extreme, we can simply distribute single copies of the tuples across the machine. This makes 'out' simple, but requires more work to find suitable tuples for 'rd' and 'in'. Schemes involving the hashing of tuple fields can help here, while a more sophisticated approach might try to analyse the usage of tuple space made by particular processes and distribute tuples accordingly.

1.4.4 Annotated Declarative Languages

The abstract nature of the declarative paradigms leaves the compiler and run-time system with much (perhaps too much) work to do in attempting to produce efficient parallel computations. Consequently, another popular approach is to build new languages with declarative foundations but augmented with either annotations or new syntactic constructs and semantics in order to make the programmer shoulder a larger part of the burden. The extent to which the resulting languages should be considered declarative with parallel features or vice-versa varies substantially from one to another.

In the functional context, the simplest help we may offer is in annotating to indicate strict arguments (Burton 1984) (thereby taking care of the issue of safety). More substantially, we could annotate function bodies to indicate which sub-expressions should be considered as candidates for parallel evaluation (thereby helping tackle the granularity problem). A more dramatic approach is to require the programmer to specify the complete process structure, including the logical communication channels implied by the corresponding function-argument graph. This approach is proposed by the language Caliban (Kelly 1989), making it effectively a 'functional occam' in which processes behave declaratively. Finally, some schemes (e.g. Hudak 1986) even require the programmer to specify process placement by annotating expressions with processor numbers indexed with respect to the current processor (effectively allowing directives such as 'reduce this sub-expression on the right neighbour of the current processor').

Amendments to logic languages focus on the elimination of non-determinism and on the use of shared variables as communication channels (Gregory 1987; Shapiro 1987; Taylor 1989). The former approach leads to 'committed choice' languages in which only one path through the possible search space is pursued (under the programmer's control), while annotating shared variables as 'read-only' by one process and 'writeable' by another allows them to be considered as inter-process channels, with synchronisation introduced by requiring the reading process to suspend while awaiting assignment to a read-only variable. The committed choice aspect of such languages means that once bound a variable cannot be rebound, corresponding to the idea that a message cannot be 'unsent'. In a final shift of responsibility from system to programmer, languages such as Strand (Foster and Taylor 1990) require the programmer to indicate process placement by means of annotations to sub-goals indicating placement relative to the current processor.

1.5 Ad-Hoc Paradigms

Our final category contains systems which have been constructed around specific architectural types and which offer few features by way of abstraction beyond those provided by the sequential languages in which they are embedded. In particular they offer no real support in the areas of problem partitioning and scheduling and provide only the support for data sharing and synchronisation which is present in the architecture itself (e.g. Seitz 1985). The sequential base language (typically C or Fortran) is augmented with a library of communications primitives which rely upon the existence of direct physical links between communicating processors. Any indirect communication (and associated routing and buffering) must be handled explicitly in the program. Rather in the manner of conventional machine code programming, such systems allow optimal exploitation of specific architectures (at least for relatively straightforward problems) at the expense of portability and often of verifiability. Early multiprocessor implementations of occam followed this route, restricting the connectivity between processes on different processors to that of the physical network. However, although entirely responsible for partition and scheduling, the programmer at least had the well defined semantics of occam to fall back on as an aid to ensuring correctness.

1.6　Overview and Trends

What conclusions can we draw from the preceding survey? Where does the future of parallel programming lie? It seems apparent that no matter which approach is taken, the programmer who hopes to make effective use of parallelism will be required to formulate solutions within a more constrained style than is allowed if a sequential implementation is envisaged. The degree to which this style is made explicit in the language varies widely between our categories, but the underlying message seems to be that there is no such thing as free parallelism. Of course, we may also observe that the design of efficient sequential programs is also difficult. Is the parallel case really harder, or is it just that we are less familiar with its tricks and short-cuts?

In close analogy with the history of sequential computing, we may also observe an increasing practical emphasis on the higher level abstractions as a means of encouraging program portability. When such models are underpinned by sound semantics, the possibility of user guided transformation becomes realistic (Darlington *et al.* 1990). Furthermore, such models often encompass useful mechanisms for abstracting those parallel tricks which we do understand for re-use (e.g. as second order functions in a functional language). We expect to see the continued emergence of such methods as our understanding of their associated implementation techniques evolves (see Chapters 2 and 5 and also Axford 1991; Cole 1989; Danelutto *et al.* 1991; Darlington *et al.* 1990; Foster and Stevens 1990; Kelly 1989; McKeown *et al.* 1990; Skillicorn 1990). Meanwhile, it would be foolish to assume that any one of the paradigms will eventually dominate. Just as different sequential paradigms seem more appropriate to different tasks, so we should expect the various approaches to parallelism to persist.

References

Alt H., Hagerup T., Melhorn K. and Preparata F. P. (1987) Simulation of Idealized Parallel Computers on More Realistic Ones. *SIAM Journal of Computing*, **16**(5) pp. 808–835.

Augustsson L. and Johnsson T. (1989) Parallel Graph Reduction with the (ν, G)-machine. In *Proc. of the Fourth International Conference on Functional Programming Languages and Computer Architecture*, pp. 202–213.

Axford T. (1991) An Abstract Model for Parallel Programming. *Computer Science Research Report CSR–91–5*, University of Birmingham.

Bird R. and Wadler P. (1988) *Introduction to Functional Programming*. Prentice Hall.

Blelloch G. E. (1990) *Vector Models for Data Parallel Computing*. MIT Press.

Burton F. W. (1984) Annotations to Control Parallelism and Reduction Order in the Distributed Evaluation of Functional Programs. *ACM TOPLAS*, **6**(2) pp. 159–174.

Carriero N. and Gelernter D. (1989) Linda in Context. *Communications of the ACM*, **32**(4) pp. 444–458.

Carriero N. and Gelernter D. (1990) *How to write Parallel Programs*. MIT Press.

Cole M. I. (1989) *Algorithmic Skeletons: Structured Management of Parallel Computation*. Pitman.

Crammond J. (1985) A Comparative Study of Unification Algorithms in Or-parallel Execution of Logic Languages. In *Proc. of the IEEE Conference on Parallel Processing*.

Danelutto M., Pelagatti S. and Vanneschi M. (1991) Parallel Programming with Algorithmic Motifs. *Report HPL–PSC–91–16*, Hewlett-Packard Pisa Science Center.

Darlington J., Khoshnevisan H., McLoughlin L. M. J., Perry N., Pull H. M., Sephton K. M. and While R. L. (1989) An Introduction to the Flagship Programming Environment. In *CONPAR 88*, pp. 108–115.

Darlington J., Reeve M. and Wright S. (1990) Programming Parallel Computer Systems using Functional Languages and Program Transformation. In *Parallel Computing 89*. Elsevier Science Publishers (North Holland).

Foster I. and Stevens R. (1990) Parallel Programming with Algorithmic Motifs. *Preprint MCS–P124–0190*, Argonne National Laboratory.

Foster I. and Taylor S. (1990) *Strand: New Concepts in Parallel Programming*. Prentice-Hall.

Gibbons A. and Rytter W. (1988) *Efficient Parallel Algorithms*. Cambridge University Press.

Goldberg B. (1988) Multiprocessor Execution of Functional Programs. *International Journal of Parallel Programming*, **17**(5) pp. 425–473.

Gottlieb A., Grishman R., Kruskal C. P., McAuliffe K. P., Rudolph L. and Snir M. (1983) The NYU Ultracomputer—Designing an MIMD Shared Memory Parallel Computer. *IEEE Transactions on Computers*, **C–32**(2) pp. 175–189.

Gregory S. (1987) *Parallel Logic Programming in PARLOG*. Addison-Wesley.

Hillis W. D. (1985) *The Connection Machine*. MIT Press.

Hoare C. A. R. (1978) Communicating Sequential Processes. *CACM*, **21**(8) pp. 666–677.

Hudak P. (1986) Para-Functional Programming. *IEEE Computer*, pp. 60–70, August .

Inmos Ltd. (1988) *Occam 2 Reference Manual*. Prentice-Hall International.

Iverson K. E. (1962) *A Programming Language*. John Wiley.

Karp R. M. and Ramachandran V. (1990) Parallel Algorithms for Shared Memory Machines. In J. van Leeuwen, editor, *Handbook of Theoretical Computer Science*, pp. 870–941. Elsevier Science Publishers.

Kelly P. (1989) *Functional Programming for Loosely Coupled Multiprocessors*. Pitman.

McKeown G. P., Rayward-Smith V. J. and Turpin H. J. (1990) Branch-and-Bound as a Higher Order Function. *School of Information Systems Internal Report SYS–C90–03*, University of East Anglia.

Melhorn K. and Vishkin U. (1984) Randomized and Deterministic Simulation of PRAMs by Parallel Machines with Restricted Granularity of Parallel Memories. *Acta Informatica*, **21** pp. 339–374.

Metcalf M. and Reid J. (1990) *Fortran 90 Explained*. Oxford University Press.

Nicolau A., Padua D., Gelernter D. and Gross T. (1991) *Advances in Languages and Compilers for Parallel Processing*. Pitman.

Parkinson D. and Litt J. (1990) *Massively Parallel Computing with the DAP*. Pitman.

Perrot R. H. (1979) A Language for Array and Vector Processors. *ACM Transaction on Programming Languages and Systems*, **1**(2) pp. 177–195.

Peyton-Jones S. L. (1987) *The Implementation of Functional Programming Languages*. Prentice Hall.

Peyton-Jones S. L. (1989) Parallel Implementations of Functional Programming Languages. *Computer Journal*, **32**(2) pp. 175–186.

Sarkar V. (1989) *Partitioning and Scheduling Parallel Programs for Multiprocessing*. Pitman.

Seitz C. L. (1985) The Cosmic Cube. *CACM*, **28**(1) pp. 22–33.

Shapiro E. (1987) *Concurrent Prolog: Collected Papers*. MIT Press.

Skillicorn D. B. (1990) Architecture-Independent Parallel Computation. *Computer*, **23**(12) pp. 38–51.

Snyder L. (1986) Type Architectures, Shared Memory and the Corollary of Modest Potential. *Annual Review of Computer Science*, , **1** pp. 289–317.

Szymanski B. K. (1991) *Parallel Functional Languages and Compilers*. ACM Press.

Taylor S. (1989) *Parallel Logic Programming Techniques*. Prentice-Hall.

Valiant L. G. (1983) Optimality of a Two Phase Strategy for Routing in Interconnection Networks. *IEEE Transactions on Computers*, **C–32**(9) pp. 861–863.

Valiant L. G. (1990) General Purpose Parallel Architectures. In J. van Leeuwen, editor, *Handbook of Theoretical Computer Science*, pp. 944–971. Elsevier Science Publishers.

Wolfe M. (1989) *Optimising Supercompilers for Supercomputers*. Pitman.

CHAPTER 2

The Divide-and-Conquer Paradigm as a Basis for Parallel Language Design

Tom Axford

University of Birmingham, UK

2.1 Introduction

2.1.1 A Brief History

Since the earliest days of computer programming, algorithms have been known which follow the so-called 'divide-and-conquer' method. Many of the best and most widely used computer algorithms are these divide-and-conquer (d-c) algorithms. The binary search algorithm and Hoare's quicksort are two very well known examples. There are many, many others for all types of computing problems, both simple and complicated, both general and specialised.

Earlier references to d-c in the programming literature do not attempt to formalise it (e.g. Aho *et al.* 1974), but simply use the idea informally to help with the explanation of an algorithm. In (Horowitz and Sahni 1978), on the other hand, d-c is described as a *control abstraction* and a program for it is given in a Pascal-like pseudocode, calling four other procedures to supply the details needed to implement any specific algorithm within the general d-c family. Although Horowitz and Sahni gave this completely general d-c program, they never thereafter used it as a program, but simply as a model to be copied when constructing programs for more specific d-c algorithms. More recently, many other authors have followed a similar approach and formally defined the d-c paradigm as an algorithm which can be discussed in its own right, rather than considering only particular instances of it.

D-c algorithms have long been recognised as highly suitable for parallel implementation because the sub-problems generated can be solved independently and hence in parallel. Many d-c algorithms generate very large numbers of sub-problems in typical real applications and hence can potentially be implemented with a high degree of parallelism. There have been various proposals for parallel architectures designed specifically for the d-c family of algorithms (Peters 1981; Preparata and Vuillemin 1981; Horowitz and Zorat 1983; McBurney and Sleep 1987) and the suggestion of basing a parallel programming language on the d-c paradigm appears to have first been made over ten years ago (Preparata and Vuillemin 1981), although the first, and so far only, language to have been developed in detail appeared quite recently (Mou 1990). Much of the recent work on using the general d-c paradigm for parallel processing has been done in the context of functional programming languages (Mou and Hudak 1988; Cole 1989; Kelly 1989; Mou 1990; Rabhi and Manson 1990), which are perceived to have some fundamental advantages over conventional procedural languages. In an influential and well known paper (Backus 1978), John Backus argued strongly in favour of functional languages instead of the conventional procedural languages, and these arguments have been echoed many times by more recent authors (e.g. Hudak 1989; Hughes 1989). The main reasons for preferring functional languages are discussed later.

2.1.2   The Divide-and-Conquer Paradigm

The general d-c paradigm may be described informally as follows: partition the problem into separate, smaller sub-problems, all of which are essentially the same but with different data; find solutions to these sub-problems; and then combine these solutions into a solution for the whole. This approach is used recursively, so that the sub-problems may themselves be further partitioned into smaller sub-problems and so on, until the sub-problems obtained are so small that a solution to each is easy to find directly.

In a Pascal-like pseudocode, the general d-c paradigm may be defined more formally as:

```
procedure divcon(p)
begin
    if simple(p)
    then
        return solve(p)
    else
        (p1,...,pn) := divide(p);
        return combine(divcon(p1),...,divcon(pn))
    fi
end
```

where *p* is the problem data and *p1,...,pn* are the problem data for each of the sub-problems into which the problem is partitioned. The function *simple* tests to see if the problem is sufficiently simple that it can be solved directly (by the function *solve*) in which case the partitioning need not be done. The function *divide* partitions the problem into *n* smaller problems, while the function *combine* takes the solutions to the *n* sub-problems and combines them into a solution to the whole problem.

2.2   Divide-and-Conquer in Procedural Languages

2.2.1   What is Wrong with Existing Languages?

Procedural languages such as Fortran, Pascal, C, Ada, etc. are well-established and highly successful vehicles for programming conventional computers based on the von Neumann machine architecture, which is an essentially sequential, uniprocessor architecture. The only really common use of parallel processing is for a few very specialised and heavily used functions of computer operating systems for which dedicated hardware is built (e.g. peripheral device controllers for handling input and output in parallel with other processing).

General purpose parallel computer systems have been readily available for at least a decade now. It is widely appreciated that these computers offer far superior price/performance ratios than conventional uniprocessor machines, yet they are used far less. As Geoffrey Fox has said:

> ... essentially all successful reasonable performance uses of parallel machines have used explicit user decomposition which is low level and machine dependent. We expect that we must find more portable attractive methods if parallel computers are to take over from the conventional architectures (Fox 1989)

And on the same theme, D. B. Skillicorn says:

> It is scarcely surprising that most potential users have avoided making the transition to parallel machines and development of parallel software. ... The solution to these problems lies in finding concurrent programming languages or programming models that are architecture independent. (Skillicorn 1991)

Many others have expressed similar views (e.g. Pancake 1991). Indeed, it is widely recognised that conventional programming languages have proved disappointingly unsuitable for automatic parallelisation, and that new languages are needed. Attempts to extend existing languages have been highly architecture dependent, and they also fail to solve the problem that the underlying language is fundamentally sequential in nature. This means that programs must use the (machine-dependent) language extensions to achieve worthwhile parallel speedups. Existing programs (entirely within the base language) fail to benefit unless they are substantially redesigned and rewritten.

Why are conventional languages so difficult to implement efficiently in parallel? A major reason for this is the type of control structures used in these languages. In essence, the control structures can usually all be reduced to three fundamental types:

1. Simple sequencing of statements (denoted by the semicolon in Pascal).

2. The **if...then...else...** construct (indeed, this construct is itself unnecessary, as it is easy to represent it in terms of the other two. That is going

to extremes, however, as no real programming language limits the programmer so severely. We wish to consider the simplest set of constructs that allow us to retain the essential flavour and style of most conventional programs).

3. The **while...do...** construct.

There are no obvious opportunities for parallelism in any of these. The compiler is totally unable to extract any parallelism from a program by simply looking at its overall control structure; it must inevitably look deeper (and that is very much more difficult to do).

2.2.2   A Divide-and-Conquer Construct

In this section we define a control structure that implements the d-c paradigm. This control structure can be used as an alternative to the usual **while...do...** construct (or similar loop constructs). Any computable function can be programmed in a conventional procedural language using only three control structures: sequencing, conditional statements (**if...then...else...**) and d-c statements (i.e. the new construct). A proof of this statement is not given here, although it is not difficult to construct a proof: essentially all that is required is to show that any instance of the old **while...do...** statement can be programmed in terms of the new d-c construct (of course, some further assumptions are needed about the programming language, but nothing out of the usual).

Definition of the Divide-and-Conquer Construct

A general d-c algorithm could be defined in the form of a procedure as in Section 2.1.2. This requires four user-defined procedures to specify the details. A less cumbersome way to do it is to define a new control construct, in a similar vein to the traditional control constructs such as **if, for, while, case,** etc. The d-c construct that we define here has five parts to it, separated and bounded by six keywords: **divcon, test, solve, ordivide, andcombine** and **nocvid** (the last terminates the complete construct and is **divcon** written backwards).

The general form is:

> **divcon** $P_1\{v_1, ..., v_m/\}$
> **test** $B\{v_1, ..., v_m\}$
> **solve** $P_2\{w_1, ..., w_n/v_1, ..., v_m\}$
> **ordivide** $P_3\{v'_1, ..., v'_m, v''_1, ..., v''_m/v_1, ..., v_m\}$
> **andcombine** $P_4\{w_1, ..., w_n/w'_1, ..., w'_n, w''_1, ..., w''_n\}$
> **nocvid**

where $P_i\{x, ..., y/u, ..., v\}$ denotes any piece of program (consisting of complete statements and constructs only) that computes the variables $x, ..., y$ from the variables $u, ..., v$ (and from constants and any other variables, such as global variables, that are in scope at the beginning of the **divcon** construct);

and $B\{x, ..., y\}$ denotes any boolean expression using the variables $x, ..., y$ (and constants and global variables).

It is important that no side effects are permitted within the d-c construct; i.e. the code for B does not update any variables at all (whether local or global) and the code for $P_i\{x, ..., y/u, ..., v\}$ does not update any variables other than the $x, ..., y$ explicitly mentioned.

The meaning of this construct is defined by the equivalent program:

> **procedure** *divcon(*in $v_1, ..., v_m$, **out** $w_1, ...w_n$)
> **begin**
> > **if** $B\{v_1, ..., v_m\}$ **then** $P_2\{w_1, ..., w_n/v_1, ..., v_m\}$
> > **else**
> > > $P_3\{v'_1, ..., v'_m, v''_1, ..., v''_m/v_1, ..., v_m\};$
> > > $divcon(v'_1, ..., v'_m, w'_1, ..., w'_n);$
> > > $divcon(v''_1, ..., v''_m, w''_1, ..., w''_n);$
> > > $P_4\{w_1, ..., w_n/w'_1, ..., w'_n, w''_1, ..., w''_n\};$
> > **fi**
> **end;**
> $P_1\{v_1, ..., v_m/\};$
> $divcon(v_1, ..., v_m, w_1, ..., w_n);$

where the two recursive calls of *divcon* (between P_3 and P_4) can be taken in either order (or concurrently).

So, the construct

> **divcon** P_1 **test** B **solve** P_2 **ordivide** P_3 **andcombine** P_4 **nocvid**

can be understood as follows:

P_1 is code to declare and initialise the variables $v_1, ..., v_m$ that will be used in the first phase of the d-c method. B is a conditional expression (in terms of $v_1, ..., v_m$). If this expression is true, the problem will not be sub-divided further, but a solution found directly using P_2, which is the code to compute the result (represented by the variables $w_1, ..., w_n$) directly from $v_1, ..., v_m$. If the conditional expression B is false, however, the problem is sub-divided into two parts using P_3, which is code to compute $v'_1, ..., v'_m$ and $v''_1, ..., v''_m$ from $v_1, ..., v_m$. The singly primed variables represent one sub-problem, while the doubly primed variables represent the second sub-problem. The two sub-problems are themselves solved using the same d-c approach applied recursively. The results of these two sub-problems will be represented by $w'_1, ..., w'_n$ and $w''_1, ..., w''_n$ respectively. The final result is computed by P_4, which is code to compute $w_1, ..., w_n$ from $w'_1, ..., w'_n$ and $w''_1, ..., w''_n$, i.e. to combine the solutions to the two sub-problems.

Two very simple examples serve to illustrate the style of programming required by the **divcon** construct. Both examples are problems that would normally be programmed very easily with a single loop using **for** or **while** loops.

Example: Summation

A program to add up an array of numbers (i.e. $a[1] + ... + a[n]$) can be written as a single application of the d-c construct, in which the variables i and j denote the first and last index values of the array to be summed (initially 1 and n), and *sum* denotes the sum of the array (the result of the computation):

```
divcon
        i := 1;
        j := n;
test
        i = j
solve
        sum := a[i]
ordivide
        i' := i;
        j' := (i+j) DIV 2;
        i" := j' + 1;
        j" := j
andcombine
        sum := sum' + sum"
nocvid
```

The first part (after **divcon**) defines the initial values of the variables i and j which will be used as working variables in the d-c process. The next part (after **test**) is a boolean expression that tests if the problem is sufficiently simple that it can be solved directly without the need for partitioning. In this example, no partitioning is done if the array contains only one element.

The third part (after **solve**) solves the problem in the case that no partitioning is to be carried out. This occurs only if the array contains just one element, so the sum is that element. The new variable *sum* is defined in terms of i and j (and any constants or globals, such as the array a which is treated as a global variable as far as the d-c construct is concerned), and *sum* will eventually contain the final result, i.e. the sum of all elements of the array. The fourth part (after **ordivide**) divides the problem into exactly two sub-problems: i', j', i'' and j'' are computed from i and j. The pair (i', j') is the data for one sub-problem (summing the bottom half of the array), while (i'', j'') is the data for the other sub-problem (summing the top half of the array).

The final part (after **andcombine**) combines the results of the two sub-problems to give the result of the whole problem: *sum* is computed from *sum'* and *sum''*, which denote the results of the two sub-problems. In this example, the sum of the whole array is simply the sum of the sums of the two halves of the array.

Another Example: Find the First Zero

A program to find the first zero in the array $a[1..n]$ can also be implemented as
a single simple application of the d-c construct. The variables i and j denote the
first and last index values, as before. The result is denoted by two variables:
found is true if a zero element is found, and false if all elements of the array
are non-zero; *posn* is the index of the first zero element if there is one (*posn* is
undefined if all elements are non-zero).

```
divcon
        i := 1;
        j := n;
test
        i = j
solve
        found := (a[i] = 0);
        posn := i;
ordivide
        i' := i;
        j' := (i+j) DIV 2;
        i'' := j' + 1;
        j'' := j
andcombine
        found := found' or found'';
        posn := if found' then posn' else posn'' fi
nocvid
```

This program works by dividing the array into two halves and solving for each
half separately. If the array contains only one element, however, no sub-division
is necessary and the result can be computed directly simply by looking at the
only element to see if it is zero. The solutions to the two halves of the array
are combined by (i) taking the logical OR of the two values of *found*, and (ii)
if a zero was found in the first half then it is the first zero in the whole array,
otherwise the zero found in the second half is the first zero of the whole array.

Discussion

To the author's knowledge, no programming language with this d-c construct
(or anything very similar) has ever been implemented. A purely sequential
implementation of the d-c construct is likely to be less efficient than equivalent
loop constructs, although not seriously so. This is due to the somewhat greater
overheads associated with the d-c method. In the two examples given in the
previous section, if n is the number of elements in the array, then the execution
time is $O(n)$ for serial execution, which is the same as for the normal loop
implementations, the only difference being that the constant multiplicative
factor may be somewhat greater for d-c than for simple loops.

For a parallel implementation, however, the gains should be striking. A shared-memory (PRAM) architecture is required. Whenever the problem is sub-divided in two by the d-c algorithm, the two sub-problems can then be solved in parallel on separate processors. Although they use their own versions of the working variables, all global variables (i.e. all variables in scope at the beginning of the **divcon** construct) are readable by all the sub-problems generated, so the obvious way to implement these variables is in shared memory. The potential parallelism is limited only by the size of the problem. In practice, the optimum amount of parallelism is likely to be considerably less than the maximum, but the important point is that it is easy to implement the program with any specified degree of parallelism (up to the maximum) without requiring help from the programmer.

For our two simple examples, if the maximum degree of parallelism is implemented (which requires $O(n)$ processors), the execution time is $O(\log n)$, a speedup of $O(n/(\log n))$. Of course, the speedup depends on the problem, but many common processing problems are subject to speedups of this order. If n is very large, the parallelism will be limited by the number of available processors (say p), but the speedup will typically be approximately p in these cases (i.e. close to the ideal speedup).

It should be completely feasible to implement a programming language containing the d-c construct described above (or an equivalent construct). The most difficult part is in prohibiting side effects within the d-c construct. Although it is possible to leave this entirely to the programmer's self-discipline, that would permit the development of unsafe and hence potentially unreliable software.

Another dilemma is whether or not to include conventional loop constructs (such as **for, while,** etc.) also. It is not necessary for these to be included; all computable functions can be programmed without them. Nevertheless, for a small minority of programming problems it may be inconvenient and cumbersome if they are not available at all and absolutely everything has to be programmed in terms of only conditional statements and d-c.

On the other hand, if conventional loop constructs are retained in the language, most programmers are so familiar with using loops that they will probably continue to use them rather than make the effort to learn the techniques for using d-c instead. If this happens, programs will continue to be written in the same old way and will completely fail to obtain the benefits of the parallelism available through d-c.

The ideal situation would be to have a language which provided d-c constructs in addition to the usual loops, but for the great majority of programs to be written using solely d-c and conditionals. These programs could then be implemented easily and efficiently either in the usual serial form on uniprocessors, or in parallel on shared memory multiprocessors.

2.3 Divide-and-Conquer in Functional Languages

2.3.1 Are Functional Languages Inherently More Parallel Than Procedural Languages?

Procedural language programs consist of a sequence of operations on the state of the machine. There are three main features of such languages that hinder their parallel implementation. The first is that any program is a sequence of operations and the programmer has to impose an order on the operations in his program, whether or not that order is logically necessary. It is very difficult for the compiler to deduce that some statements must be kept in the order written whereas other statements may be safely re-ordered.

The second feature that causes difficulties is the state itself. A construct such as the d-c construct introduced earlier allows the execution sequence to be split into two parallel sequences, but it does not split the state. In general, both execution sequences can access all variables in scope when the d-c construct is entered. Indeed, procedural languages routinely allow communication between different routines to occur via global or outer-block variables. Programmers are used to exploiting this facility and hence tend to think of shared variables as a natural and familiar means of communication. Most theoretical work on parallel algorithms has been in conventional procedural languages and it is no coincidence that far more work has been done on shared memory algorithms than on message passing algorithms. It is this fundamental rôle of state in procedural languages that makes shared memory parallelism seem easier and more natural.

Thirdly, the fact that most procedural languages permit side effects (e.g. global and outer-block variables may be updated by any procedure) makes it practically impossible for a compiler to change the order of execution of many parts of a program that would otherwise be perfectly safe to change. Hence automatic parallel implementation is very difficult to achieve.

Functional programming languages, on the other hand, have rather different characteristics. The programmer does not have to impose a sequential order on all operations whether or not it is required. In a functional program, there is no explicit order of operations. The compiler must deduce any necessary sequencing from the data dependencies (e.g. if x depends on y, then the computation of y must precede the computation of x).

Furthermore, there is no state, as such, in a functional language and hence there is less bias towards shared memory parallelism. Message passing approaches seem to fit comfortably into a functional language framework.[1]

Because of these potential advantages, and the more general benefits of functional programming (Hudak 1989; Hughes 1989), most of the recent work

---

[1] Most functional languages retain some bias towards shared memory, however, because outer block variables are usually accessible within function definitions and other program structures. As variables in functional programs cannot be changed after definition, access to outer block variables is equivalent to read-only shared memory.

on the use of d-c as a general procedure or function which can be used as a building block in program construction has been done in a functional language context (Burton and Sleep 1981; Mou and Hudak 1988; Cole 1989; Kelly 1989; Mou 1990; Rabhi 1990). We follow the same approach here, and the remainder of this chapter discusses d-c as a basis for parallelism in functional languages. This is not to say that a similar approach cannot be used to achieve parallelism in procedural languages. Any functional language program can be translated into a procedural language program, and usually the overall style and structure of the original program can be largely preserved (just as it is possible to translate a Fortran program into Pascal, and the result still has a recognisably Fortran-like style).

Functional programming languages do not include loop constructs such as **while...do...**, nor do they allow any explicit sequencing of operations, although there is implicit sequencing arising from the data dependencies. The programmer defines the control structure of his program by using **if...then...else...** expressions and recursive function definitions.[2] Whether or not parallelism can be easily found in such programs depends very much on the particular recursive definitions used. The d-c paradigm often permits a high degree of parallelism and can very often be used as a convenient alternative to the more familiar programming structures.

2.3.2   A Simple Introduction to Functional Languages

For the benefit of readers who are not familiar with modern lazy functional programming languages such as Miranda (Turner 1985) and Haskell (Hudak *et al.* 1991), we informally introduce a very simple functional language, sufficient for the purposes of illustration in the rest of this chapter. The language is approximately a common subset of Miranda and Haskell, but a few small differences have been introduced to achieve a high degree of simplicity, clarity and generality. The language is untyped and the pattern-matching features of Miranda and Haskell have been omitted.

Expressions

Expressions in the language can be constructed in the following ways:

1. As the **application** of a function to a single argument (in functional languages a function always has exactly one argument, hence the brackets around the argument can be omitted):

   $f x$

---

[2]Strictly speaking, not even recursion is necessary. Any program can be written solely in terms of non-recursive function definition, provided higher order functions are allowed (the so-called Y combinator of combinatory logic can be used to simulate recursion). It is conventional to use recursion, however. In any event, the same arguments would apply to the use of the Y combinator or any other way of representing recursion.

denotes $f(x)$ in mathematics. Functions can give other functions as their results, and these may be applied to further arguments in turn (the default bracketing is from the left):

$$f\,x\,y\,z \qquad \text{means} \qquad ((f\,x)\,y)\,z$$

i.e. f is applied to its argument x, the result (another function) is applied to y and the result of that (yet another function) is applied to z. This is normally how we represent a function which would conventionally be defined to have three arguments.

In general, all of f, x, y and z may themselves be expressions (bracketed if necessary).

2. As a **function definition** of the form:

$$x \to e$$

where x denotes any identifier and is the formal argument of the function, and e may be any expression (which usually involves the formal argument x). The whole expression denotes the function f where $f(x) = e$ in mathematics. It is a way of defining a function without having to give it a name. Default bracketing of these expressions is from the right, so:

$$x \to y \to x + y \qquad \text{means} \qquad x \to (y \to x + y)$$

which defines a function which takes an argument x and gives as its result another function, which takes an argument y and gives as its result the sum of x and y.

3. As a **conventional expression** with infix operators, **if** expressions, brackets, etc. For example:

if $x > 0$ **then** $x + 6$ **else** $2 * (x - 6)$ **fi**

has its usual meaning. Prefix function application always binds more tightly than infix operators, so:

$$f\,x + g\,y \qquad \text{means:} \qquad (f\,x) + (g\,y)$$

and both denote the mathematical expression $f(x) + g(x)$. The infix expression $x + y$ is interpreted as equivalent to $(+)\,x\,y$ where $(+)$ denotes the prefix form of $+$: $(+) = x \to y \to x + y$.

4. As a **tuple** containing two or more components which are separated by commas and enclosed in brackets, e.g.

$$(x+y, f\,a)$$

is the tuple whose first component is $(x+y)$ and whose second component is $(f\,a)$.

5. As a **where** clause which is of the form: any expression followed by the keyword **where**, followed by a list of definitions. These definitions are local to the expression. For example:

$x + y$ **where** $x = 3; y = 7$ equals $3 + 7$

$f = x \rightarrow g$ **where** $g = y \rightarrow a$ equals $f = x \rightarrow y \rightarrow a$

Program layout (i.e. the indentation) is used to indicate the extent of the list of definitions in a **where** clause.

Definitions

Expressions are used to construct **definition statements** which have the form:

$x = e;$

where x denotes any identifier and e is any expression. Although definitions are superficially like assignment statements in procedural languages, the identifiers in functional languages do not denote variables, and once an identifier has been defined it cannot be redefined to a new value (unless the identifier is being used in a different scope or context, in which case it denotes a logically distinct object). For this reason, definitions may be written in any order and this order does not indicate the order in which they will be evaluated.

Tuples of identifiers are allowed on the left-hand side of definition statements:

$(x,y,z) = e;$ is equivalent to $x = e\ 1; y = e\ 2; z = e\ 3;$

Tuples are treated as equivalent to partial functions over the integers, so that any tuple applied to the integer i gives the i-th component of that tuple, e.g. $(a,b,c)3$ is equivalent to c.

Function Definitions

Functions can be defined using the definition statements already introduced. For example:

$factorial = n \rightarrow$ **if** $n=0$ **then** 1 **else** $n * factorial(n-1)$ **fi**

An alternative syntactic form of function definition statement is permitted, which many programmers prefer:

$f\,x = ...$ is synonymous with $f = x \rightarrow ...$

$f\,x\,y = ...$ is synonymous with $f = x \rightarrow y \rightarrow ...$

and so on. For example, the same factorial function definition may be written:

$factorial\ n =$ **if** $n=0$ **then** 1 **else** $n * factorial(n-1)$ **fi**

and function composition $(.)$ may be defined in any of the following ways:

$(.) = f \rightarrow g \rightarrow x \rightarrow f(g\ x)$

$(.) f g x = f(g\ x)$

$f . g = x \rightarrow f(g\ x)$

$(f . g) x = f(g\ x)$

2.3.3   A Function for Divide-and-Conquer

A divide-and-conquer function is easy to define in functional programming languages. The first proposal to use such a function as a general basis for parallelism was made by Burton and Sleep (1981), and others have developed and extended this approach. Following the approach we used in the procedural language context, a simplified form of the general d-c algorithm is used in which the sub-division is always into exactly two parts at a time. Of course, this entails no significant loss of generality as subdivision into any number of parts is possible; it simply requires more steps in which to do it.

Using our simple functional language, we define the d-c function *divcon* as follows:

> *divcon simple solve divide combine data =*
> **if** *simple data* **then** *solve data* **else** *combine result1 result2* **fi**
> > **where**
> > *result1 = divcon simple solve divide combine data1;*
> > *result2 = divcon simple solve divide combine data2;*
> > *(data1, data2) = divide data;*

The function *divcon* takes five arguments: the first three are unary functions (*simple, solve* and *divide*); the fourth is a binary function (*combine*); and the last (*data*) is a data object of arbitrary complexity. This last argument is the data for the particular problem to be solved using the d-c method. The first four arguments effectively define the particular d-c algorithm to be used, and correspond to the four functions with the same names introduced in the procedural form of d-c in Section 2.1.2.

The parallel implementation of *divcon* in a functional language is not fundamentally different from its parallel implementation in a procedural language. Both rely on the property that the two sub-problems can be solved independently and in parallel (*result1* and *result2* in the above program).

2.4   Lists with Divide-and-Conquer

The basic data structure used in most functional languages is the list, following the lead set by Lisp in the 1960s (McCarthy 1978a, 1978b). The conventional definition of a list makes use of the primitive functions *head, tail, cons* and *null*[3] and is well-suited to a particular style of programming which is illustrated by the following programs taken from the Haskell Report (Hudak *et al.* 1991) (with syntax changes to remove the pattern matching) which define: *s++t*, the

---

[3]*head s* is the first element of the list *s*; *tail s* is the list obtained by removing the first element of *s*; *cons x s* (more commonly written *x : s*) is the new list obtained by adding the item *x* to the beginning of *s* (so *x* is the head of the new list); *null s* is true if the list *s* is empty and false otherwise.

list formed by concatenating two lists *s* and *t*; and *s!!i*, the *i*-th element of the list *s*:

$$s ++ t = \textbf{if } null\ s \textbf{ then } t \textbf{ else } head\ s : (tail\ s ++ t) \textbf{ fi}$$

$$s\ !!\ i = \textbf{if } i = 0 \textbf{ then } head\ s \textbf{ else } tail\ s\ !!\ (i-1) \textbf{ fi}$$

This style of programming in which operations on a list are defined recursively by operating on the head of the list and recursively applying the definition to the tail of the list does not contain any obvious parallelism. Furthermore, the primitive functions *head, tail, cons* and *null* are not well suited to use with the d-c paradigm. An alternative set of primitive functions for lists has been proposed (Axford and Joy 1991) which is well suited to use with d-c algorithms and easily parallelisable. These primitives are described in the next section.

2.4.1 A Divide-and-Conquer Model of Lists

The model contains six primitives (the first is a constant, the rest are functions): (i) *[]* is the empty list; (ii) *singleton x* (usually written as *[x]*) is the list containing the single element *x*; (iii) *s++t* is the list formed by concatenating lists *s* and *t*; (iv) *split s* is a pair of lists obtained by partitioning the list *s* into two non-empty parts (*s* must contain at least two elements); (v) *#s* is the number of elements in the list *s*; and (vi) *element s* is the only element in the list *s* (defined only if *s* contains exactly one element).

The algebraic specification of this model is:

$$\#[] = 0$$

$$\#[x] = 1$$

$$\#(s ++ t) = \#s + \#t$$

$$element\ [x] = x$$

$$s ++ [] = s$$

$$s ++ (t ++ u) = (s ++ t) ++ u$$

$$split\ ([x] ++ [y]) = ([x],[y])$$

$\#u \geq 2$ **and** $split\ u = (s,t)$ **implies:**
$$s ++ t = u,\ \#s \geq 1,\ \#t \geq 1$$

In this model of lists, the primitive operation for building up lists is concatenation (instead of *cons* in the traditional model), and the primitive function for breaking down lists is *split* (instead of *tail*).

2.4.2 Representation in the Computer

Suppose that a list is represented as a binary tree. Elements of the list are stored in the leaves of the tree, each leaf node containing exactly one element. No elements are stored in branch nodes, but each branch node contains the size of the list and two pointers: the left one points to the first part of the list, while the right points to the second part of the list. Ideally these two parts of the list should be approximately equal in length (i.e. the tree should be balanced), but that is not a requirement for correctness of the representation, it affects only the performance.

The representations of the empty list (*[ ]*), a singleton list (*[a]*), and a list of two elements (*[a,b]*) are, respectively:

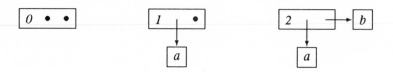

The symbol • denotes a nil pointer and occurs only in lists containing either no elements or just a single element.

Two alternative representations of the list *[a,b,c]* are:

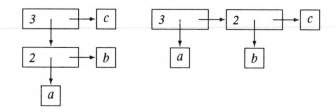

For lists of three or more elements, the representation is not unique, but several different structures are possible. All are equally valid and will give exactly the same results, although the performance of a program may depend upon how well-balanced the representation is.

The above representation allows straightforward and obvious implementations of all five primitive functions. All execute in constant time, irrespective of the lengths of the lists involved. None of these primitives offers any scope for parallelism, however. That comes with their use in d-c programs and, in particular, their use in the implementation of higher-level list-processing functions.

This representation of lists is somewhat less efficient than the traditional representation in typical circumstances on sequential von Neumann architectures. The binary tree structure uses more pointers than a simple chain (up to twice as many) and this requires more memory space and increases the execution time for typical list-processing operations. This loss of efficiency (of up to a factor of two in both memory space and execution time) in the sequential implementation is insignificant in comparison to the gains that can be made with parallel implementation.

2.4.3 Some Examples of Use

Some Higher-Level List Functions

Programs involving lists rarely need to use the primitive functions directly. Most things can be done much more easily using general high-level list-processing functions. One such function that is very widely applicable is *map*, defined as:

$$map\ f\ [x1,...,xn]\ =\ [f\ x1,\ ...,\ f\ xn]$$

This function may be programmed in the d-c style as:

map f s =
    if *#s=1* **then** *[f(element s)]* **else** *map f s1* **++** *map f s2* **fi**
        where *(s1,s2)* = *split s;*

or, using the function *divcon* explicitly:

map f s = *divcon single solve split (++) s* **where**
        single s = *(#s=1);*
        solve s = *[f(element s)]*

Another very useful function is *reduce* which has the following definition, in which $\oplus$ denotes any infix operator:

$$reduce\ (\oplus)\ [x1,...,xn]\ =\ x1 \oplus x2 \oplus ... \oplus xn$$

The operator $\oplus$ must be associative for the result to be uniquely defined. If the list is empty, the result is undefined. If the list contains only one element, the result is that element.

The function *reduce* may be programmed using d-c:

reduce f s =
    if *#s=1* **then** *element s* **else** *f(reduce f s1)(reduce f s2)* **fi**
        where *(s1,s2)* = *split s;*

A third very useful high-level function is *filter*, which finds the sublist obtained by removing all elements which do not satisfy a given condition. It can be programmed using d-c (this time using *divcon* explicitly):

filter p s = *divcon single solve split (++) s* **where**
        single s = *(#s=1);*
        solve s = **if** *p(element s)* **then** *s* **else** *[]* **fi**

All three functions, *map, reduce* and *filter*, have sequential execution times of $O(n)$ provided that the list primitives execute in constant time. In a parallel implementation on a shared memory architecture, if the d-c parallelism is fully used, all three have execution times of $O(\log n)$, requiring $O(n)$ processors.

Many other common functions of lists have similar performance characteristics, with similarly large parallel speedup. In practice, if n is much larger than the number of available processors p, then p will limit the parallelism rather than n, and the speedup will be close to the ideal (i.e. close to p).

Other Applications

Instead of programming in terms of the primitive list-processing functions directly, it is much better to use high-level functions such as *map, reduce* and *filter*. All of these functions can be implemented using either the traditional *head, tail* and *cons* primitives or the new primitives, or any other suitable set of primitives for that matter. Furthermore, they may be implemented on many different computer architectures, using many different types of list representation. Programs that are written in terms of these high-level functions alone are thus immediately portable across all these different methods of implementation, both serial and parallel.

A good example of a problem that can be easily programmed purely in terms of high-level list functions is the ray tracing problem described next.

Example: Ray Tracing

Suppose we have a list of objects and a list of rays (representing physical objects and light rays in three dimensional space). The problem is to compute the first impact of each ray on an object.

Each ray is represented as a starting point and a direction. An impact with an object is simply represented as a distance, which is the distance the ray travels from its starting point before it hits the object. The first impact for a given ray is then the impact represented by the minimum distance. We do not go into the details of how objects are represented, or how the point of impact of a ray with an object is computed, but concentrate solely on the overall structure of the program.

One possible solution is as follows, expressed as a program for the function *impacts* which takes two arguments, a list of rays and a list of objects, and gives a list of impacts as its result (this list is in the same order as the input list of rays):

> *impacts rays objects* = *map firstimpact rays* **where**
> $\qquad$ *firstimpact ray* = *reduce min (map impact objects)*
> $\qquad\qquad$ **where**
> $\qquad\qquad$ *impact object* = <distance travelled by *ray*
> $\qquad\qquad\qquad\qquad\qquad\qquad\qquad$ to hit *object*>;
> $\qquad\qquad$ *min x y* = **if** $x \le y$ **then** *x* **else** *y* **fi**

The program uses *map* to apply the function *firstimpact* to each separate ray in the list; *firstimpact* being a function which takes a single ray as its argument and finds the first impact of that ray with any of the objects. That is done by first using *map* to apply the function *impact* to each separate object. The function *impact* takes a single object as its argument and finds the impact of the current ray with that object. When the list of impacts of one ray with all the objects has been found, the function *reduce min* is used to find the first impact of that ray.

The outermost structure of this program is an application of *map* to the list of rays. This will take $O(\log n)$ time for a fully parallel implementation (on

$O(n)$ processors), where n is the number of rays. The computation of the first impact of a single ray is an application of *map* to all the objects, followed by an application of *reduce* to all the objects, both of which take $O(\log m)$ time in a fully parallel implementation (on $O(n)$ processors), where m is the number of objects. Thus the execution time of the complete program is $O(\log m + \log n)$ for a fully parallel implementation (on $O(m \times n)$ processors).

2.5 Arrays with Divide-and-Conquer

The common use of arrays in programming languages typically provides very little to support parallelism. The primitive operations on arrays are accesses to single elements of the array, the element being identified by its index. For d-c style programming, operations are required which can partition arrays and put them back together again. Such operations should ideally be defined as primitive operations, so that the implementor is free to implement them in whatever manner is most efficient.

The programming language Divacon adopts just such an approach, using a functional language as the framework.

2.5.1 Divacon

Divacon (Mou 1990) is a parallel functional programming language which has been designed particularly for architectures with parallel processor networks in the form of a hypercube or closely related topologies such as the butterfly, perfect shuffle or cube-connected-cycles. Divacon has been implemented in prototype form on the Connection Machine at Yale.

Overview of the Language

Divacon is a functional programming language which is conventional in its basic data types (integer, float, character and boolean) and operators ($+, -, *, \div, >$, $<$, etc.). The main data structures are tuples and arrays. Tuples are used statically (there are no operators to change the size of tuples), while arrays are dynamic and can be divided into smaller arrays or combined to form larger arrays (these are the operations needed for d-c programming). In fact, Divacon tuples are more like conventional arrays in many respects (except that their syntax is not at all the same), while Divacon arrays are really much more sophisticated data structures than normal arrays, having primitives which operate on the array as a whole, unlike the primitives for normal arrays which operate on single elements only.

In addition to the operations required by d-c algorithms for sub-dividing and re-combining arrays, some 'communication' operations are also provided on arrays. These permute the elements of an array according to a specified rule (defined in terms of the array indices only, not the values of the elements).

For example, a simple rule would be to move the i-th element to the $(i+1)$-th position (modulo the size of the array).

An analog of the function *map* on lists (as defined in Section 2.4.3) is defined as a primitive function on both tuples and arrays. In Divacon this function is called 'distribution' and is denoted by the symbol *!* (used as a prefix operator).

The other main feature of Divacon is the divide-and-conquer function itself. Three forms are provided, all of which are variations on the general d-c paradigm.

It would occupy too much space to attempt a full description of Divacon here. The basic data types and operators, and many other basic features of the language, are much like those in many other programming languages. The important new features of the language are its treatment of arrays (effectively a new type of data structure) and the d-c functions.

In the description which follows, many details of the syntax of Divacon are glossed over as they are not relevant to the present discussion. In some examples, the syntax has been changed to bring it more into line with our earlier examples and to avoid the need to describe the syntax of Divacon in detail.

Arrays in Divacon

An array is a function $A : I \rightarrow V$, where I is a set of integer tuples called its index set, and V its indexed set. Each index (i.e. each element of I) has the form $(i_0, ..., i_{m-1})$ where $0 \leq i_j < S_j$ for all $0 \leq j \leq m - 1$. This defines an m-dimensional array of size S_j in the j-th dimension. Primitive functions are provided in Divacon to give the size and shape of any array (hence requiring that this information be stored as part of the array).

Divide Functions   Several primitive functions are provided to partition arrays into smaller arrays. The binary divide function d_b m divides a vector (i.e. a one-dimensional array) into two subvectors, the first consisting of the first m elements of the original, and the second consisting of the remainder. The left-right divide function d_{lr} is a special case of this for which $m = n/2$, where n is the number of elements in the vector. The even-odd divide function d_{eo} divides a vector into two subvectors, the first consisting of the even-numbered elements, while the second consists of the odd-numbered elements (the order of the elements otherwise remains unchanged).

For example,

$$d_b \ 4 \ [a \ b \ c \ d \ e] = ([a \ b \ c \ d], \ [e \ f])$$
$$d_{lr} \ [a \ b \ c \ d \ e \ f] = ([a \ b \ c], \ [d \ e \ f])$$
$$d_{eo} \ [a \ b \ c \ d \ e \ f] = ([a \ c \ e], \ [b \ d \ f])$$

Similar operators are defined for multi-dimensional arrays also. These operators to sub-divide arrays into smaller arrays are fundamental to the use of d-c algorithms on data which is represented in the form of arrays. All of these division operators are 'polymorphic' in the sense that the way in which the array is sub-divided is determined purely by its size and shape, never on its contents.

Combine Functions   Primitive combine functions are the inverse of the primitive divide functions. The combine functions c_{lr} and c_{eo} are defined to be the inverses of d_{lr} and d_{eo}, respectively.

So, for example,

$$c_{lr} \ ([a \ b \ c], \ [d \ e \ f]) = [a \ b \ c \ d \ e \ f]$$
$$c_{eo} \ ([a \ b \ c], \ [d \ e \ f]) = [a \ d \ b \ e \ c \ f]$$

Similar operators are defined for multi-dimensional arrays also. All the combine operators are polymorphic in the same sense as the divide operators.

Distribution Operator   The distribution operator $!$ is equivalent to the function *map* introduced in Section 2.4.3. So, for a function f, the new function $!f$ applies f separately to each element of an array (or tuple).

For example,

$$!+ \ [(3,5) \ (1,4) \ (2,2)] = [8 \ 5 \ 4]$$
$$!* \ [(3,5) \ (1,4) \ (2,2)] = [15 \ 4 \ 4]$$

(Notice that Divacon regards $+$, $*$, etc. as prefix operators without enclosing them in brackets, and the argument of each is a single pair of numbers: they are not curried; e.g. $+(3,4)=7$.)

As with most of the Divacon primitives, the distribution operator can be used with multi-dimensional arrays also.

Communications Operator   This operator (denoted $\#$) permutes the elements of an array. It is applied to a function which specifies the particular permutation required. If f is a function, and $[a_1...a_n]$ is an array, then $\#f \ [a_1...a_n] = [(a_1, a_f \ 1)...(a_n, a_f \ n)]$. This definition permits not only permutations of the array, but also the broadcasting of a single element into all positions in the array, etc.

For example, if *const m x* $= m$, and *mirr i* $= -i$, (interpreted modulo the size of the array),

$$\#(const \ 0)[3 \ 5 \ 7 \ 9 \ 11] = [(3,3) \ (5,3) \ (7,3) \ (9,3) \ (11,3)]$$
$$\#mirr \ [3 \ 5 \ 7 \ 9 \ 11] = [(3,11) \ (5,9) \ (7,7) \ (9,5) \ (11,3)]$$

and so $\#(const \ m)$ broadcasts element m to all positions in the array, while $\#mirr$ defines a mirror image reflection about the centre of the array.

Divide-and-Conquer Functions

Divacon defines three versions of the general d-c paradigm.

The function *Naive-PDC* can be defined as follows:

> *Naive-PDC (divide, combine, simple, solve) = f*
> **where** *f data* = **if** *simple data* **then** *solve data*
> **else** *(combine . !f . divide) data* **fi**

(where . denotes function composition). The arguments to *Naive-PDC* are in a different order to those used in the d-c functions introduced earlier in this chapter, but are otherwise very similar except that the function *divide* partitions the problem not just into two sub-problems, but into an array (of arbitrary size) of sub-problems, and *combine* combines an array of answers to those sub-problems.

Another d-c function ('parallel d-c') is defined as follows:

> *PDC (divide, combine, pre, post, simple, solve) = f*
> **where** *f data* = **if** *simple data* **then** *solve data*
> **else** *(combine . post . !f . pre) data* **fi**

The functions *pre* and *post* have been included to provide pre-adjustment and post-adjustment of the data within the d-c construct.

A third form of the general d-c paradigm is provided. It imposes a linear order on the recursive operations and hence is called sequential divide-and-conquer:

> *SDC (divide, combine, $(\mu_0, ..., \mu_{k-2})$, simple, solve) = f*
> **where** *f data* = **if** *simple data* **then** *solve data*
> **else** *(combine . h . divide) data* **fi**
> **where** $h\ (x_0,...,x_{k-1}) = (y_0,...,y_{k-1})$
> **where**
> $y_0 = f\,x_0;$
> $y_1 = (f . \mu_0)\ (x_1,y_0);$
> . . .
> $y_{k-1} = (f . \mu_{k-2})\ (x_{k-1},y_{k-2})$

The function *divide* partitions the problem into *k* parts, and the functions $\mu_0, ..., \mu_{k-2}$ define the way in which the solutions to these *k* parts interact.

This sequential d-c function is more general than the d-c algorithms previously discussed, but it has the disadvantage of requiring the sub-problems to be solved in a specified order (the solution to each must be available before the next in the sequence can be solved), hence no structural parallelism is possible. The only possible parallelism is within the separate sub-problems (or by transforming the program into one of the more restricted forms of d-c).

Implementation of Divacon

A prototype implementation of Divacon has been constructed in *Lisp on the Connection Machine at Yale University. The implementation uses both data parallelism and control parallelism. Arrays are distributed over the available processors and hence represented in a highly parallel form. Operations on arrays, such as their creation, the divide, combine and index translation operations can then be performed in parallel. All these operations take constant time provided the size of the array does not exceed the number of processors.

The parallel d-c functions are also implemented in parallel, and provide the only source of control parallelism.

It is non-trivial to implement Divacon in *Lisp, which is essentially a data-parallel language and provides only flat parallel data structures. It is not entirely clear how much the implementation has constrained the performance and design of Divacon, and whether or not its limitations are due to the implementation language or to the underlying machine architecture. Later implementations have been completed for the MasPar MP–1 (2D mesh architecture) and WaveTracer DTC (3D mesh), both massively parallel machines using bit-processors.

Programming in Divacon

The d-c model provided by Divacon is both powerful and convenient. A very wide range of applications can be readily programmed with the constructs provided giving programs which execute in parallel with high performance. Some examples of such problems, which have all been programmed in Divacon, are polynomial evaluation, matrix multiplication, monotonic sort, FFT and banded linear systems.

The divide and combine functions for arrays in Divacon are all polymorphic in the sense defined earlier, i.e. the subdivision of an array is independent of the values of the elements of the array. This means that only 'polymorphic' d-c algorithms can be programmed easily. This rules out an important class of d-c algorithms which use non-polymorphic divide and combine operations. A simple example is the divide operation in quicksort, in which the array is split into two parts, one part being all elements less than a guessed median value, and the other part being the remainder. Clearly, this operation depends not on the indices of the elements, but on their values.

2.6 Sets and Mappings with Divide-and-Conquer

The basic concepts of set theory mathematics are widely used in computer science, yet relatively few programming languages have incorporated sets as basic data structures. The best known procedural language based on sets is SETL, which has been in use for some two decades and has had some success as a very high-level language for general software development and prototyping (Schwartz *et al.* 1986). It has remained very much a minor language, however, and does not seriously compete with the major programming languages.

The use of set theory in program specification is much more universal. The major formal specification methods, such as Z (Spivey 1989) and VDM (Jones 1986), are based on set theory mathematics, and use sets and relations as basic data structures out of which everything else is constructed. Experience with these methods has shown that set theory provides a powerful and convenient way in which the programmer can express the organisation of his data. Recently there has been increasing interest in representing sets and relations directly in programming languages so that they can be used as tools for further software development (North 1990; Treadway 1990; Marino and Succi 1991).

The common programming languages have not found it appropriate to include sets and relations as basic data structures because they are rather more difficult and somewhat less efficient to implement than arrays or lists, and their conceptual advantages were felt to be less important than maximising performance. Furthermore, why change something that has worked well for years?

The balance of advantage *is* changing, however. Not only are sets and relations mathematically more powerful and elegant than arrays and lists, but they probably offer easier and more efficient parallel implementations. In the rest of this section, we introduce the idea of using mappings as basic data structures and some primitive functions for operating on such structures using the d-c style of programming (Axford 1991).

For simplicity, we choose to define a single type of data structure only (although it is quite possible to provide other types of structures as well, or instead). The chosen structure is called a 'mapping' and corresponds to a finite partial function in set theory mathematics. It is a convenient choice because it is reasonably easy to implement efficiently on most computer architectures, whether serial or parallel, and it is convenient to use in programming. Sets need not be provided separately as they are very easily represented in terms of mappings. Sequences and arrays are simply special cases of mappings. Relations are more general, but they are also more difficult to implement efficiently, so they appear to be a less attractive choice as the basic type of structure. This is an area in which there is rather limited experience, however, particularly of parallel implementations for general-purpose use, and a great deal more research is needed to find optimum solutions.

2.6.1　Primitive Functions of Mappings

Any abstract data structure can be defined by giving an algebraic specification of the properties of the primitive functions which operate on it. In choosing a suitable set of primitive functions, mathematicians will usually give priority to simplicity and elegance in the specification itself. Here, we are more concerned with choosing a set of primitive functions that are convenient for the programmer and can be efficiently implemented on both serial and parallel architectures. The set of primitives is not intended to be the simplest or most elegant from a mathematical viewpoint. Furthermore, a complete formal specification is rather long and tedious, and not necessary for our purposes here, so the primitives will be defined more informally for brevity and ease of understanding.

A mapping may be thought of as a data structure which represents a function of a single argument over a finite domain. The representation stores the function values for all possible argument values. We will refer to the argument value as the *key* and to the corresponding function value as its *attribute*. The set of keys is the domain of the function, each key being 'mapped' to its attribute.

The following functions are considered as primitives:

CreateMap	*MapUnion*	*ForAll*
IntegerSequence	*MapSize*	*MapMap*
ApplyMap	*SubMap*	*ReduceMap*
MapIntersection	*IsIn*	*ReduceList*
MapDifference	*ForOne*	

Each is now defined in turn.

CreateMap

This takes a single argument which is a tuple[4] of pairs. The pairs are the key-attribute pairs which define the mapping. If any duplicate keys are present in the tuple, then the result is undefined. The order in which the pairs occur is irrelevant. For example:

> *CreateMap ((3,6),(1,4),(0,1))*
> *CreateMap ((1,4),(0,1),(3,6))*

each represent the mapping $\{0\mapsto1,\ 1\mapsto4,\ 3\mapsto6\}$.

We will usually write $\{(x1,y1),...,(xn,yn)\}$ as a shorthand for *CreateMap ((x1,y1),...,(xn,yn))*.

IntegerSequence

This function takes two arguments, both integers, and creates the mapping whose domain is the sequence of integers from the first argument to the second. The attribute values created are the same as the keys. The infix operator form *m..n* will normally be used in preference to the prefix function *(IntegerSequence m n)*. So, for example:

> *3..7*   has the value $\{(3,3),(4,4),(5,5),(6,6),(7,7)\}$

ApplyMap

This operation applies the mapping as if it were a function, i.e. it maps a given key to its attribute. It takes two arguments, the first is a mapping and the second is the key value to be mapped. Thus *(ApplyMap s k)* is the attribute associated with the key *k* in the mapping *s*. For example:

> *ApplyMap $\{(3,6),(1,4),(0,1)\}$ 1*   has the value *4*.

---

[4]We assume that tuple structures already exist in whatever programming language we are using. If not, it is quite straightforward, if more tedious and less efficient, to build up mappings of any size by adding elements one at a time.

MapIntersection

This function takes the set intersection of the domains of two mappings as the domain of the resultant mapping. The attribute of a given key in the result is a function of the attributes of that key in the two argument mappings. The first argument is this function, the second and third arguments are the two mappings. So, *(MapIntersection f s t)* is the mapping in which every key, x, that is present in this mapping is also a key in s and t; and if *(x,y)* is in s and *(x,z)* is in t, then *(x, f y z)* is in *(MapIntersection f s t)*. For example:

$\quad$ *MapIntersection (*) {(1,7),(2,45),(3,2)} {(0,3),(1,2),(3,3)}*

has the value *{(1,14),(3,6)}*.

MapDifference

This function takes two mappings as its arguments. The difference of the domains of the two mappings is the domain of the result. The attributes remain unchanged. In other words, *(MapDifference s t)* is the mapping which contains all the elements of s for which the keys differ from all keys in t. For example:

$\quad$ *MapDifference {(1,7),(2,45),(3,2),(4,7)} {(0,3),(1,2),(3,3),(4,4)}*

has the value *{(2,45)}*.

MapUnion

This function takes the set union of the domains of two mappings as the domain of the resultant mapping. The attribute of a given key in the result is the attribute of that key in the argument in which it occurs (if it occurs in only one), otherwise it is a function of the two attributes if that key occurs in both. MapUnion is applied to three arguments: the first is a binary function, the second and third are mappings. For example:

$\quad$ *MapUnion (*) {(1,7),(2,45),(3,2),(4,7)} {(0,3),(1,2),(3,3),(4,4)}*

has the value *{(0,3),(1,14),(2,45),(3,6),(4,28)}*.

MapSize

The size of a mapping is the number of elements it contains. For example:

$\quad$ *MapSize {(1,7),(2,45),(3,24),(7,17)}* $\qquad$ has the value *4*.

SubMap

This function gives the sub-mapping that consists of all elements of a given mapping whose keys satisfy a given condition. It takes two arguments, the first is a unary function (the condition), while the second is the mapping. For example:

SubMap (x→x>2) {(1,7),(2,45),(3,24),(4,17)}

has the value *{(3,24),(4,17)}*, i.e. all elements with keys greater than 2.

IsIn

This function tests if a given value is present in the domain of a given mapping. It takes two arguments, the first is the key value to be tested for, the second is the mapping in which the key is to be looked for. The result is a boolean. For example:

IsIn 3 {(1,7),(2,45),(3,24),(4,17)}

has the value *true* because the mapping contains the key *3*.

ForOne

This function tests whether or not a given condition is true for at least one member of the domain of the mapping. It takes two arguments: the first is a monadic function whose result is a boolean and the second is a mapping; the result is a boolean.

ForOne (x→x>40) {(1,7),(2,45),(3,24),(4,17)}

has the value *false* because there is no key greater than *40*.

ForAll

This is the similar function which tests whether or not the condition is true for all values in the domain of the mapping. For example:

ForAll (x→x>0) {(1,7),(2,45),(3,24),(4,17)}

has the value *true* because every key in the mapping is greater than zero.

MapMap

This function performs the same operation on every element in a mapping. So, *(MapMap f s)* is the mapping in which every element of the form *(x,y)* in *s* becomes *(x, f x y)* in the result. The first argument is a binary function, the second is a mapping and the result is a mapping. For example:

*MapMap (x→y→2*y) {(1,7),(2,45),(3,24),(4,17)}*

has the value *{(1,14),(2,90),(3,48),(4,34)}*, i.e. every attribute is doubled.

ReduceMap

This function reduces a mapping to a single value by applying a binary function repeatedly to all the elements. It takes four arguments: the first and second are binary functions, the third is any type of object, and the fourth is a mapping. *(ReduceMap f g z s)* is defined to have the value z if s is empty; it has the value *(g x y)* if s contains only one element $s=\{(x,y)\}$; and in all other cases (i.e. two or more elements in s) it is defined recursively to be

f (ReduceMap f g z s1) (ReduceMap f g z s2)

where $s1$ and $s2$ are any two parts of the mapping s, such that both of the following equations are true (where f is any function):

MapIntersection f s1 s2 $= \{\}$
MapUnion f s1 s2 $= s$

The first argument, f, is not significant in either case.

Proof Obligations   *ReduceMap* is a powerful and useful function, but it has the rather dangerous property of being potentially non-deterministic, i.e. its definition is sometimes ambiguous, and when evaluated in different ways it can lead to different results. To avoid non-determinism (i.e. to ensure that its value is always uniquely defined), the function *(ReduceMap f g z s)* should only be used when f is commutative and associative, and when z is an identity element of f, i.e. when all three of the following equalities hold:

f a b = f b a
f (f a b) c = f a (f b c)
f z a = a

for all a, b and c. These must be regarded as proof obligations on the programmer whenever he or she uses *ReduceMap*. If these properties are not satisfied by the arguments of *ReduceMap*, then the results may be non-deterministic. Non-deterministic functions are dangerous because they are not referentially transparent and hence program analysis is made very much more difficult and counter-intuitive. For example, we can no longer make the apparently obvious statement that:

ReduceMap f g z s = ReduceMap f g z s

simply because *ReduceMap* may give different values on different occasions even though all its arguments remain the same.

Another way to avoid this problem is to tighten up the specification by defining precisely the order of evaluation. In the functional language Haskell, the function *foldr* is similar to *ReduceMap* except that the order of evaluation is precisely defined. This is certainly a simple solution which avoids the problem of non-determinism, but it also makes general parallel implementation very much more difficult and less efficient, so we rule it out on those grounds. The definition of *ReduceMap* given above allows the implementor a great deal of freedom to choose the most efficient order of evaluation, at the expense of putting an obligation on the programmer to make sure that he or she uses *ReduceMap* only when its arguments have the required properties. If the programmer chooses to ignore this obligation, then he has only himself to blame if either (i) some implementations of his program give incorrect results, or (ii) mathematical analysis of the program is misleading.

With the increasing trend towards more formal analysis and verification of programs, and with the increasing availability of software tools to aid the programmer at this task, imposing such proof obligations on the programmer is likely to become much more acceptable in the future. The great majority of programmers today would probably regard any such proof obligations as thoroughly undesirable, but these attitudes are slowly changing. Already, it is widely recognised that formal methods of program analysis and proof are becoming essential in some specialised fields of programming such as concurrency and safety-critical systems. The wider use of formal methods will inevitably follow.

ReduceList

Suppose it is required to reduce a mapping with a function that is non-commutative. In this case it is necessary to define the order in which the elements are enumerated, although the order in which the reduction operations are carried out may still be left undefined provided the function is associative. Suppose that the data is in the form of a list and the order of elements in the list is specified to be the order of enumeration for the reduction.

In this context, a list *[x1,...,xn]* is taken to be simply a shorthand for the mapping $\{(1,x1),...,(n,xn)\}$ with the domain *1,...,n*. The function *(ReduceList f g z s)* is defined to be equal to *z* if *s* is empty; to *(g x)* if *s* is the list *[x]* (i.e. the list containing the single element *x*); and in all other cases it is defined recursively as:

$$f \, (ReduceList \, f \, g \, z \, s1) \, (ReduceList \, f \, g \, z \, s2)$$

where *s1* and *s2* are any two non-empty lists such that *s* is equal to the concatenation of *s1* and *s2* (in that order).

Proof Obligations The requirements on the arguments of *ReduceList* for the result to be deterministic are less stringent than before. They are simply that *f* be associative:

$$f \, (f \, a \, b) \, c \, = \, f \, a \, (f \, b \, c)$$

2.6.2 Some Applications

Ray Tracing

Consider the ray tracing problem introduced earlier in Section 2.4.3. The data consists of a set of objects and a set of rays (e.g. representing light rays) and we wish to find the first impact of each ray on an object. Suppose the data is supplied in the form of two mappings. In both cases, the domain of the mapping (i.e. the set of keys) is unimportant and may simply be the integers from 1 to n. In the first mapping (called *rays*) the attributes are the rays, while in the second mapping (called *objects*) the attributes are the objects. The required result is another mapping over the same domain as *rays*, but with the attributes being the first impacts of the corresponding rays. An impact is represented by the distance along the ray to that impact (a ray is assumed to have a starting point, and the distance may be infinite if there is no impact). We do not concern ourselves with the details of the representation of the individual rays and objects.

Using a similar d-c approach to that adopted in Section 2.4.3, we first subdivide the problem by sub-dividing the set of rays, and then in the computation of the first impact of a given ray, we sub-divide further by sub-dividing the set of objects. A function to do this can be programmed as follows:

> *impacts rays objects* = *MapMap firstimpact rays* **where**
>   *firstimpact key ray* = *mindistance* **where**
>   *mindistance* = *ReduceMap min impact infinity objects;*
>   *impact key object* = <distance travelled
>   by *ray* to hit *object*>;
>   *min x y* = **if** $x \leq y$ **then** *x* **else** *y* **fi**;
>   *infinity* = <value to represent no impact>;

This program uses *MapMap* to apply the function *firstimpact* to every ray (a parallel implementation could do these computations concurrently). The definition of *firstimpact* then uses *ReduceMap* to apply the function *impact* to every object to find the impact of the current ray with one object, and to reduce these with the function *min* to obtain the first impact (the first impact being the impact at the minimum distance). Note that the definition of *impact* defines *object* to be an argument (along with *key*, which is not used), but *ray* is a free variable (i.e. it is accessed directly as an outer-block variable rather than passed to the function as an argument).

An alternative style of program divides the objects first:

> *impacts rays objects* =
>   *ReduceMap (MapUnion min) getimpacts* {} *objects*
>   **where**
>   *min x y* = **if** $x \leq y$ **then** *x* **else** *y* **fi**;
>   *getimpacts key object* = *MapMap findimpact rays*
>   **where**
>   *findimpact key ray* = <distance travelled
>   by *ray* to hit *object*>;

The structure of this program is rather different. Firstly, *ReduceMap* is used to apply the function *getimpacts* to each object separately (a parallel implementation could do these concurrently). The function *getimpacts* computes the list of impacts of all the rays with a single object (the argument). This is done by using *MapMap* to apply *findimpact* to each ray separately, where *findimpact* is very similar to *impact* in the previous program, but its second argument this time is *ray*, and *object* is a free variable. It computes the impact of a single ray on a single object. Finally, the function *ReduceMap* (in the first line of the program) combines all the sets of impacts, using *(MapUnion min)*, which picks out the first impact for each ray.

Both solutions offer a great deal of parallelism if the primitive functions *MapMap* and *ReduceMap* can be implemented in parallel. Such parallel implementation is discussed further in Section 2.6.3.

In both programs, the use of *ReduceMap* gives rise to proof obligations. It is not difficult to show that the arguments of *ReduceMap* have the required properties so the results are uniquely determined.

Hoare's Quicksort Algorithm

Hoare's 'quicksort' algorithm can be programmed using the general d-c function as shown below. For simplicity we assume that the file contains at least one element and that all elements are different. The input data is assumed to be in the form of a list represented as a mapping with the domain *1,...,n*.

The program below uses the function *divcon*. The final argument of *divcon* is a triple *(m,n,s)* and the program sorts the segment of the sequence *s* between index values *m* and *n*.

```
sort s  =  divcon simple solve divide (MapUnion f)
                                        (1, MapSize s, s) where
        simple (m,n,s)  =  (m=n);
        solve (m,n,s)  =  s;
        divide (m,n,s)  =  ((m,p,s1),(p+1,n,s2)) where
              median  =  if first<second then first else second fi;
              first  =  ApplyMap s m;
              second  =  ApplyMap s (m+1);
              s1  =  belowmedian m n;
              s2  =  abovemedian m n;
              belowmedian m n  =
                   if m>n then {}
                   elsf y≤median then MapUnion f {(m,y)} rest
                   else rest fi where
                        rest  =  belowmedian (m+1) n;
                        y  =  ApplyMap s m;
              abovemedian m n  =
                   if m>n then {}
                   elsf y>x then MapUnion f {(n,y)} rest
```

> **else** *rest* **fi where**
>   *rest = abovemedian m (n–1);*
>   *y = ApplyMap s n;*
> *p = m + MapSize s1 – 1;*

In this program, the expressions *(belowmedian m n)* and *(abovemedian m n)* give sequences of all the elements of the segment of *s* from *m* to *n* which are below or above the median, respectively. The former sequence begins at index *m*, while the latter sequence ends at index *n*. The estimation of a median value is very crude, being simply the smaller of the first two items in the file. (It is easy to improve this.) The combine phase of the d-c algorithm uses *(MapUnion f)* to rebuild the file (*f* is irrelevant).

For example, if the initial data is {*(1,4),(2,3),(3,7),(4,2),(5,6)*} (representing the list *[4,3,7,2,6]*) then the value used for *median* will be *3*, and the subfiles *s1* and *s2* will be computed to be {*(1,3),(2,2)*} and {*(3,4),(4,7),(5,6)*}, respectively (representing the lists *[3,2]* and *[4,7,6]*). Notice that the definitions of *belowmedian* and *abovemedian* are recursive because the algorithm used to compute them is inherently sequential (there is no obvious way in which the computation of these functions may benefit from parallel processing).

Convex Hull

The convex hull of a set of points in a plane is the smallest enclosing convex polygon. For example, the set of eleven points shown in the diagram has the convex hull indicated:

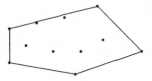

A program for finding the convex hull of a set of points can be written using a divide-and-conquer algorithm, using the fact that it is relatively easy to combine two non-overlapping convex polygons into a single convex polygon which encloses the original two (Preparata & Hong 1977). (This involves much less work than combining two overlapping convex polygons.) We can take advantage of this if we divide the original set of points into non-overlapping subsets. An easy way to do this is to order the points in order of their X-coordinates (and if two points have equal X-coordinates, they are ordered on their Y-coordinates). If this ordered list of points is now divided into sublists, each sublist will have a convex hull which does not overlap the convex hull of any other sublist.

For the example given above, the points can be divided into two sets with convex hulls as shown:

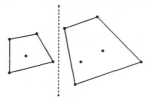

A suitable program to do this is shown below. The data is assumed to be in the form of a list of points (represented as a mapping called *points*) which have already been ordered in the required manner.

> *convexhull points* = *ReduceList f g z points* **where**
>   *f hull1 hull2* = <compute the convex hull
>   which encloses both *hull1* and *hull2*>;
>   *g point* = <the polygon consisting of that one point>;
>   *z* = <the empty polygon>;

We can easily represent a polygon as an ordered list of points (its vertices) and it is easy to write programs for the functions *g* and *z*. The function *f* which combines two convex hulls is harder, but the details do not really matter here, as we have sufficient of the structure of the program to show the potential for parallel implementation.

A more efficient program is likely to result if we combine the sorting and the convex hull computation into a single application of d-c. This is easy to do, using the quicksort algorithm for the sorting:

> *convexhull points* = *polygon* **where**
>   *polygon* = *divcon lengthOne I divide combine points;*
>   *lengthOne s* = *(MapSize s = 1);*
>   *I x = x;*
>   *divide s* = *(s1,s2)* **where**
>   *s1* = *SubMap p1 s;*
>   *p1 x* = *(x≤median);*
>   *s2* = *SubMap p2 s;*
>   *p2 x* = *(x>median);*
>   *median* = <estimate the median value>;
>   *combine p1 p2* = <convex hull enclosing *p1* and *p2*>;

In this single application of *divcon*, the function *divide* effectively does the sorting, while *combine* does the convex hull computation.

The fact that this quite complex problem can be solved by a program which is in the form of a single application of *divcon* (with all the detailed computation captured in the arguments) illustrates well the power of the abstract d-c function. A good compiler could reasonably be expected to obtain quite large-grain parallelism from such a program. In the next section, the problems of implementation are discussed further.

2.6.3 Parallel Implementation of Mappings

The primitive functions of mappings can all be implemented in the divide-and-conquer style if mappings are represented as trees, hash tables, or other representation that allows partitioning into two approximately equal parts, preferably in constant time (i.e. independently of the number of elements in the mapping). This allows efficient parallel implementation on multiprocessor shared memory (i.e. P-RAM) architectures, just as can be done for lists.

Mappings are also well suited to parallel implementation on distributed architectures (without shared memory). In this case, the representation of a mapping may be distributed across the local memories of the processors, a different part of the mapping being stored on each processor. This approach is essentially like a very high-level SIMD architecture. Real SIMD machines available at present are generally SIMD at a low level, with each machine instruction being executed simultaneously by all processors on their own local data. In a distributed mapping implementation, the synchronisation is at a much higher level, with each high-level primitive function being executed simultaneously by all processors. For example, consider the following situation.

Suppose the mapping is distributed across all available processors by using a hashing function of the keys. For P processors, an element with key k is assigned to processor i, where $i=hash\ k$ and the function *hash* gives values in the range $1,...,P$ for all arguments. The primitive functions of mappings can be implemented as follows.

ApplyMap

For the function application *ApplyMap s k*, the only part of the mapping that is required is that containing the key k, which resides on processor number *hash k*. All other processors can remain idle.

The execution time for such a parallel distributed implementation is the sum of the following: (i) the time to compute *hash k* on the master processor; (ii) the time to communicate with processor number *hash k*; (iii) the time to evaluate *ApplyMap s k* on the part of the mapping held on processor number *hash k*; and (iv) the time to return the result to the master processor.

MapIntersection

The function application *MapIntersection f s t* can be implemented by concurrently evaluating *MapIntersection f si ti* on each processor, where *si* and *ti* are those parts of the mappings *s* and *t*, respectively, which reside on processor number *i*.

The execution time is the sum of: (i) the time to broadcast the instruction to all processors; and (ii) the maximum time to compute *MapIntersection f si ti* (the maximum being taken across all values of the processor number, *i*). Assuming that the sequential processing of *MapIntersection* takes $O(n)$ time (where n is the total number of elements in the mappings *s* and *t*), and that the broadcast communication time is constant, then the overall parallel execution time is $O(n/P)$, or constant if P (the number of processors) is $O(n)$, provided that the hashing function distributes the mapping evenly across the processors. There is insufficient practical experience to be able to say at present just how well this method will work in typical computing applications. Some form of dynamic load balancing may well be needed in addition to the distribution of load provided by the hashing function.

The functions *MapDifference* and *MapUnion* can be implemented similarly, with similar execution time behaviour.

MapSize

The function application *MapSize s* can be implemented by evaluating *MapSize si* concurrently on each processor and then summing all the individual results. The time taken is the sum of:

1. the time to broadcast the instruction to all processors;

2. the maximum time to compute *MapSize si*; and

3. the time to sum the results.

We assume that the broadcast time is constant, and that the time to compute *MapSize* sequentially is at most $O(n)$, although it may be constant in many implementations. Either way, for an even distribution of the mapping and $O(n)$ processors, the parallel time for 2. is approximately constant. The time for the final step (summing the results obtained on the individual processors) is $O(\log n)$, which dominates the timing, hence the total parallel execution time of *MapSize* is also $O(\log n)$.

Discussion

The other primitive functions can be implemented in similar ways using d-c algorithms to achieve high parallelism. No programming language has yet been constructed using this approach, so the advantages of mappings as basic data structures for parallel programming are essentially speculative at the present time. Many practical problems remain to be overcome, but the approach looks extremely promising nevertheless.

2.7   Summary and Conclusions

The development of parallel software is floundering (Pancake 1991) because there is no universal, architecture-independent model of programming. The purpose of this chapter has been to suggest that there is a promising case for considering the divide-and-conquer paradigm as a fundamental design principle with which to guide the design of both control structures and data structures for new models of programming. The important characteristic of d-c is that it contains inherent parallelism in a very abstract and general form, and hence is not dependent upon a particular architecture.

Procedural Languages and Side Effects

A d-c control construct for procedural languages can be defined and, although unfamiliar, is not difficult to use. This construct can replace the usual loop constructs (such as **for** and **while** statements) and permits obvious control parallelism. The main problem with such a construct is that side effects must be prohibited for it to be well defined. Very few procedural languages prohibit side effects and hence neither users nor compiler writers are attuned to the side-effect-free style of programming, although it has for long been advocated as good practice in most circumstances. Of course, as freedom from side effects is required only within the d-c construct itself, the rest of the program need not be side-effects free. It would be possible, although unsafe, to put the d-c construct into any normal procedural language and rely solely on the self-discipline of the programmer to avoid side effects.

The toleration of side effects by most procedural languages is itself a serious problem for automatic parallel implementation (as it is an equally serious problem for other automatic optimisations, for the same reasons). The admission of side effects means effectively that an implementation model must be defined as part of the language definition (because without it the side effects are meaningless). This requires all implementations of the language to be consistent with this model, thus severely constraining the implementor. A good principle for any architecture-independent language to follow is to allow as much freedom as possible on methods of implementation.

Functional Languages

Pure functional languages are completely free of side effects, so they avoid one of the main weaknesses of the common procedural languages (although it is true that procedural languages without side effects do exist, but they are not in common use). Functional languages also make it easy to use functions which operate on other functions and generate yet further functions as their results. They provide a very convenient medium in which to define and use d-c functions. A further major advantage for parallel programming is that they do not require the programmer to put all program instructions into sequential order, whether or not the ordering of instructions is logically necessary.

In addition, functional languages provide a very much better medium for program analysis; for example, that required when verifying the proof obligations imposed by the use of non-deterministic functions such as *reduce* and *ReduceMap*.

Nevertheless, it is not essential to use a functional language. Most of the discussion and examples given in this chapter have been in the context of a functional programming language, but there is nothing that could not be done in a similar way in a procedural language, albeit with more difficulty. A functional language gives a number of benefits, not least being greater simplicity and elegance.

Control Parallelism v. Data Parallelism

Many authors have defined d-c procedures or functions, and their inherent control parallelism is well known: whenever the problem is divided into independent sub-problems, the solutions to those sub-problems may be obtained concurrently. The overall efficiency of a d-c algorithm is very dependent upon the efficiency with which the problem can be divided into parts, and the efficiency with which the solutions to the parts can be combined to give the overall solution. Often, large data structures are required to represent the problem data and/or the problem solution. Efficient ways of dividing and combining these data structures are then needed—rather than the efficient single-element operations required for the traditional iterative loop style of programming which we are all so used to.

Hence, the effective use of the control parallelism available in d-c programs often depends very much on having suitable data structures available, ones for which very fast divide and combine operations can be implemented. In this chapter, three classes of data structure have been considered from this point of view: lists, arrays and mappings.

Lists

By choosing a different set of primitive functions to the usual *head, tail* and *cons,* we can define lists in such a way that divide and combine operations are primitives, and everything else is programmed in terms of these. This opens the way to very efficient parallel implementation of many d-c programs using lists on shared memory (PRAM) architectures. As yet, no general programming language has been constructed using this approach, so experience of this style of parallel programming is extremely limited. The method looks very promising, however, particularly for those types of programming languages which have traditionally used lists as their primary data structures, but further investigation is required.

The efficient implementation of lists for d-c programs on distributed (non-shared-memory) architectures is an even more open question, although there appears to be no obvious reason why it should not be possible to find satisfactory representations on these architectures also. More research is needed to determine if this is a useful approach. Of course, lists can easily be represented as 1-D arrays and we can look for suitable ways of defining and representing arrays instead.

Arrays

Arrays can be re-defined to provide divide and combine operations as primitives instead of (or as well as) the usual single-element access operations of conventional arrays. George Mou's work on the language Divacon has shown how this can be done in a way suited to data parallel implementation of the arrays on hypercube and related architectures. Divacon is the first (and so far, only) language whose primary aim is to provide a medium in which programs with a high degree of d-c parallelism can be easily and conveniently written. While the language is quite general-purpose, it is targeted at certain hypercube architectures (in particular, the Connection Machine); and the limitations of the target architecture have somewhat constrained the language design. Nevertheless, Divacon supports efficient programs for the simpler types of d-c algorithms and has demonstrated that these can be programmed simply and conveniently, and then processed fully automatically to achieve very high degrees of parallelism.

Arrays in Divacon are defined with divide and combine primitives, as well as operations to permute the elements in specified ways (dependent only on the array indices, not on the values of the elements) or copy some elements to other positions in the array (again specified only by position, not by value). Such permute and copy operations are needed for the data-parallel representation used in the Connection Machine implementation of Divacon, but would not be necessary if a shared memory implementation was being used instead. On the other hand, Divacon provides no simple and efficient way of dividing (or combining) arrays by content, as needed in programming the quicksort algorithm, for example. This limitation is again due to the need to be able to implement the language easily and efficiently on the Connection Machine.

Sets and Mappings

Probably the most attractive data structures for an architecture-independent programming language are sets and mappings. They are powerful and convenient for the programmer, as evidenced by their very widespread use in specification languages, for which convenience of use is much more important than convenience of implementation. Of course, data structures based on sets and mappings are more complicated and less efficient to implement sequentially than conventional arrays or lists. The situation changes markedly for parallel implementation, however, for which absence of a specified order to the elements of sets and mappings gives a useful additional degree of freedom to the implementation.

Many programs in conventional languages use lists or arrays to represent data for which the order is unspecified and irrelevant to the computation. Imposing an unnecessary order on the elements of such data structures imposes unnecessary constraints on the optimisations which the language processor can carry out (unless it has a very deep knowledge of the mathematical properties of the program, which is typically far beyond the abilities of any normal compiler or interpreter). The use of sets and mappings avoids the need to impose arbitrary order where it is unnecessary. A parallel implementation then has the freedom to handle the elements in whatever order happens to be convenient, which is often much more efficient than having to maintain a strict ordering.

Data structures based on sets and mappings are very much less common in programming languages than arrays and lists (data structures such as Pascal sets are not counted as they are so limited in practice as to be essentially useless for the type of programming advocated here), so there is relatively little experience of their use, even with sequential implementations. No general purpose programming language yet exists which uses sets or mappings as its basic type of data structure and which is suitable for writing programs with a high degree of d-c parallelism. This is an area in which much more research is needed: to find the best set of primitives, to investigate the ease of programming with these primitives and to find efficient implementations of the primitives on a variety of architectures.

References

Aho A. V., Hopcroft J. E. and Ullman J. D. (1974) *The Design and Analysis of Computer Algorithms*. Addison-Wesley.

Axford T. (1991) An Abstract Model for Parallel Programming. *Research Report CSR–91–5*, School of Computer Science, University of Birmingham, Birmingham B15 2TT.

Axford T. and Joy M. (1991) List Processing in Parallel. *Research Report CSR–91–8*, School of Computer Science, University of Birmingham, Birmingham B15 2TT.

Backus J. (1978) Can Programming Be Liberated from the von Neumann Style? A Functional Style and Its Algebra of Programs. *Comm. ACM*, **21**(8), pp. 613–41.

Burton F. W. and Sleep M. R. (1981) Executing Functional Programs on a Virtual Tree of Processors. In *Proc. Conf. Functional Programming Languages and Computer Architecture*, New Hampshire, pp. 187–94.

Cole M. (1989) *Algorithmic Skeletons: Structured Management of Parallel Computation*. Pitman.

Fox G. C. (1989) Parallel Computing comes of Age: Supercomputer Level Parallel Computations at Caltech. *Concurrency: Practice and Experience*, **1**(1), pp. 63–103.

Horowitz E. and Sahni S. (1978) *Fundamentals of Computer Algorithms*. Pitman.

Horowitz E. and Zorat A. (1983) Divide-and-Conquer for Parallel Processing. *IEEE Trans. Computers,* **C–32**(6), pp. 582–5.

Hudak P. (1989) Conception, Evolution and Application of Functional Programming Languages. *ACM Computing Surveys,* **21**(3), pp. 359–411.

Hudak P., Peyton-Jones S., Walder P., Boutel B., Fairbairn J., Fasel J., Guzm'an M.M., Hammond K., Hughes J., Johnson T., Kieburtz R., Nikhil R., Partain W. and Peterson J. (1991) Report on the Programming Language Haskell—A Non-Strict, Purely Functional Language—Version 1.1, Yale University, Department of Computer Science.

Hughes J. (1989) Why Functional Programming Matters. *Computer J.* **32**(2), pp. 98–107.

Jones C. B. (1986) *Systematic Software Development Using VDM.* Prentice-Hall.

Kelly P. (1989) *Functional Programming of Loosely-Coupled Multiprocessors.* Pitman.

McBurney D. L. and Sleep M. R. (1987) Transputer-Based Experiments with the ZAPP Architecture. In *Proc. PARLE 1987,* published as *Lecture Notes in Computer Science* (ed. de Bakker *et al.*) **258**, pp. 242–59, Springer-Verlag.

McCarthy J. (1978a) A Micro-Manual for Lisp—Not the Whole Truth. *ACM SIGPLAN Notices* **13**(8), pp. 215–16.

McCarthy J. (1978b) History of Lisp. *ACM SIGPLAN Notices* **13**(8), pp. 217–23.

Marino G. and Succi G. (1991) Functional Programming with Bags. *Internal Report,* DIST, Università di Genova, via Opera Pia 11a, 16145 Genova, Italy.

Mou Z. G. (1990) Divacon: A Parallel Language for Scientific Computing Based on Divide-and-Conquer. In *Proc. 3rd Symp. Frontiers Massively Parallel Computation,* IEEE.

Mou Z. G. and Hudak P. (1988) An Algebraic Model for Divide-and-Conquer and Its Parallelism. *J. Supercomputing,* **2**, pp. 257–78.

North N. D. (1990) An Implementation of Sets and Maps as Miranda Abstract Data Types. *NPL Report DITC 162/90,* National Physical Laboratory, Teddington, TW11 0LW.

Pancake C. M. (1991) Software Support for Parallel Computing: Where Are We Headed? *Comm. ACM,* **34**(11), pp. 52–64.

Peters F. J. (1981) Tree Machines and Divide-and-Conquer Algorithms. In *CONPAR81; Proc. Conf. Analysing Problem-Classes Parallel Computing*, pp. 25–36.

Preparata F. P. and Hong S. J. (1977) Convex Hulls of Finite Sets of Points in Two and Three Dimensions. *Comm. ACM*, **20**(2), pp. 87–93.

Preparata F. P. and Vuillemin J. (1981) The Cube-Connected Cycles: A Versatile Network for Parallel Computation. *Comm. ACM,* **24**(5), pp. 300–9.

Rabhi F. A. and Manson G. A. (1990) Experimenting with Divide-and-Conquer Algorithms on a Parallel Graph Reduction Machine. *Research Report CS–90–2*, Department of Computer Science, University of Sheffield, Sheffield S10 2TN, U.K.

Schwartz J. T., Dewar R. B. K., Dubinsky E. and Schonberg E. (1986) *Programming with Sets: An Introduction to SETL*. Springer-Verlag.

Skillicorn D. B. (1991) Practical Concurrent Programming for Parallel Machines. *Computer J.*, **34**(4), pp. 302–10.

Spivey J. M. (1989) *The Z Notation: A Reference Manual*. Prentice-Hall.

Treadway P. L. (1990) The Use of Sets as an Application Programming Technique. *ACM SIGPLAN Notices*, **25**(5), pp. 103–16.

Turner D. A. (1985) Miranda: A Non-Strict Functional Language with Polymorphic Types. In *Proc. 1985 Conf. Functional Programming Languages and Computer Architecture*, published as *Lecture Notes in Computer Science* **201**, pp. 1–16, Springer-Verlag.

CHAPTER 3

The Neural Network Paradigm

Dean Shumsheruddin
University of Birmingham, UK

3.1   Introduction

This chapter provides an introduction to neural networks and discusses how they can be used as parallel computers.

A neural network is an information processing system consisting of a collection of simple processing units, termed nodes or neurons, connected together by links with varying strengths or weights. Input signals are fed into input units in the network and signals are propagated between the processing units through the weighted links. Output units give rise to the output signals from the network. There are many different types of neural network, with many different kinds of processing units, signals and network topologies.

In general, neural networks are not programmed in the way that conventional computers are. Instead, they acquire the ability to perform a specific information processing task during a learning or training process, which sets the strengths of the links between the processing units.

Research into neural networks has a long history. In the early 1940s McCulloch and Pitts proved that neural networks could, in principle, perform arbitrary arithmetic and logical computations (McCulloch and Pitts 1943). In the late 1940s Donald Hebb studied neural network models of conditioning and developed the first learning rule for neural networks (Hebb 1949). In the late 1950s Frank Rosenblatt and his colleagues developed the first practical neurocomputer, which they called the Perceptron (Rosenblatt 1958).

There was considerable progress in the field during the 1960s. However, this came to an abrupt end with the work of Minsky and Papert (1969). They showed that the Perceptron had very limited powers of computation. They proved that perceptrons were unable to compute the exclusive OR of two binary input signals. Unfortunately they implied that these limitations applied to all neural networks and this view became generally accepted.

During the 1970s only a handful of researchers continued to study neural networks. However, over the course of the 1980s there was an explosion of interest in neural networks. This was primarily a result of the realisation that the limitations of perceptrons were not shared by more complex networks, and the development of new learning mechanisms including error-backpropagation.

Today a great deal of research into the theory and applications of a vast range of neural networks is in progress, and there is every reason to expect that this will continue to be the case.

For many applications, neural networks have important advantages over conventional parallel computing systems. In addition to the ability to learn to perform a task, neural networks can offer a considerable degree of fault-tolerance. Their fine-grained parallelism enables them to solve many problems in applications such as image processing much faster than conventional systems. However, they also have a number of limitations not found in conventional parallel computers and more research is needed to overcome these limitations.

Section 3.2 contains an overview of biological neural networks. It discusses their basic structure and mechanism of operation. Animal nervous systems provided the original inspiration for research into artificial neural networks and developments in neurobiology and neural computation have proceeded hand in hand since then. Many mechanisms employed in artificial neural networks are based on mechanisms discovered in neurobiology, and techniques developed for computer simulation of artificial neural networks have been successfully employed in the investigation of animal nervous systems.

Section 3.3 presents an overview of artificial neural networks. It begins with a discussion of the essential features of artificial neural networks and then goes on to discuss two well known examples in more detail. These include the multi-layer backpropagation network which can be trained with the delta-rule and backpropagation algorithm.

Section 3.4 discusses the computer simulation of neural networks. Most investigations into neural networks, and many practical applications, are carried out by means of computer simulation. Neural networks are usually simulated with a general purpose package or a special program written in a conventional programming language running on a conventional computer. However, techniques have also been developed to simulate neural networks on parallel computers including transputer networks and connection machines.

Section 3.5 presents an overview of current developments in the field of hardware neural networks. In order to obtain the full potential speed of operation of neural networks, they must be implemented in hardware. This section discusses the basic principles of implementing neural networks in hardware and examines two important examples.

Finally, Section 3.6 draws some conclusions and attempts to predict some likely future developments in the area of neural networks for parallel computation.

This chapter concentrates on the general principles of neural networks. Chapter 14 describes applications of neural networks in the important area of robot control. For a more comprehensive introduction to neural networks see Khanna (1990) or Hecht-Nielsen (1990).

3.2 Biological Neural Networks

Animal nervous systems are control systems. Their inputs are signals from the animal's sense organs and their outputs are signals to the animal's muscles and

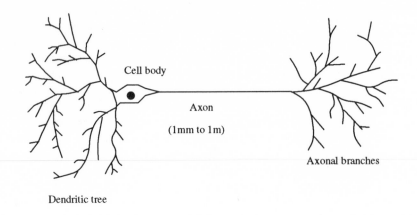

Cell body

Axon

(1mm to 1m)

Axonal branches

Dendritic tree

Figure 3.1: The structure of a typical cortical neuron

glands. Nervous systems have evolved to increase the probability of survival of the animals they control and their genes. At a microscopic level, mammalian nervous systems are very similar. However, at a macroscopic level they differ greatly in size and complexity.

The human nervous system is the most complex and sophisticated. It consists of the central nervous system, which includes the brain and spinal cord, and the peripheral nervous system, which includes the peripheral nerves and nerve ganglia. The peripheral nervous system carries out very little information processing. Its main purpose is to transmit information from sense organs to the central nervous system and from the central nervous system to muscles and glands (Holmes 1990).

At a macroscopic scale, the brain and spinal cord consist of grey and white matter. The grey matter is organised into various nuclei and a thin layer over the surface of the brain, termed the cortex. The white matter fills the spaces between the cortex and nuclei. At a microscopic scale the brain and spinal cord consist of neurons and supporting cells termed neuroglia. The human central nervous system contains approximately 4×10^{10} neurons.

Although they vary greatly in their sizes and shapes, all neurons have the same basic structure. Figure 3.1 shows the structure of a typical cortical neuron. It consists of a cell body, a tree of fine processes called dendrites, and a process called an axon, which branches out into another tree of fine processes. The cell body is typically 10–20 microns across and contains a nucleus, like almost all other cells. The dendritic processes are typically 1 micron in diameter, but they may spread out over a region a few millimetres across. The axon is typically 1–2 microns in diameter, but may be up to a meter long. The axonal branches may also spread over a region a few millimetres wide.

Neurons are connected together at specialised junctions called synapses.

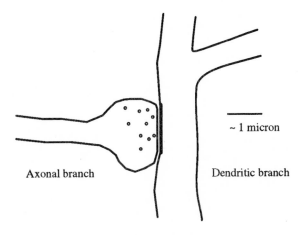

~ 1 micron

Axonal branch

Dendritic branch

Figure 3.2: The structure of a typical cortical synapse

Figure 3.2 shows a typical synapse in the central nervous system. The synapse connects the end of an axonal branch of one neuron to a dendritic process of a second neuron. At the end of the axonal branch is a presynaptic terminal. This contains minute granules of neurotransmitter. Many different substances are used as transmitters in the nervous system including, glutamate, acetylcholine, noradrenaline and γ-amino butyric acid. The cell membrane of the dendrite is thickened at the synapse. This postsynaptic membrane contains complex receptor molecules which can bind to molecules of neurotransmitter, and can also change the permeability of the postsynaptic membrane to various ions.

A typical cortical neuron makes contacts with of the order of 10^3 other neurons through its axonal branches and receives contacts from of the order of 10^3 other neurons onto its dendritic tree. Although the majority of synapses are like the one shown in Figure 3.2, there are also some synapses between dendrites (dendro-dendritic synapses) and some synapses from one axonal process onto another (axo-axonal synapses). There are also a number of synapses between neurons which do not use chemical neurotransmitters (electrical synapses).

Neurons operate by generating and transmitting electrical signals termed action potentials. Normally, there is a steady voltage across the cell membrane of a neuron, termed its resting potential. The inside of the cell is normally around $-70\,\text{mV}$ relative to the potential outside the cell. Figure 3.3 shows how this voltage is generated. The cell membrane is a good insulator and is normally fairly permeable to potassium ions but impermeable to sodium ions. The inside of the cell has a high concentration of potassium ions and a low concentration of sodium ions, compared to the extracellular fluid. These concentrations are maintained, in the long term, by active pumping of ions across the membrane. When the cell is at rest, a relatively small number of potassium ions diffuse out

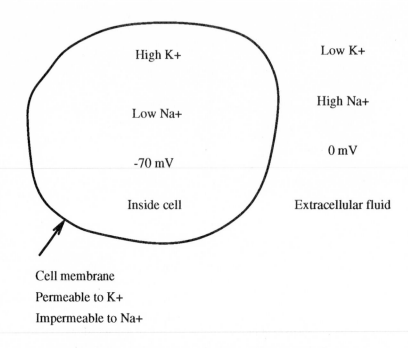

High K+   Low K+

Low Na+   High Na+

                                     0 mV
-70 mV

Inside cell   Extracellular fluid

Cell membrane
Permeable to K+
Impermeable to Na+

Figure 3.3: The origin of a neuron's resting potential

of the cell, down the concentration gradient. As a result the potential in the cell falls until it is negative enough to attract sufficient potassium ions back into the cell to balance the loss by diffusion.

Figure 3.4 shows an action potential. The graph shows the voltage inside the cell body, compared to the extracellular fluid, over the course of about 5 milliseconds. During the action potential, the voltage rises rapidly to around +10 mV and then returns more slowly to the resting level. The electrical activity of the neuron is generated by a relatively complex process. The cell membrane contains large molecules which are sensitive to the voltage across it and serve as channels for sodium ions to flow through it. Normally these channels are closed.

However, if the voltage inside the cell rises above a threshold at about −50 mV, the channels open. This allows a sudden inflow of sodium ions into the cell, down both a concentration and potential gradient. The influx of sodium ions causes the potential in the cell to rise rapidly to about +10 mV. When the potential reaches 0 mV, the channels close again and remain closed. Potassium ions then diffuse out of the cell relatively slowly, until the potential returns to −70 mV. At this potential, the sodium channels reset themselves so that they can open again next time the voltage inside the cell rises.

Action potentials usually start in the cell body. When the voltage inside the

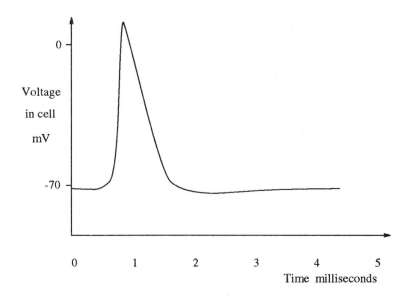

Figure 3.4: An action potential

cell body rises, current flows into the first part of the axon causing the voltage inside it to rise above the −50 mV threshold. This causes an action potential in the first part of the axon, which triggers an action potential in the next part of the axon, and so on. As a result, the action potential travels down the cell's axon at a speed of 1–10 metres per second. Neurons tend to generate action potentials, or fire, in complex patterns of bursts with long gaps in between. The fastest neurons can fire at a rate of up to 500 Hz during a burst.

Synapses transmit signals from one neuron to another. Some synapses are excitatory and some are inhibitory. At an excitatory synapse, firing of the presynaptic cell makes the postsynaptic cell more likely to fire. At an inhibitory synapse firing of the presynaptic cell makes the postsynaptic cell less likely to fire.

When an action potential reaches a presynaptic terminal of an axon, it causes a few of the granules of neurotransmitter to fuse with the presynaptic membrane. Some of the transmitter molecules diffuse across the synaptic gap and bind to the receptors on the postsynaptic membrane. At an excitatory synapse the receptors that have bound transmitter molecules increase the permeability of the postsynaptic membrane to sodium ions and raise the potential inside the dendrite. At an inhibitory synapse receptors change the membrane's permeability to other ions (including chloride) and reduce the potential inside the dendrite. Potentials generated inside the dendrites at synapses spread by passive conduction into the cell body. If there is sufficient net excitatory input, the cell will fire.

Although the basic mechanism of operation of individual neurons is now reasonably well understood, most of the details remain to be discovered. Our knowledge of how networks of neurons operate together on a larger scale is fragmentary. However, there are many interesting theories in this area. One of the most impressive is Edelman's theory of neuronal group selection (Edelman 1987).

3.3   Artificial Neural Networks

An artificial neural network is an information processing system consisting of a set of simple processing units, known as neurons, connected together by a set of links with varying strengths. Artificial neural networks may be implemented in hardware, or simulated on a conventional serial or parallel computer.

3.3.1   Basic Principles

There are many different types of neural network, with different kinds of neurons, different types of signals, and different patterns of connectivity. However, the vast majority fit into the general framework for neural networks, or parallel distributed processing systems, proposed by Rumelhart *et al.* (1986a). According to this general framework, a neural network consists of the following components:

1. A set of processing units.

2. A state of activation.

3. An output function for each unit.

4. A pattern of connectivity between units.

5. A propagation rule for propagating patterns of activities through the network of connectivities.

6. An activation rule for combining the inputs impinging on a unit with the current state of that unit to produce a new level of activation for the unit.

7. A learning rule whereby patterns of connectivity are modified by experience.

8. An environment within which the system must operate.

All neural networks have a set of processing units. These units operate in parallel, either synchronously, as in the case of most computer-simulated networks, or asynchronously, like biological neural networks. Each unit receives inputs from one or more of its neighbours, computes an output value, and sends it to one or more of its neighbours. Input units receive signals from the environment, and output units send signals to the environment. The computational power of units is usually very limited, and they calculate their outputs entirely on the basis of locally available information.

Units have a state of activation, which varies with time. Different networks allow different sets of activation values for their units. Activation levels may be discrete or continuous, bounded or unbounded. Many networks employ binary units having 0 and 1 as the only two possible levels of activation. Other networks allow integer-valued or real-valued activation levels.

Each unit has an output function, which maps its current activation level to its output signal. In a simple network the identity function is used. In more sophisticated models, a threshold function or a stochastic function may be used.

The units of the network are connected together by a set of links. Each link has a strength, or weight. Links are usually unidirectional, as in the case of biological systems, but may be bi-directional. Some networks employ more than one kind of link. The strength of the links in the network may be pre-determined fixed values, but they are usually set by a learning algorithm.

The rule of propagation and the activation rule specify how the inputs to the unit are combined to produce its activation level. In many networks, the activation of a node is the sum of the products of its inputs and the weight of their links:

$$A_i(t+1) = \sum_{j=1}^{n} A_j w_{ij}$$

where A_i is the activity of the ith node, and w_{ij} is the weight of the link from node j to node i.

The learning rule specifies how the weights of the connections in the network are to be adjusted during the learning process. During learning, the weights are usually adjusted in a large number of small steps. The simplest learning rule is Hebb's rule, which was mentioned in the introduction. This rule states that if two units are active simultaneously, the strength of the link between them should be increased (Hebb 1949). This rule is not sufficiently powerful for many problems, and most modern networks use the a version of the delta-rule:

$$\Delta w_{ij} = \eta(T_i - A_i)A_j$$

where Δw_{ij} is the change to be made to w_{ij} during one learning step, η is a learning rate constant and T_i is the target value for the activity of the ith node (Widrow and Hoff 1960; Rumelhart *et al.* 1986b).

3.3.2 Backpropagation Networks

Backpropagation networks, or multi-layer perceptrons, are the commonest type of artificial neural network. They can learn an arbitrary mapping between input patterns and output patterns (Rumelhart *et al.* 1986b). They are being used for a wide range of pattern matching and classification applications.

Figure 3.5 shows the structure of a typical backpropagation network. The nodes of a backpropagation network are arranged in three or more layers. There

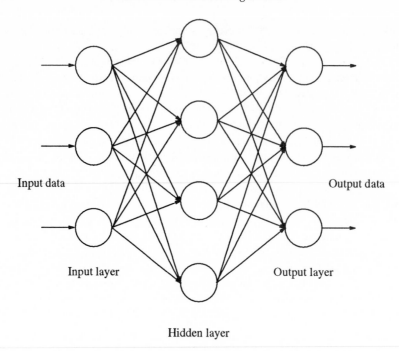

Input data

Output data

Input layer

Output layer

Hidden layer

Figure 3.5: A typical three layer backpropagation network

may be an arbitrary number of nodes in each layer. All input signals are fed into the first layer. There are connections from each node in the first layer to all the nodes in the second layer. There are also connections from each node in the second layer to all the nodes in the third layer. There are no connections in the reverse direction and no connections between the nodes within a layer. All output signals are taken from the third layer. The nodes in the second and third layers have internal thresholds associated with them. The nodes in the network have real-valued activation levels, which may be either positive or negative. The weights of the connections are also real-valued, and may be positive or negative.

During normal operation, input patterns are used to set the activation level of the input nodes. Activity is then propagated from the input nodes to the hidden nodes. Finally, activity is propagated from the hidden nodes to the output nodes. Propagation takes place according to the following equations:

$$net_i = \sum_{j=1}^{n} a_j w_{ij}$$

$$a_i = \sigma(net_i - \theta_i)$$

where net_i is the net input to the ith node in the current layer, n is the number of nodes in the previous layer and w_{ij} is the strength of the connection from the jth node to the ith node. a_i is the activity of the ith node, θ_i is its threshold and σ is the sigmoid or logistic function:

$$\sigma(a) = \frac{1}{1 + e^{-a}}$$

In order to train the network to perform a particular task, the thresholds of the nodes and the strengths of the links are initially set to small random values. Then the network is trained with a set of training cases. Each case contains an input pattern and the corresponding desired output pattern. It may be necessary to run through the complete set of training cases many times in order to achieve good performance.

In order to train the network on a single case, the input pattern is fed into the first layer of the network and activity is propagated to the hidden layer. Then activity is propagated from the hidden layer to the output layer. The actual activities of the output nodes are then compared with the target activities in the training case.

The weights of the links from the hidden layer to the output layer are adjusted according to the delta rule:

$$\Delta w_{ij} = \eta \delta_i a_j$$

where Δw_{ij} is the change to be made in the weight of the link from the jth node to the ith node, η is a learning rate constant between zero and one, a_j is the activity of the jth node and

$$\delta_i = a_i(1 - a_i)(t_i - a_i)$$

where a_i is the actual activity of the ith node and t_i is the target activity for it.

Then the weights of the links from the input nodes to the hidden nodes are adjusted according to the same rule:

$$\Delta w_{ij} = \eta \delta_i a_j$$

However, for the hidden nodes the values of δ_i are calculated by propagating the error signals backwards from the output layer according to the following formula:

$$\delta_i = a_i(1 - a_i) \sum_k \delta_k w_{ik}$$

where δ_k is the value of δ_i for the kth output node.

The node thresholds are adjusted in a similar way by treating them as weights of links from a constant-valued input. For many applications, the number of learning trials required may be reduced by adding a momentum term to the weight changes according to the following formula:

$$\Delta w_{ij} = \eta \delta_i a_j + \alpha \Delta' w_{ij}$$

where α is a momentum constant between zero and one, and $\Delta'w_{ij}$ is the value of Δw_{ij} for the previous learning trial.

Backpropagation networks have been applied to a very wide range of problems. They have proved to be particularly useful for classification problems, such as medical diagnosis. In a typical medical diagnostic application, a network is trained to diagnose a number of diseases on the basis of a set of symptoms. One input node is used to represent each possible symptom. Input nodes are normally set to an activity level of 1 or 0 to indicate the presence or absence of their symptom in the case being diagnosed. One output node is used to represent each of the possible diseases. When the network is working correctly, the output node corresponding to the correct disease will have an activity of 1 and all the others will have an activity of 0.

The network is trained on a set of cases containing the pattern of symptoms and correct disease for each case. Normally the training examples are taken from the records of a large set of real patients. However, for many applications, backpropagation networks can also be trained with sets of examples generated by a conventional computer program.

3.3.3   Kohonen Networks

Kohonen networks are a type of self-organising feature map (Kohonen 1984). They are being applied to vector quantisation and cluster analysis problems, especially in fields such as image processing.

They were originally inspired by the neural maps found in the brain. In the visual system for example, an animal's visual field is continuously mapped onto a region of its cortex. Neurons close together in the visual cortex process information from neighbouring points in the visual field. However, this mapping is not linear as a greater area of cortex is devoted to processing information from the centre of the visual field than its periphery.

Figure 3.6 shows the structure of a typical Kohonen network. It consists of a linear array of inputs and normally a rectangular array of map nodes. There may be an arbitrary number of input nodes and an arbitrary number of map nodes. There are connections from each input node to every map node. There are also bi-directional connections between neighbouring map nodes. The activities of the nodes and strengths of the links are all real-valued.

During normal operation, a pattern of input signals is applied to the N input nodes and activity is propagated to the map nodes. Each map node calculates the Euclidean distance between the point it represents in the input vector space and the point corresponding to the actual input vector according to the following formula:

$$d_i = \sum_{j=1}^{N}(x_j - w_{ij})^2$$

where d_i is the distance for the ith map node, x_j is the activity of the jth input node and w_{ij} is the strength of the link from the jth input to the ith map node.

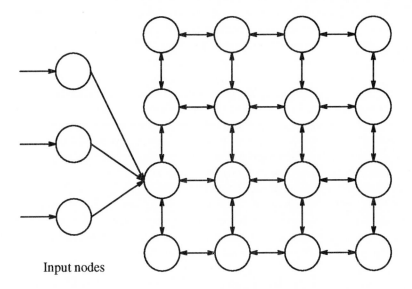

Input nodes

Feature map nodes

Figure 3.6: A Kohonen network. For clarity, the links from the input nodes to just one of the map nodes are shown. There are similar links to all the other nodes in the map.

The map nodes then inhibit each other through the links between them. As a result of this lateral inhibition, the node representing the point closest to the actual input vector generates an output of 1. All the other map nodes generate an output of 0.

The network is trained on a sample set of input vectors. No target output values are needed as the network is self-organising. The strengths of the links from input nodes to map nodes are initially set to small random values. Each map node has a neighbourhood of surrounding nodes in the map. This is initially set to a large, normally square, region of the map and gradually reduced in size over the course of the learning process until it contains just one node.

During a single learning trial, an input vector is applied to the input nodes and the map node representing the closest point to the actual input is selected, as during normal operation. Then the strengths of the links from the input nodes to the selected node, and all of the nodes in its neighbourhood, are updated according to the following formula:

$$\Delta w_{ij} = \eta(x_j - w_{ij})$$

where Δw_{ij} is the change to be made in the strength of the link from the jth input node to the ith map node, x_j is the value of the jth input and w_{ij} is the current weight of the link. η is the learning rate constant. This is initially set to 1.0 and decays slowly to 0 over the course of learning. It is the combination of the shrinking neighbourhoods and decaying learning rate constant that is responsible for the self-organisation of the network.

Kohonen networks have been applied to a wide range of vector quantisation problems. A good example is the compression of 24-bit colour images to 8-bit colour images. A 24-bit colour image has 8-bits for the intensity each of the three primary colours, red, green and blue, for each pixel. An 8-bit colour image has an 8-bit code for the colour of each pixel, and a look-up table containing the intensities of red, green and blue for each of the 256 possible codes.

A Kohonen network with three input nodes and 256 map nodes is used for this application. The input nodes represent the intensities of the three primary colours of a pixel. Each map node represents a code value. The weights from the input nodes to a map node represent the colour corresponding to its code value. The network is first trained on a sample of the pixels in the image. During this process it learns an optimal set of 256 colours to use in the compressed image. Then the values in the look-up table are taken from the weights in the network, and each pixel is fed through the network to generate the code corresponding to the colour closest to it.

3.4 Computer Simulation of Neural Networks

At present, the vast majority of neural networks are simulated on conventional, serial or parallel, computers. Computer simulations are ideal for research into neural networks as they can be developed very quickly and cheaply. They are very flexible and make it easy to experiment with alternative network structures, activation functions and learning algorithms. They also allow easy collection and analysis of data on the behaviour and performance of the network.

The main disadvantage of simulating a neural network on a conventional computer is that it runs much more slowly than it would do if implemented in VLSI hardware. However, computer simulations of neural networks are fast enough for many practical applications such as classification problems, including medical diagnosis.

Networks are usually simulated with a general purpose package or a special program written in a conventional programming language. Most general purpose neural network simulation packages, and special purpose simulation programs, are designed to run on serial computers. However, an increasing number are being developed to run on coarse-grained and even fine-grained parallel computers.

3.4.1 Neural Network Simulation Packages

Using a general purpose neural network simulation package to simulate a network has a number of advantages over developing a special purpose program. Using a package is easier and faster than writing a program. Most packages provide good user interfaces and debugging tools for network simulations, and most packages include thoroughly tested and debugged library routines for simulating common types of network such as the backpropagation network. However, writing a special purpose program to simulate a network allows greater flexibility and enables simulation of arbitrary designs of networks. Specially written programs can be optimised for a particular type of network and almost always run faster than simulations developed using a package. However, packages are ideal for developing prototype neural network simulations for practical applications.

The Rochester Connectionist Simulator was one of the first general purpose neural network simulation packages. It was developed at the University of Rochester and is in the public domain. It is written in the C programming language. It runs under the UNIX operating system on a range of machines including Sun and DEC workstations. Versions are also available for a number of microcomputers. The system provides an extensive set of library routines which can be combined with a small user-written program to simulate a complex network. It also provides a graphical interface to display the operation of the network and plot graphs of outputs and errors during learning (Feldman *et al.* 1988).

Aspirin/MIGRAINES is a neural network simulation environment developed by the MITRE Corporation. It is now available free of charge. It consists of a code generator that builds neural network simulations by reading a network description, written in a language called Aspirin, and generates a C program to simulate the network. An interface called MIGRAINES is provided to export data from the neural network to a graphical display. It is primarily designed to simulate backpropagation networks (Leighton and Wieland 1991).

PlaNet is a neural network simulator designed to run on a workstation running the UNIX operating system and the X window system. It was developed at the University of Colorado at Boulder and is also available free. PlaNet reads a network description written in a high-level neural network description language and then simulates the specified network. It provides sophisticated graphical display facilities and an automated network training mechanism. It has a modular structure which allows the user to replace parts of the system with his or her own programs. This makes it especially suitable for experimenting with new network designs (Miyata 1990).

There are now a very large number of commercial neural network simulation packages available. Most of them run on microcomputers and provide similar facilities to the three simulators discussed above.

3.4.2 Simulation on Conventional Computers

Most special purpose neural network simulation programs, and most neural network simulation packages, are written in conventional programming languages. C is probably the most popular language for neural network simulation at present, although object-oriented programming languages such as C++ are rapidly gaining popularity.

In order to simulate a backpropagation network, for example, a program employs data structures to represent the current state of the network. It employs an algorithm to generate the next state of the network from the current state of the network, and the current state of the network's input signals. The state of the network, including the activities and thresholds of all the nodes and strength of all the connections, is normally stored in one and two-dimensional arrays of 32 or 64-bit floating point numbers. The simulation algorithm normally updates the states of all the nodes in the network synchronously.

In some networks, for example Boltzmann networks, the inputs to a neuron determine the probability of it taking on a particular activation value rather than determining the activation value directly. Simulations of this type of network rely on pseudo-random number generation algorithms to determine the states of nodes. It is also possible to simulate networks with nodes that operate asynchronously by using pseudo-random numbers, but such simulations tend to run extremely slowly and are rarely used (see McClelland and Rumelhart 1988).

3.4.3   Simulation on Parallel Computers

Now that parallel computers are becoming widely available, they are frequently being used to simulate neural networks. Most types of neural network simulation can be parallelised very effectively. The majority of parallel simulations of neural networks have employed coarse-grained parallel computers, such as multi-processor UNIX minicomputers, networks of workstations, or small transputer networks. However, a number of simulations have been developed for fine-grained parallel computers such as the connection machine.

Simulating a backpropagation network on a coarse-grained parallel computer is relatively straightforward. Simulations of backpropagation networks take a long time to train. In order to parallelise the training process, the training algorithm needs to be modified to employ batch-mode learning. Normally, the weights of every link in a backpropagation network are updated during the processing of each training case. In batch-mode learning, the weights are not modified on each training case, but the weight changes for each case are accumulated. When the complete set of cases has been processed, the weights are modified by the accumulated changes.

Figure 3.7 shows how a network of workstations can be used to run a parallel simulation of a backpropagation network. A master copy of the simulation program runs on one machine and slave copies of the program run on the other machines. The set of training cases is divided equally between the slave copies of the program.

The master program generates the initial random set of weights and thresholds for the network. It broadcasts this set of weights and thresholds to the slave programs. Each slave program initialises its network with these values and runs through its share of the training cases, accumulating the weight changes. The

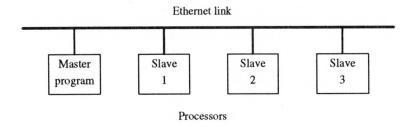

Figure 3.7: Simulating a neural network on a network of processors

slave programs then send their accumulated weight changes back to the master program. The master program adds the weight changes from each of the slaves together and applies the result to the initial random values for the weights and thresholds. It then broadcasts the new weight and threshold values to the slave processors for their next cycle through the training cases. This process is completely equivalent to batch-mode learning on a single processor.

Simulating a backpropagation network on a fine-grained parallel computer such as a connection machine presents more difficult problems. The CM-1, the original connection machine, has 64K processors. Belloch and Rosenberg (1987) were among the first to simulate a backpropagation network on a connection machine. They used one processor to simulate each unit in the network, one processor to simulate each outgoing weight from a unit, and one processor to simulate each incoming weight.

During normal operation, the processors simulating units in one layer send their activities to their outgoing weight processors. The outgoing weight processors multiply the activities by the strengths of their links and send the results to the incoming weight processors of the next layer. Incoming weight processors add all their incoming values together and feed them to the unit processors for their layer of the network. This arrangement allows all the layers to be simulated in parallel with very fast throughput of data. However, the speed of this type of simulation tends to be limited by inter-processor communication delays. In addition, using connection machines for neural network simulation has a relatively poor cost-performance ratio compared to using transputer networks.

3.5 Hardware Neural Networks

Neural networks implemented in hardware run many times faster than computer simulations of them. Although computer simulations are fast enough for many applications, neural networks need to be implemented in hardware to exploit their full potential.

A wide range of hardware implementations of neural networks is currently being developed. At one end of the spectrum, there are von-Neumann machines optimised for the simulation of neural networks. They are extremely flexible, but are relatively slow. In the middle of the spectrum there are parallel digital computers with processors designed to simulate nodes and local memories designed to hold connection weights. These are considerably faster, although they are less flexible. At the other end of the spectrum there are biologically inspired analog VLSI circuits designed to implement one particular network structure. These are extremely fast but are inflexible.

There are a number of hardware neural network implementations based on optical computers. However, the majority of hardware neural networks are based on digital electronic circuits and a smaller number are based on analog or hybrid technologies.

3.5.1 The CNAPS Neurocomputer

The CNAPS Neurocomputer chip, developed by Adaptive Solutions, is a good example of a parallel digital computer based system (McCartor 1991). The chip contains 64 processors running at 25 MHz. Each processor has a 32-bit adder, 9-bit by 16-bit multiplier, shifter, logic unit, 32 16-bit registers and 4096 bytes of local memory. All the processors receive the same instruction, but may execute it conditionally.

Input and output take place over 8-bit input and output busses connected to all the processors. The output bus is connected to the input bus so that one processor can broadcast signals to all the others. Multiple chips can easily be connected together to build a larger network. A sequencer chip controls instruction flow, input and output. An eight chip configuration on one board can update 2.3 billion connections per second (CPS) in learning mode and process 9.6 billion CPS in feed forward mode.

McCartor (1991) developed an efficient implementation of a backpropagation network using an array of eight CNAPS chips. The network has 1900 inputs, 500 hidden nodes and 12 outputs. Weights were updated after each input and no momentum was used. 16-bit integers were used to represent real weights between -8.0 and +8.0. Errors were represented with 8-bit signed integers. The sigmoid output function was implemented as an 8-bit by 256 look-up table. The basic backpropagation network structure was extended by adding reverse connections from the output layer nodes to the hidden layer nodes.· These reverse links were used to propagate error signals backwards from the output layer to the hidden layer during training. Their weights were updated in parallel with the corresponding forward links.

The performance of the network compared favourably with that of similar networks implemented on several other neurocomputer chips. In feed-forward mode, the network ran more than 10,000 times faster than a software simulation running on a Sun 3.

3.5.2 The Silicon Retina

The Silicon Retina is an analog hardware neural network based on the structure of the biological neural network of the eye. It is designed to be used as the front end of an intelligent vision system (Mead and Mahowald 1988, Mead 1989).

The human retina contains a multi-layer network of neurons. The outermost layer contains rod and cone cells which detect light. The next layer contains horizontal cells, which are connected to many rods and cones and transmit information horizontally through the retina. The next layer contains bipolar cells which are connected to a small number of rods and cones, and transmit information inwards. The rods and cones have special triad synapses connecting them to both horizontal and bipolar cells. The next layer contains amacrine cells which also transmit information horizontally. The innermost layer contains ganglion cells which collect information from a relatively wide area and transmit it to the brain via the optic nerve.

The Silicon Retina contains a hexagonal network of photoreceptors with a logarithmic response to light, simulating rods and cones. It contains a horizontal network of resistors in a hexagonal grid, simulating the horizontal cell layer. It also contains bipolar units made up of two amplifiers, simulating bipolar cells and ganglion cells.

The Silicon Retina is designed to work in the same way as a biological retina. The photoreceptors measure the intensity of light at a point in the image projected onto the retina. Their logarithmic response enables them to operate over a very wide range of lighting levels without generating unreasonable signal levels. The horizontal resistor network calculates an average intensity over the circular region surrounding each photoreceptor. The bipolar units measure the difference between the intensity at a single photoreceptor and the average intensity of the surrounding region calculated by the resistive network.

The Silicon Retina exhibits many of the characteristics of a real retina, including its response to points of light, static objects and moving edges. However it operates many times faster than a real retina.

3.6   Future Developments

Since the resurgence of interest in neural networks in the early 1980s remarkable progress has been made in the field. At the same time there has been rapid growth and development in the field of parallel computing in general. There is every reason to believe that rapid progress will continue to be made in these areas of computer science over the next decade.

In the short term it seems likely that there will be many more practical applications of neural networks, especially in fields where rapid progress is being made at the present time. These include vision and image processing, handwritten character recognition, speech synthesis and recognition, and control and navigation.

In the medium term, it seems likely that theoretical advances in neurobiology and artificial neural networks will continue. At the same time hardware implementations of neural networks will become much faster, cheaper and more widely available.

In the long term, it seems inevitable that the general trend towards parallel computing will continue, and that neural networks will occupy a significant proportion of the market for information processing systems.

References

Belloch G. and Rosenberg C. R. (1987) Network learning on the connection machine. In *Proc. of the Tenth International Joint Conference on Artificial Intelligence, Milan, Italy*, pp. 323–326.

Edelman G. M. (1987) *Neural Darwinism*. Basic Books, New York.

Feldman J. A., Fanty M. A. and Goddard H. N. (1988) Computing with structured neural networks. *Computer* **21**, pp. 91–103.

Hebb D. (1949) *The Organization of Behaviour*. Wiley, New York.

Hecht-Nielsen R. (1990) *Neurocomputing*. Addison-Wesley.

Holmes O. (1990) *Human Neurophysiology*. Unwin Hyman.

Khanna T. (1990) *Foundations of Neural Networks*. Addison-Wesley.

Kohonen T. (1984) *Self-Organization and Associative Memory*. Springer-Verlag, Berlin.

Leighton R. and Wieland A. (1991) *The Aspirin/MIGRAINES Software Tools User's Manual*. MITRE Corporation, Washington.

McCartor H. (1991) Back propagation implementation on the Adaptive Solutions CNAPS neurocomputer chip. In *Advances in Neural Information Processing Systems 3*, edited by Lippmann R. P., Moody J. E and Touretzky D. S. pp. 1028–1031. Morgan Kaufmann.

McClelland J. L. and Rumelhart D. E. (1988) *Explorations in Parallel Distributed Processing*. MIT Press, Cambridge MA.

McCulloch W. S., and Pitts W. (1943) A logical calculus of the ideas immanent in nervous activity. *Bulletin of Mathematical Biology*, **5**, pp. 115–133.

Mead C. A. (1989) *Analog VLSI and Neural Systems*. Addison-Wesley, Reading, MA.

Mead C. A. and Mahowald M. A. (1988) A silicon model of early visual processing. *Neural Networks*, **1**, pp. 91–97.

Minsky M. and Papert S. (1969) *Perceptrons*. MIT Press, Cambridge MA.

Miyata Y. (1990) *A User's Guide to PlaNet Version 5.6, A Tool for Construct-ing, Running, and Looking into a PDP Network.* Computer Science Department, University of Colorado at Boulder, Boulder CO.

Rosenblatt F. (1958) The perceptron: A probabilistic model for information storage and organization in the brain. *Psychological Review,* **65**, pp. 386–408.

Rumelhart D. E., Hinton G. E. and McClelland J. L. (1986a) A general framework for parallel distributed processing. In *Parallel Distributed Processing*, edited by Rumelhart D. E. and McClelland J. L. MIT Press, Cambridge MA.

Rumelhart D. E., Hinton G. E. and Williams R. J.. (1986b) Learning internal representations by error propagation. In *Parallel Distributed Processing*, edited by Rumelhart D. E. and McClelland J. L. MIT Press, Cambridge MA.

Widrow G. and Hoff M. E. (1960) Adaptive switching circuits. *Institute of Radio Engineers, Western Electronic Show and Convention, Convention Record, Part 4,* pp. 96–104.

CHAPTER 4

Problem-Driven versus Hardware-Driven Approaches to the Design of Concurrent Programs

David W. Bustard and James S. Weston
University of Ulster, Coleraine, UK

4.1 Introduction

This chapter is about the design of concurrent programs. Its premise is that there are essentially two approaches to consider: problem-driven design, in which emphasis is placed on representing the concurrency inherent in a problem, and hardware-driven design, in which the main concern is to make best use of the hardware available for parallel processing. The chapter discusses and illustrates these approaches and their inter-relationship.

Some authors tend to use the terms concurrent program and parallel program interchangeably when referring to software in which two or more activities may proceed at the same time, e.g. Gehani and McGettrick (1988). The view taken here, however, is that the terms are different and, more importantly, that they largely imply different approaches to software design. The following definitions are offered:

- a sequential program specifies sequential execution of a list of statements; its execution is called a process;

- a concurrent program specifies two or more sequential programs that may be executed concurrently as parallel processes (Andrews and Schneider 1983); and

- a parallel program is a concurrent program that is designed for execution on parallel hardware.

Thus, concurrent program is a generic term used to describe any program involving actual or potential parallel behaviour whereas parallel programs form a sub-class of concurrent program, where each program in the sub-class is designed for execution in specific parallel processing environments (Bustard 1990).

Distributed programs form an important sub-class of parallel programs. They may be defined as follows:

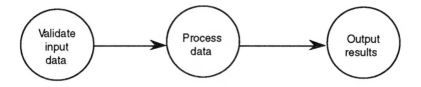

Figure 4.1: General data processing model

- a distributed program is a parallel program designed for execution on a network of autonomous processors that do not share main memory (Bal *et al.* 1989).

A concurrent program is primarily a coherent unit of software. If two or more pieces of communicating software run in parallel, the result is a concurrent program when the pieces form a conceptual whole; otherwise, the situation is viewed as programs communicating through an agreed protocol. The communicating programs do, however, constitute a concurrent system (or parallel system or distributed system, as appropriate).

All programs are potentially concurrent in the sense that their design may identify components capable of parallel execution. Figure 4.1, for example, suggests a structure that might be found in almost any piece of software.

Software of this form is potentially concurrent because the input of data and the output of results can both be buffered to enable either activity to proceed in parallel with the main computation. This type of concurrency is common in many application programs. Usually, however, it is implemented in the operating system on top of which the program runs, rather than being expressed explicitly within the program.

A concurrent program can be described as inherently concurrent if it is constructed to control or model physical systems that involve parallel activity. Examples include real-time systems that have parallel inputs or simulation systems that describe real-world parallel activities such as aircraft activity at an airport or device activity on a computer processor board.

In this article, the design of potentially or inherently concurrent programs will be described as problem-driven, indicating that the designer essentially responds to the basic nature of the system involved in order to reflect its inherent structure. The hardware-driven approach to the design of parallel programs is much more proactive, with the designer actively seeking to introduce parallel activity to make the best use of available parallel hardware; i.e., to use parallelism to enhance program performance.

The problem-driven and hardware-driven approaches to design are discussed directly in later sections. Before that, however, there is an introduction to the main example used to illustrate the discussion in general and an examination of the life-cycle model that encompasses both design techniques. In particular, there is a detailed discussion of the ways in which requirements might be expressed since such descriptions form the basis of design. Indeed, there is really no clear demarcation between the expression of requirements and the development of software to meet those requirements.

4.2   Spelling Checker Example

It was felt that the distinction between the problem-driven and hardware-driven approaches to design would be best illustrated by using the same example throughout, allowing any differences that appear to be more easily attributed to variations in the design techniques. Furthermore, it seemed desirable that any example chosen should be readily understood so that its presentation would not distract the reader from the general design discussion. The example itself needed to be one that had some inherent concurrency and also scope for optimisation through parallelism, thereby enabling the use of each design approach to be justified separately.

In the end a spelling checker was chosen because of its simplicity and the fact that it has a processing speed requirement that might benefit from a parallel solution. A spelling checker is typically an integral part of a word-processing package but it is still possible to consider the operation of the checker in isolation as it is usually separated from the basic text editing function.

4.3   The Software Life Cycle

Concurrent and sequential programs are developed and maintained in much the same way. Each goes through a life cycle in which the requirements of a client are defined and software implemented to meet those requirements. Implementation entails the design of a suitable software structure, the realisation of that structure in a programming notation, and the validation and verification of the resulting system with respect to the defined requirements.

4.4   Requirements Modelling

Requirements are classified (Ince 1988) as either:

- functional, defining the functions that a system must support; or

- non-functional, defining constraints on an implementation, such as the use of a particular programming language or the need to achieve defined levels of performance.

Thus, in the case of a spelling checker, one main function exists: to identify all words in a given piece of text that are absent from an on-line dictionary. Secondary operations, such as a function to add new words to the dictionary might also be defined. An obvious performance constraint is that the checking rate for a piece of text should be up to some pre-defined word-per-minute level.

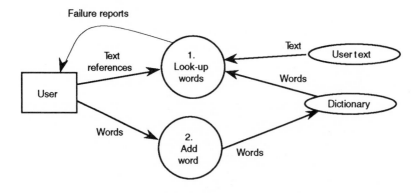

Figure 4.2: Spelling checker: dataflow diagram

Requirements are first expressed in natural language, perhaps as a series of numbered statements. These are initially informal and possibly incomplete, inconsistent, and ambiguous. Deriving a software design from the requirements will help to identify and clear up many of the problems present but some shortcomings may only be discovered at code, or test, or even delivery time. It is fairly obvious (Davis 1990) that the earlier a problem is detected in the software life cycle the cheaper it is to fix, so there is a strong argument for spending additional time trying to identify and remove problems in requirements before design commences. This is achieved by constructing one or more models of the requirements that enable various aspects of their definitions to be analysed in detail. A few of the common techniques are now considered, although in the space available there is only room to give a flavour of each approach.

4.4.1 Dataflow Diagrams

One of the oldest and certainly the most widely used technique for describing requirements is the dataflow diagram (Yourdon 1989). Figure 4.2 shows a dataflow diagram (DFD) for the spelling checker.

Dataflow diagrams are graphs whose nodes represent processes, data stores, and terminators; and whose arcs represent flows of data. Processes, shown as circles, or 'bubbles' in the diagram, represent system functions. These transform inputs into outputs. Data stores, shown here as ellipses, represent information generated or required by processes. Terminators, shown as boxes, represent the external entities with which the system interacts.

A dataflow description is inherently concurrent in that it permits parallel data transformations. In the above example, for instance, it allows for the possibility that some words can be looked up in the dictionary while others are added to the dictionary (subject to mutual exclusion constraints). Its main advantage is that

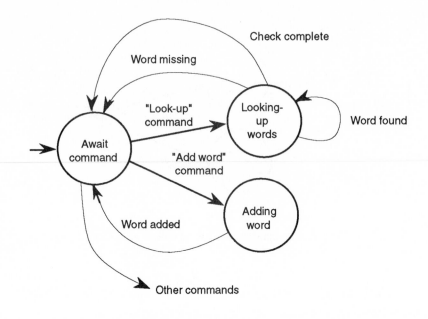

Figure 4.3: Spelling checker: state transition diagram

it provides a useful overview of the information flow in a computing system and so gives a map around which to build a detailed requirements narrative.

4.4.2 Finite State Machines

A number of other requirements definition techniques focus on the description of system behaviour. The essence of these techniques is that a program is characterised by its current state which may be changed by the occurrence of an event—a state transition. One of the most commonly used techniques is the finite state machine description (Davis 1988). It can be expressed as a state transition matrix or a state transition diagram, the latter being illustrated in Figure 4.3 for the spelling checker.

The circles show possible system states and the arcs identify events that cause transitions among the states. Thus the spelling checker is defined to be in one of three states: waiting for a command, looking-up words in the dictionary or adding a new word to it. Most events cause a change of state, the one exception being that the look-up continues after a word is found in the dictionary.

A finite state machine describes the behaviour of a single process. Communicating finite state machines (Shields 1989) or their recent variant, statecharts (Harel 1988) are required to describe concurrent processes.

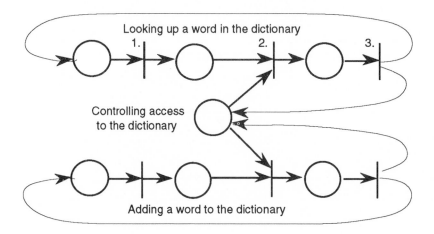

Figure 4.4: Spelling checker: Petri net diagram

4.4.3 Petri Nets

Petri nets (Peterson 1981; Reisig 1985) can be used to represent and simulate the conditions under which state transitions occur in cases where the identity of individual states is not relevant to the description. Figure 4.4, for example, shows a Petri net model of the spelling checker in which the checking of a word in the dictionary and the addition of a word to the dictionary can proceed independently. To maintain the integrity of the dictionary only one operation at a time may be performed on it.

There are several varieties of Petri net. In a condition/event net, events, shown as vertical bars, are linked by conditions, shown as circles. In the diagram, for example, the top sequence of events, from left to right, represents:

- the arrival of a request to look-up a word in the dictionary;

- permission to use the dictionary; and

- the completion of the dictionary operation.

Each event has a set of input and a set of output conditions. An event occurs when all of its input conditions are satisfied. For example, access to the dictionary is given if there is a request to use the dictionary and no other operation currently has access to it. After an event, the connected output conditions are enabled.

A Petri net can be marked with tokens to identify the conditions that are initially satisfied. Applying the event transition rules then allows the tokens to be advanced from input to output conditions. In this way the behaviour defined by the net can be simulated and the possible event sequences explored. Figure

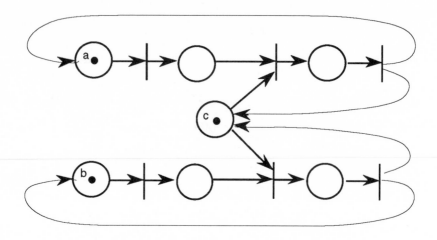

Figure 4.5: Spelling checker: marked Petri net diagram

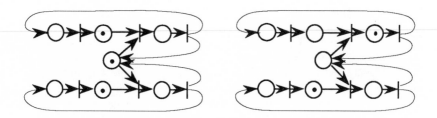

Figure 4.6: Spelling checker: Petri net diagram

4.5, for example, shows an initial marking in which the look-up (a) and add word (b) events are enabled and the dictionary is available for use (c).

Figure 4.6 then shows two successive steps in the simulation of the net in Figure 4.5, identifying the progression of the dictionary look-up and add word operations. The diagrams show that although both operations attempt to gain access to the dictionary at the same time only one actually succeeds—the outcome being non-deterministic.

A Petri net has an equivalent mathematical description that can be analysed more formally.

4.4.4 Temporal Logic

Temporal logic (Pneuli 1986; Galton 1987) is another formal approach to behaviour description. Its emphasis is on the safety and liveness properties of programs (Lamport 1989). A safety property is an invariant of a computation;

that is, a condition that is true of all program states. One example is absence of deadlock, i.e., there must not be a program state in which a process awaits an event that cannot occur. A liveness property identifies what a program must do; it defines what will happen (eventually) in a computation. One example is the required property that any program operation that starts will eventually terminate.

4.4.5 Process Algebra

The use of process algebra is another formal approach to the description of concurrent programs. The general technique is to define a program and its environment in terms of communicating processes (Hoare 1985; Brinksma 1988; Milner 1989). Each process is described by the events or actions in which it is involved. There are, in general, three types of event specified:

- an event internal to a single process;

- a synchronisation (communication) between one process and one or more other processes; and

- a synchronisation between one or more processes and the environment of the system described.

Process descriptions compose events to constrain their order of occurrence. Typically, processes are defined in a recursive fashion. For example, the action of a spelling checker might be described as an infinite sequence of either look-up or add word events, as follows:

spelling_checker: (lookup [] addword);
                     spelling_checker

The precise notation and set of operators available for process descriptions vary from one notation to another, but each supports the same basic approach to requirements definition. Mostly the notations are text-based but graphical equivalents have been developed in some cases, e.g. Bustard *et al.* (1988a) and ISO (1990).

4.4.6 Axiomatic Specification

Another approach to requirements description is to focus on the semantics of the functions that are needed. Axiomatic specification (Gries 1981) defines the meaning of a function as a post-condition predicate over the inputs and outputs of the function. In addition a pre-condition defines the necessary initial state of the inputs for the correct operation of the function. In the case of the spelling checker the add word function needs a pre-condition that the word is not already present in the dictionary and a post-condition indicating that the supplied word is present in the modified dictionary. Standard mathematical concepts and related operators can be used in forming the predicates. An axiomatic description of the add word function might take the following form:

function	add word
inputs	word, dictionary
outputs	dictionary
pre-condition	word $\notin$ dictionary
post-condition	word $\in$ dictionary

4.4.7 Algebraic Specification

Algebraic specification, e.g. Erhig and Mahr (1985), is concerned with types rather than individual functions; the functions are operations associated with a type. For the spelling checker the dictionary is a type and the functions to look up a word in the dictionary and add a new word are operations of that type. A set of equations define the inter-relationship of the functions and in doing so give them meaning. For example, using a variation of the notation suggested in Sommerville (1989) a specification of the dictionary might take the following form:

```
type    dictionarytype
        sort    dictionary
        import  word, Boolean
        operations
                create → dictionary
                lookup (word, dictionary) → Boolean
                addword (word, dictionary) → dictionary
        equations
                forall w, w1: word; d: dictionary
                        lookup (w, create) = false;
                        lookup (addword (w, d), w1) =
                                if w = w1 then true else lookup (w1, d)
endtype dictionarytype
```

The points to note here are:

- a distinction is made between the values, or sort of each type and its overall definition;

- one type definition can build on another by importing it;

- operation definitions introduce a function name, and define the sort of its inputs and result—this is the function signature;

- a function 'create' has been introduced so that the equations that give meaning to the dictionary functions can be fully defined;

```
  ┌─ DictionaryDefinition ──────────────────┐
  │  Dictionary: P WORD                      │
  │ ─────────────────────────────           │
  │  #Dictionary < 5000                      │
  └──────────────────────────────────────────┘
```

Figure 4.7: Z schema for dictionary definition

- the first equation states that the create operation produces a dictionary such that all attempts to look up any word in it will fail, i.e. an empty dictionary; and

- the second equation defines the dictionary 'lookup' recursively, returning true if the given word has been inserted through 'addword'.

4.4.8 Model-Based Specification

Model-based specification is yet another formal approach to requirements definition. The basic technique involves describing a program state using well understood mathematical entities such as sets and functions and then defining the required program functions in terms of their effect on the model state. The two best-known examples of this approach are VDM (Jones 1986) and Z (Hayes 1987; Ince 1988; Diller 1990). As an illustration of the approach consider the Z notation.

A Z specification is a collection of schemas. A schema has three components: a name by which it can be referenced, a signature introducing variables and finally a predicate relating values of the variables to each other and to global variables in other schemas. A schema is usually presented in a graphical form as shown in Figure 4.7. The signature in this example defines the dictionary as a powerset of words, where a word would have its own separate schema—probably defining each as a sequence of characters. The predicate indicates that the dictionary will contain no more than 5000 words.

Given this definition of a dictionary the look-up operation can then be defined as shown in Figure 4.8.

Some notation needs to be explained. The reference to DictionaryDefinition means that the schema definition for the directory is imported and the preceding Ξ indicates that the dictionary is not affected by the look-up operation. The ? qualifier to the variable w means that the variable is an input and the ! qualifier to found, means that it is an output. The predicate indicates (rather predictably) that the value of found! is determined by whether or not the supplied word w? is in the dictionary. The add word operation can be defined in a similar way as shown in Figure 4.9. The use of the Δ before DictionaryDefinition indicates

```
┌─ Lookup ─────────────────────────────┐
│ Ξ DictionaryDefinition               │
│ w?: WORD                             │
│ found!: Boolean                      │
│ ──────────────────────────────────  │
│ found! = w? ∈ Dictionary             │
└──────────────────────────────────────┘
```

Figure 4.8: Z schema for the dictionary look-up operation

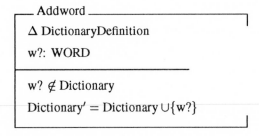

```
┌─ Addword ─────────────────────────────┐
│ Δ DictionaryDefinition                │
│ w?: WORD                              │
│ ───────────────────────────────────  │
│ w? ∉ Dictionary                       │
│ Dictionary' = Dictionary ∪{w?}        │
└───────────────────────────────────────┘
```

Figure 4.9: Z schema for the dictionary add word operation

that the dictionary is modified.

4.5 Choosing a Requirements Description Technique

The tour of the common techniques that are available for making requirements more precise is now almost complete and reveals that the program developer is faced with a bewildering choice. It is clearly impractical to apply all of these techniques in any particular instance and really not desirable anyway because some of them tend to serve the same purpose.

In making a choice it may help to note that the techniques largely fall into three classes:

1. Those concerned with defining structure: identifying the components of a requirement and their inter-relationship. The construction of dataflow diagrams falls under this heading as does object-oriented analysis—a discussion of which has been delayed to the next section.

2. Those concerned with defining behaviour: identifying the temporal ordering of some actions or events with respect to others. The relevant

techniques include finite state machines, statecharts, Petri nets, temporal logic and process algebra.

3. Those concerned with defining semantics: identifying precisely what each function means. Relevant techniques include axiomatic specification, algebraic specification and model-based specification.

These three system views are inter-related but sufficiently distinct to make it practical to consider using one technique from each group during program development. The inter-relationship may in fact be an advantage as it can provide a cross-check on the adequacy of each description.

Another factor to consider when selecting a requirement description technique is its implications for software design. The ideal technique will permit a smooth transition from specification to design and from design to coding, with a fully defined procedure for making each transition. Some research work has been done on developing ways of refining an initial, formal statement of requirements into code through a series of a correctness-preserving transformations (Jones 1986) but so far such techniques have had little impact on day-to-day practice. Instead, it has been common to use techniques and notations at specification, design and code levels that are reasonably sympathetic to each other. One well-known approach to the implementation of data processing programs, for example, is to start with dataflow diagrams, develop them into structure charts, then into pseudocode and finally into a programming notation such as COBOL (Yourdon 1989). The links between the steps are, however, largely informal and thorough inspections are required to ensure consistency.

Another, more recent technique involves an object-oriented approach to analysis, design and programming (Booch 1991). This technique seems to offer several advantages and is discussed in the next section as the single recommended problem-driven approach to concurrent program design. Before that, however, we need to take a brief time-out to consider how hardware-driven software design fits into the software development picture developed so far.

Hardware-driven design was defined to be an approach that concentrated on making good use of available parallel hardware. Thus, the need for such an approach tends to emerge from requirements analysis as a non-functional performance requirement. By implication, the use of parallel hardware should have no influence on the software functions required or on the perceived external behaviour of the software. Consequently, all of the techniques for modelling requirements that have been discussed above really only support problem-driven analysis. This is perhaps to be expected because performance requirements are easily stated and understood. Once software design starts, however, performance needs become significant and will have a major impact on design structure. The next two sections consider the problem-driven and hardware-driven approaches to design separately, followed by a short section that attempts to bring the two discussions together.

4.6   Problem-Driven Software Design

Many statements have been made about the general nature of design and a good discussion of the subject can be found in Booch (1991). In broad terms, design is widely perceived as an engineering activity with both a scientific and artistic component. Petroski describes the process thus: 'The conception of a design for a new structure can involve as much a leap of the imagination and as much a synthesis of experience and knowledge as any artist is required to bring to his canvas or paper. And once that design is articulated by the engineer as artist, it must be analysed by the engineer as scientist in as rigorous an application of the scientific method as any scientist must make' (Petroski 1985).

The main conclusion to be drawn from this definition is that it is impossible to define a software development method that will unfailingly lead the software engineer down a path from requirements to implementation. This does not mean, however, that the would-be designer is without guidance; as Brooks puts it in his 'silver bullet' paper (Brooks 1987), 'there is no royal road, but there is a road'.

Surprisingly perhaps, there is also some consensus on the nature of the road. The object-oriented approach to software development is now enjoying almost universal acclaim as the preferred way to analyse, design and represent software. The technique is not new as it is effectively what Dahl was describing in the late 60s/early 70s (Dahl *et al.* 1972) and which he incorporated in the programming language Simula (Birtwistle *et al.* 1973). It is also, in essence, the technique promoted by Parnas as modular design (Parnas 1972)—an information-hiding approach to program structure. This relatively long history tends to suggest that the basic approach has some intrinsic value.

An object represents an individual, identifiable item, unit or entity, either real or abstract, with a well-defined role in the problem domain. This definition is not related specifically to software and means that the object view of systems is equally applicable to any phase of system development . For a spelling checker, the identifiable objects include the user who invokes the checking operation, the text on which the check is performed and the dictionary against which the check is made.

Graphical notations have been proposed for the documentation of a collection of objects (Booch 1991). These serve to identify the objects and summarise their interaction. Figure 4.10 shows a very simple form of object diagram (Bustard *et al.* 1988b) for the word processing system of which the spelling checker is a part. Here objects are nodes in the graph and arcs used to link those that are inter-dependent.

This diagram gives a neat summary of the main components of the system, on top of which can be added full descriptions of the objects and their links. Some graphical representations enable further information to be attached to the summary diagrams (Booch 1991). For example, distinctions can be made between different types of link such as an indication that one object uses another or instantiates another or inherits definitions from another. Diagrams may also

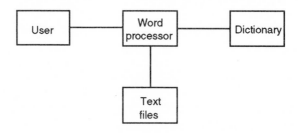

Figure 4.10: Object diagram for a word processor system

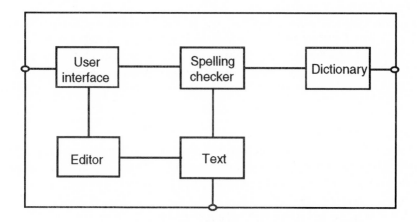

Figure 4.11: Object diagram for a word processor program

distinguish between objects and object types, known as classes.

The object-oriented approach to description is hierarchical in that it allows each object in a system to be described in the same way as the system itself. For example, the word processor software might be designed as the collection of objects shown in Figure 4.11.

The purpose of the objects in this diagram is largely self-evident:

- the user interface object deals with all user interaction;

- the spelling checker and editor objects deal with the spell checking and editing functions;

- the text object maintains the current text being developed; and

- the dictionary object maintains the dictionary.

This diagram is essentially a peek inside the word processor object box in Figure 4.10.

At some point, the semantics of the objects and their inter-connections have to be fully defined but initially it often helps to sketch out object diagrams informally and develop these through several levels of hierarchy. In this way the design can be explored and developed quickly. When a design seems satisfactory its definition is documented precisely. If the target implementation language is object-like then it can be used to define the object communication mechanism and the data involved (Booch 1991). The behaviour and semantics of objects can be documented using any of the requirements description techniques discussed earlier. As a rough rule of thumb, object-oriented design stops when the key abstractions are simple enough to require no further decomposition. More detailed design might then be documented in the form of pseudocode.

The object-oriented design technique is ideal for the development of concurrent programs because the objects identified have autonomy and so may operate in parallel. When translating a design into code the potential for concurrency may or may not be preserved depending on the type of target programming language chosen. A language supporting concurrency is desirable if it is known that the program may execute on parallel hardware. It may also be desirable if the problem is inherently concurrent and so requires concurrency features to express the solution adequately.

For programs that need to make good use of available hardware it will be necessary to take account of the characteristics of that hardware at a much earlier stage. This approach to design is now considered.

4.7   Hardware-Driven Software Design

In today's world many problems exist which require over one thousand times the computing power of the most powerful sequential Von Neumann computers. Such problems include three-dimensional fluid flow calculations, real-time simulations of complex systems, speech and image processing, and intelligent robots. Since these problems are inherently concurrent, they may be tackled using computers with multiple processors operating in parallel (Decegama 1989). Other problems having solutions which are most naturally expressed as static or dynamic networks of communicating processes yield further opportunities to explore the use of parallelism. For example, control systems for industrial plants may be decomposed into sets of data collection, data integration and filtering processes, decision making processes and processes which are responsible for implementing the decisions taken.

Problem-driven design largely ignores the characteristics of the target hardware although it does in effect assume a Von Neumann model in the design of individual processes. The universality of the Von Neumann model means that the structure of a sequential algorithm designed for the solution of a given problem is, to a large extent, independent of the particular computer upon which it is implemented. There are, however, other standard models to consider that involve more than one processing element. Flynn's taxonomy (Flynn 1966) of computer architectures partitions the class of all such parallel machines as follows:

- SIMD machines contain many processors that execute the same instruction simultaneously on different data (the paradigm of lock-step parallelism); and

- MIMD machines contain many independent processors, each of which executes its own individual program. MIMD machines may be further classified according to whether each processor has access to any part of the whole machine's memory (shared memory machines) or whether the memory is distributed among the processors (distributed memory machines). In the latter case the processors are connected by a high performance switching network.

The type of parallel machine used may often have a profound impact on the algorithmic structure of software designed for these machines (Clint *et al.* 1983; Jordan 1986; Fox *et al.* 1988). Thus, a recurring theme in parallel processing has been that the topology or structure of an algorithm should match as closely as possible the topology or architecture of the target machine (Fox *et al.* 1988).

The task of solving a problem on a parallel machine may be viewed as a mapping between one complex system and another, namely, a mapping between the process structure of the software used to solve the problem and the processor structure of the computer on which the software will run. To make best use of the target hardware this mapping should be one-to-one (Muhlenbein *et al.* 1987). In order to accomplish this mapping the software must be structured in an appropriate manner. Three general strategies for identifying suitable software components have been defined (Carriero and Gelernter 1989). These are known as:

- agenda parallelism,

- result parallelism, and

- specialist parallelism.

The following sub-sections consider each strategy in turn, each being illustrated by means of appropriate solutions to one of the subproblems associated with an interactive spelling checker. The mapping of the solutions onto suitable architectures is also briefly described.

4.7.1   Agenda Parallelism

A program may be structured as a sequence of steps, or agenda of activities. With agenda parallelism, the objective is to identify simultaneous computations in as many of the agenda activities as is possible. This type of parallelism is particularly suitable for those problems which involve the same transformation (or series of transformations) of all the elements of a data domain.

Such parallelism maps directly onto SIMD machines but is also appropriate for other types of parallel processing hardware. In particular, it is suitable for a homogeneous set of worker-processors which receive instructions from and report back to a master-processor: the paradigm of master-worker parallelism (Carriero and Gelernter 1989). In the literature the terminology processor farms (Pritchard *et al.* 1987) or task farms (Trew and Wilson 1991) is also used to describe this type of architecture. In such configurations the master-processor contains the master-process which

- controls the implementation of the agenda of activities;

- produces the subtasks associated with each item on the agenda; and

- farms out the subtasks to be processed by the worker-processors.

If the subtasks which make up an item on the agenda are independent of each other they may be processed by the worker-processors in parallel. In such circumstances each worker-processor repeatedly acquires and processes a subtask until all subtasks associated with a particular item on the agenda have been completed. Thus, in general, each worker-processor helps with the current item on the agenda.

In the case of the spelling checker, consider the problem of matching the words in the text with those in the dictionary. If it is assumed that information on unmatched text words is recorded in a file, Mismatches, then the following pseudo-code represents one possible solution to the problem:

```
begin (* Checktext *)
        for each word in text
                Compare word with those in dictionary;
                        if a mismatch occurs then append details in Mismatches ;
        endfor;
Display Mismatches ;
end.
```

This solution involves four basic activities:

- the selection of the words in the text;

- the checking of the selected words;

- the construction of the file Mismatches, and

- the presentation of the file Mismatches to the user.

Observe that no information is given concerning the order in which the words are to be selected for checking or, indeed, whether more than one word is to be selected at any given time.

Agenda parallelism, utilising the architecture of a processor-farm, may be incorporated into this solution in a number of ways, one of which is outlined below:

Each of the worker-processors is first assigned a copy of the dictionary and then repeatedly performs the following activities:

1. request a word from the master-processor;

2. perform the check function, reporting any mis-match to the master-processor; and

3. indicate to the master-processor the completion of 1 and 2.

The above activities cease when all of the words of the given text have been checked. The master-processor assumes responsibility for the selection (in some order) and distribution of the words to the worker-processors, and also for the construction and display of the file Mismatches . In this case, the data domain (the text words) is distributed across the worker-processors each of which performs, asynchronously, the same functions.

4.7.2 Result Parallelism

With result parallelism the design of the algorithm is based on the structure of the solution data: the parallelism is achieved by computing sub-structures of the solution simultaneously. Other terms used to describe such parallelism are geometric parallelism (Pritchard *et al.* 1987), data parallelism (Hillis 1985), and domain decomposition (Fox 1987; Fox *et al.* 1988).

Essentially, the data domain associated with the problem solution is distributed uniformly across the available processor array, with each processor being responsible for a defined spatial area or volume. Since it is frequently the case that the elements of the result structure are not completely independent it is usual for neighbouring processors to exchange data at appropriate intervals. Thus, processors may remain idle at times while waiting for information from other processors.

For optimum domain decomposition the problems of balancing the computational workload among the processors and of minimising the communication between the processors must be addressed. For many geometrically regular problems, decompositions which adequately balance the workload usually correspond to simple geometric partitions; but for more complex problems, such as irregular simulations and in artificial intelligence applications, more sophisticated techniques for problem decomposition are required. Currently, investigation of automatic load-balancing methods is a central feature of research in this type of concurrent computation (Kramer and Muhlenbein 1987).

Result parallelism generally requires each processor to have a complete copy of the program and a subset of the data domain from which it generates a subset of the required results. The result approach to parallelism is particularly suitable for those problems which produce multiple element data structures, or which have an underlying regular geometrical structure with spatially limited interactions. A number of examples of solutions to such regular problems are to be found in the recent literature, e.g. Fox *et al.* (1988) and Weston *et al.* (1991). These cover many areas of science which use domain decomposition techniques.

Returning to the spelling checker, the problem of checking the words in a given text with those in a dictionary may also be solved using result parallelism and an array of processors. One such solution is as follows:

- Assign a copy of the dictionary to each processor and partition the given text (the data domain) into a number of distinct subsections.

- Assign each subsection to a separate processor. Each processor will then check the words in its assigned text and constructs its own file of mis-matches.

- Whenever the processors complete their check, display the files of mis-matches sequentially to the user or merge them into the file Mismatches and display as required.

Essentially, each processor performs a spelling check on its own set of data. In this solution, the partitioning of the data domain is accomplished in a relatively straightforward manner and the communication overheads are relatively small. Further, the larger the piece of text to be analysed the more efficient the solution becomes. This follows because, for each processor, the ratio of the time taken to communicate with other processors to the time taken to perform the task of checking, which may be interpreted as the fraction of the total time spent on communication, becomes smaller. Since communication is an overhead, the smaller this ratio is the better.

4.7.3   Specialist Parallelism

With specialist parallelism, also referred to as algorithmic parallelism, the solution to a problem may be conceived as a logical network in which each node executes a relatively autonomous computation and in which inter-node communication follows predictable paths. The design of the algorithm is based on components of the problem solution, each of which has a clearly identifiable purpose and each of which co-operates with the others, in parallel, to achieve the desired solution. Thus, the parallel solution is designed around an ensemble of specialists connected into a network with the parallelism resulting from the nodes being active simultaneously. Essentially, the different functions of the problem solution which are capable of being executed in parallel are identified, each is assigned to a different processor, the processors implement their respective functions in parallel, and the data flows between the processors. Hence, in contrast to the agenda and result types of parallelism, it is code rather than data that is distributed across the processors.

The specialist or algorithmic approach to parallelism is particularly suitable for problems which partition naturally into separate realms of responsibility with well defined inter-communication channels. It is also suitable for algorithms which partition into a sequence of steps which may be implemented as a pipeline. Many examples of solutions to such problems are to be found in the recent literature, e.g. Clint *et al.* (1988).

Returning yet again to the spelling checker, a very simple pipe of three processors might be used to implement the basic activities involved in the solution. These are:

1. select words from the text;

2. check words against dictionary; and

3. present mismatches to the user.

The first processor in the pipe selects a word from the given text and passes it to the second processor for checking. Upon completion of the check, any mismatch is passed to the third processor for presentation to the user. This process is repeated, with all of the processors performing their individual activities in parallel, until all of the words of text have been checked.

4.7.4 Hardware Driven Design Procedure

As discussed under problem-driven design, the creative aspect of design, in general, means that any design procedure will really only amount to a few guidelines and a collection of standard options to explore. In the case of hardware-driven design, at present, there are the three basic design approaches to consider (as discussed above) and some additional advice on how to take advantage of these approaches.

The first practical consideration is the starting point of the software development. The software may be written from scratch but with a substantial volume of sequential software already in existence it is often necessary to base a parallel program on an existing sequential program. In this latter case, the first step is to examine the existing algorithm for those sections of the program which are independent of each other and for those loops which contain independent iterations. The order of computation may then be reorganised to create or increase the size and the number of independent tasks. Finally, the derived solution must be mapped onto the target machine. The approach here is essentially agenda parallelism, where the sequential algorithm is fanned out into parallel activity at appropriate points. It is highly unlikely that converting a sequential algorithm in this way will yield an optimal algorithm for any particular machine but it may often be the best way to preserve an earlier investment.

For new programs, the first step is to try to classify the program under the agenda, result and specialist headings. For many problems, however, the solution does not fall naturally into one of these classes and it is therefore necessary to take a hybrid approach utilising more than one technique. In practice, this is also desirable even if a solution is largely of one type. So, taking a specialist approach to a given problem, for example, it may be possible for one or more of the solution components to be implemented using the agenda or result technique.

The spelling checker provides a more concrete illustration of this point. In the result parallelism solution, discussed above, each processor in the array performs the functions of selecting and checking the words in its subsection of text and of constructing its own file of mismatches. If each processor is replaced by a cluster of three processors then a series of simple pipes can be constructed to perform these functions, thereby yielding a solution which incorporates a specialist approach to parallelism within a result approach. In general, for any given problem, all approaches to parallelism should be explored in search of the best solution.

4.8   Summary

Design is not, and indeed, cannot be a precisely defined activity. However, if armed with a repertoire of standard techniques the outcome is likely to be much more satisfactory than if approached in a haphazard manner. This chapter has identified and illustrated some of the main design techniques that are available and also discussed techniques for describing software systems at the preceding specification stage. Suggestions on how these techniques might be used in practice are also given.

The chapter draws a distinction between concurrent and parallel programming: concurrent programming is considered to be a problem-driven approach to the design of software whereas parallel programming is deemed to be driven by the characteristics of the target hardware. In concurrent programming, software components with potential for parallel execution emerge as the problem is decomposed whereas in parallel programming, components are chosen to suit the capability of the target machine for parallel processing. Both types of programming have the same initial starting point: a definition of requirements. At this stage of development the characteristics of the target hardware have an influence on performance requirements but the main concern tends to be with the nature of the problem. Thus, the same techniques for specifying requirements can be used in each case. The main techniques in common use are dataflow diagrams, finite state machines, Petri nets, temporal logic, process algebra, axiomatic specification, algebraic specification and model-based specification.

An object-oriented approach to design is highly suitable for the development of concurrent programs. For parallel programs, three main general design techniques are available: agenda, result and specialist parallelism. These can be used individually or, more typically, in combination. Specialist parallelism and object-oriented design are essentially the same approach. In both cases there is a search for objects. The difference between the two approaches is that in parallel programming the choice of objects is strongly influenced by the architecture of the target machine rather than areas of responsibility in the problem.

These specification and design techniques offer a toolbox of ideas that can usefully be employed in the construction of any concurrent or parallel program.

Acknowledgements

We are grateful to Maurice Clint for his comments on an earlier draft of this chapter. Some of the ideas presented have been developed through the SCAF-FOLD project (SERC GR/G 03700)—collaborative research involving British Aerospace, York University and the University of Ulster.

References

Andrews G. R. and Schneider F. B. (1983) Concepts and Notations for Concurrent Programming. *ACM Computing Surveys*, **15**, pp. 3–43.

Bal H. E., Steiner J. G. and Tanenbaum A. S. (1989) Programming Languages for Distributed Computing Systems. *ACM Computing Surveys*, **21**, pp. 261–322.

Birtwistle G. M., Dahl O-J., Myhraug B. and Nygaard K. (1973) *SIMULA begin*. Auerbach.

Booch G. (1991) *Object Oriented Design with applications*. Benjamin Cummings.

Brinksma E. (Ed) (1988) *Information Processing Systems OSI–LOTOS A Formal Technique Based on Temporal Ordering of Observational Behavior*. ISO IS 8807.

Brooks F. P. (1987) No Silver Bullet: Essence and Accidents of Software Engineering. *Computer*, **24**, pp. 10–19.

Bustard D. W., Norris M. T. and Orr R. A. (1988a) A Pictorial Approach to the Animation of Process-Oriented Formal Specifications. *Software Engineering Journal*, **3**, pp. 114–18.

Bustard D. W., Elder J. W. G. and Welsh J. (1988b) *Concurrent Program Structures*. Prentice-Hall International.

Bustard D. W. (1990) *Concepts of Concurrent Programming*. Curriculum Module 24, Software Engineering Institute, Carnegie Mellon University, Pittsburgh, Pa.

Carriero N. and Gelernter D. (1989) How to Write Parallel Programs: A Guide to the Perplexed. *ACM Computing Surveys*, **21**, pp. 323–57.

Clint M., Holt C., Perrot R. H. and Stewart A. (1983) The Influence of Hardware and Software Considerations on the Design of Synchronous Parallel Algorithms. *Software Practice and Experience*, **13**, pp. 961–74.

Clint M., Roantree D. and Stewart A. (1988) Towards the Construction of an Eigenvalue Engine. *Parallel Computing*, **8**, pp. 127–32.

Dahl O., Dijkstra E. W. and Hoare C. A. R. (1972) *Structured Programming*. Academic Press.

Davis A. M. (1988) A Comparison of Techniques for the Specification of External System Behavior. *Comm. ACM*, **31**, pp. 1098–115.

Davis A. M. (1990) *Software Requirements: Analysis and Specification*. Prentice-Hall, New Jersey.

Decegama A. L. (1989) *The Technology of Parallel Processing*. Prentice-Hall, New Jersey.

Diller A. (1990) *Z, An Introduction to Formal Methods*. John Wiley and Sons.

Erhig H. and Mahr B. (1985) *Fundamentals of Algebraic Specification 1*. Springer-Verlag, Berlin.

Flynn M. J. (1966) Very High-Speed Computing Systems. IEEE, **54**, pp. 1901–9.

Fox G. C. (1987) Domain Decomposition in Distributed and Shared Memory Environments. *First International Conference on Supercomputing, Athens*, pp. 1042–55.

Fox G. C., Johnson M., Lyzenga G., Otto S., Salmon J. and Walker D. (1988) *Solving Problems on Concurrent Processors*, Vol. **1**. Prentice-Hall, New Jersey.

Galton A. (Ed) (1987) *Temporal Logics and Their Applications*. Academic Press.

Gehani N. and McGettrick A. D. (Eds) (1988) *Concurrent Programming*. Addison-Wesley.

Gries D. (1981) *The Science of Programming*. Springer-Verlag.

Harel D. (1988) *On Visual Formalisms*. Comm. ACM, **31**, pp. 514–30.

Hayes E. (Ed) (1987) *Specification Case Studies*. Prentice-Hall International.

Hillis W. D. (1985) *The Connection Machine*. MIT Press.

Hoare C. A. R. (1985) *Communicating Sequential Processes.* Prentice-Hall International.

Ince D. C. (1988) *An Introduction to Discrete Mathematics and Formal System Specification.* Clarendon Press, Oxford.

ISO (1990) Proposed Draft Addendum to ISO 8807:1988 on G-LOTOS. ISO/IEC JTC1/SC21.

Jones C. B. (1986) *Systematic Software Development Using VDM.* Prentice-Hall International.

Jordan H. F. (1986) Structuring Parallel Algorithms in a MIMD, Shared Memory Environment. *Parallel Computing,* **3,** pp. 93–110.

Kramer O. and Muhlenbein H. (1987) Mapping Strategies in Message Based Multiprocessor Systems. In *PARLE (Parallel Architectures and Languages Europe)* (Ed. by de Bakker *et al.*). Springer-Verlag.

Lamport L. (1989) A Simple Approach to Specifying Concurrent Systems. *Comm. ACM,* **32,** pp. 32–45.

Milner R. (1989) *Communication and Concurrency.* Prentice-Hall International.

Muhlenbein H., Kramer O., Limburger F., Mevenkamp M. and Streitz S. (1987) Design and Rationale for MUPPET; A Programming Environment for Message Based Multiprocessors. *First International Conference on Supercomputing, Athens,* pp. 172–93.

Parnas D. L. (1972) On the Criteria to be Used in Decomposing Systems into Modules. *CACM,* **15,** pp. 1053–58.

Peterson J. (1981) *Petri Net Theory and the Modeling of Systems.* Prentice-Hall, Englewood Cliffs, N. J.

Petroski H. (1985) *To Engineer Is Human.* St. Martin's Press, New York.

Pneuli A. (1986) Applications of Temporal Logic to the Specification and Verification of Reactive Systems: A Survey of Recent Trends. In *Current Trends in Concurrency* (Ed. by J. W. de Bakker *et al.*), pp. 510–84. Springer-Verlag, New York.

Pritchard D. J., Askew C. R., Carpenter D. B., Glendinning I., Hay A. J. G and D. A. Nicole (1987) Practical Parallelism Using Transputer Arrays. In *PARLE (Practical Architectures and Languages Europe)* (Ed. by J. W. de Bakker *et al.*) Springer-Verlag.

Reisig W. (1985) *Petri Nets: An Introduction.* Springer-Verlag.

Shields M. W. (1989) *Finite State Automota*. Blackwell Scientific Publications, Oxford.

Sommerville I. (1989) *Software Engineering*, 3rd edn. Addison-Wesley.

Trew A. and Wilson G. (1991) *Past, Present, Parallel*. Springer-Verlag.

Weston J. S., Clint M. and Bleakney C. W. (1991) The Parallel Computation of Eigenvalues and Eigenvectors of Large Hermitian Matrices using the AMT DAP 510. *Concurrency: Practice and Experience*, **3**, pp. 179–85.

Yourdon E. (1989) *Modern Structured Analysis*. Prentice-Hall.

CHAPTER 5

Parallel Branch-and-Bound

G. P. McKeown, V. J. Rayward-Smith and S. A. Rush
University of East Anglia, UK

5.1 General Overview

The Branch-and-Bound paradigm is a general purpose enumerative technique for solving a wide range of problems in Combinatorial Optimisation, Operations Research and Artificial Intelligence; for a detailed discussion see Lawler and Wood (1966), Horowitz and Sahni (1978) and Ibaraki (1988). We shall introduce the Branch-and-Bound technique in general terms followed by a more formal definition using the 0/1 Knapsack problem as an example. Although the solution of this particular problem is not an especially interesting example of a Branch-and-Bound algorithm, it does provide a simple example for illustrating many of the important aspects of the Branch-and-Bound paradigm.

Branch-and-Bound would appear to be well suited to parallel processing because packets of work which could be executed independently on different processors are generated during a Branch-and-Bound algorithm. This is very encouraging as many algorithms of this type involve a considerable number of packets of work and the use of parallel processing could reduce execution times dramatically. However, the design of parallel Branch-and-Bound algorithms is not necessarily straightforward, and we shall discuss several possible strategies and show how these have been implemented on various parallel machines.

5.2 An Informal Description

Informally, the Branch-and-Bound approach can be characterised as an intelligent search for an optimal solution within a space of potential solutions to the given problem. Typically, the search space is exponentially large with respect to the size of the problem and the aim of the technique is to find the optimal solution whilst minimising the number of solutions to be explicitly considered.

During the execution of a Branch-and-Bound algorithm the search space is partitioned recursively into smaller and smaller disjoint subsets, a process known as *expansion*. Each subset generated then requires a value to be computed, corresponding to a bound on the best possible value for any solution which may be found in that section of the search space. After each iteration of the partitioning process, all subsets with a bound not better than the value for a known solution are identified and are excluded from any further part in the algorithm.

The partitioning process continues until all possible subsets have either been expanded or have been specifically excluded from further consideration. At this point, the current best solution must be at least as good as any other solution for the given problem and the algorithm terminates.

In the implementation of Branch-and-Bound algorithms the subsets of the search space are represented as *problem-states* which are generated and stored in an *active pool* of states. Associated with each problem-state is a set of potential solutions to the initial problem. The problem-states in the active pool correspond to sections of the search space which are still to be examined. Initially the active pool contains a single element, the initial problem description, which represents the whole of the search space. On termination of the algorithm the active pool is empty, indicating that the whole of the search space has either been examined or has been excluded from consideration.

The procedure described above can be characterised by a number of rules that define how the search is performed and how the search space is partitioned. These rules are known as the selection rule, the expansion rule, the branching rule and the bounding rules and are described informally as follows.

The *selection rule* is used to choose a problem-state from the active pool. After being selected the problem-state is removed from the pool and is added to the *dead pool*, consisting of problems that have been both generated and expanded.

The *expansion rule* proceeds as follows. The selected problem-state is examined to see if it provides a single feasible solution to the initial problem. If it does then the solution is considered as a possible optimal solution to the initial problem. The value generated is compared to the current best solution and the solution with the better value is preserved. If the problem-state does not provide a single solution, then it may be possible to show that it cannot possibly provide any feasible solution. In this case the problem-state is infeasible and is discarded. Otherwise, we expand the problem-state using a *branching rule* to split the problem into a number of sub-problem-states which are then added to the active pool. The newly generated problem-states represent a partitioning of the search space of the parent problem-state.

Bounding rules are used to eliminate problems which cannot lead to an optimal solution.

All Branch-and-Bound algorithms can thus be characterised by their expansion, branching, bounding and selection rules. We now present a more formal definition of Branch-and-Bound, using the 0/1 Knapsack problem as an illustrative example.

5.3   The 0/1 Knapsack Problem

In the 0/1 Knapsack problem, the objective is to maximise the possible profit which can be obtained from putting a number of items into a knapsack of known capacity. The constraints on this problem are that only whole items may be added to the knapsack and the maximum capacity of the knapsack must not be exceeded.

An instance of the 0/1 Knapsack problem consists of n items. Associated with the ith item is a volume, v_i, and a profit, p_i. Profit, p_i accrues if the ith item is included in the knapsack. For a knapsack of capacity C the problem can be formulated as follows:

$$\text{Maximise} \quad \sum_{i=1}^{n} p_i x_i$$

$$\text{subject to} \quad \sum_{i=1}^{n} v_i x_i \leq C,$$

$$x_i \in \{0, 1\}, \qquad 1 \leq i \leq n.$$

5.3.1 An Example

We now show how Branch-and-Bound might be used to solve the 0/1 Knapsack problem by considering a simple example. In this example the knapsack has a capacity of 100 and there are 4 items with the following volumes and profits.

Item	Volume	Profit
1	50	60
2	40	46
3	40	45
4	20	20

The Branch-and-Bound algorithm produces a binary search tree based on a decision as to whether or not an item should be included or excluded from the knapsack. Figure 5.1 shows the search tree generated during the sequential execution of this algorithm. Numbers inside the nodes represent the order in which problem-states are expanded while numbers outside represent bounds on the sub-problems formed. Labels on the arcs indicate whether a particular item is included or excluded from the knapsack, $+i$ indicates item i has been included while $-i$ indicates that it has been excluded. Square boxes are used to represent solution nodes, the bound on such a node being the cost corresponding to the solution.

The algorithm begins with an empty knapsack, node 1. This cannot be solved immediately so it is expanded to generate nodes 2 and 7 corresponding to the inclusion and exclusion, respectively, of item 1. The bounds on these can now be calculated using a greedy process which adds items until all of the available space is used; this process is allowed to add partial items to the knapsack. By sorting the items in descending order of profit/volume ratio, this bound provides an upper bound on the possible profit.

Node 2 has a current profit of 60 and still has 50 units of remaining space. This can hold all of item 2 and one quarter of item 3. This therefore has a bound

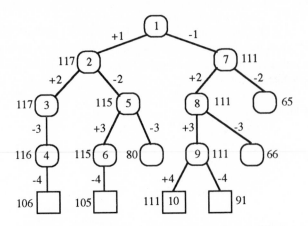

Figure 5.1: Example Branch-and-Bound search tree

of $\lfloor 60 + 46 + (\frac{1}{4} * 45) \rfloor = 117$. Node 7 has a current profit of 0 and 100 units of remaining space. It can therefore hold items 2, 3 and 4 and has a bound of $\lfloor 46 + 45 + 20 \rfloor = 111$.

We use a best-first selection rule, always selecting from the active pool that problem with the greatest bound. Thus node 2 is selected and expanded generating nodes 3 and 5 corresponding to the inclusion and exclusion, respectively, of item 2. Nodes 3 and 4 are expanded in turn but only generate 1 child apiece since, in each case, the child that includes the item exceeds the knapsack capacity.

The child generated by node 4 is actually a solution to the problem, but this fact cannot be established until the node is chosen for expansion. In this particular example it is never chosen, but is later pruned when a better solution is discovered. Nodes 5—9 are expanded in similar fashion.

Node 10 is then chosen for expansion. This node is found to be a solution node with a solution value of 111 and is the best solution so far found.

When we come to select the next node, there is no node in the active pool with a bound better than the current best and so the algorithm terminates. Our optimal solution includes items 2, 3 and 4 and has a value of 111.

5.4   A Formal Definition

We now give a formal definition of the Branch-and-Bound paradigm, using the approach given in McKeown *et al.* (1991b). Other formalisms for the paradigm have been presented in Mitten (1970), Kohler and Steiglitz (1974), Ibaraki (1978) and Kumar and Kanal (1983).

The Branch-and-Bound paradigm is a general approach for solving a wide variety of *discrete constrained optimisation problems (DCOPs)*.

In general, a constrained optimisation problem has the form:

$$Optimise \ f : S \rightarrow W$$

$$over \ F = \{s \in S \mid s \text{ is feasible}\}.$$

In a DCOP only a finite (although possibly very large) set of solutions needs to be considered in order to determine an optimal solution. All of the nodes in the Branch-and-Bound search tree correspond to DCOPs of the same type. We denote by P the problem space in which all of these DCOPs must lie.

The aim of a Branch-and-Bound algorithm is to minimise the number of elements of F which need to be explicitly enumerated. The Branch-and-Bound algorithm achieves this by combining a tree search with various strategies for pruning parts of the search tree.

S is the space of solutions for the given problem and W is the space of values for these solutions. A particular solution value, $s \in S$, is a *feasible solution* if and only if it satisfies each constraint in a specified set of constraints.

W is a totally-ordered set with ordering $\gg$ and is used as a measure indicating the optimality of a particular solution. For $w_1, w_2 \in W$,

$$w_1 \gg w_2 \Leftrightarrow (w_1 \geqslant w_2 \wedge w_1 \neq w_2).$$

For the knapsack problem, S is equal to $\{0, 1\}^n$ and $s \in S$ determines x_i values for a particular assignment of items to the knapsack. W is the set of integers, the ordering, $\geqslant$, on the set is simply $\geq$. $f(s)$ represents the profit received from the particular assignment, s. Thus, if $s = x_1 \ldots x_n$ then

$$f(s) = \sum_{i=1}^{n} p_i x_i.$$

In this example, P represents the space of problems in which an assignment has been made for some subset of the items. Each $p \in P$ has the form:

Maximise	$\sum_{i=1}^{n} p_i x_i$	
subject to	$\sum_{i=1}^{n} v_i x_i \leq C,$	
	$x_i = 0$	$i \in A$
	$x_i = 1$	$i \in B$
	$x_i = \{0, 1\},$	$i \in \{1, \ldots, n\} - A - B$
	$where A \cap B = \emptyset$	
	$and A \subseteq \{1, \ldots, n\}, B \subseteq \{1, \ldots, n\}$	

By interpreting F as a function over P we may denote P by a pair of functions (f, F), where $f : S \rightarrow W$ is the function to be optimised and $F(p) =$

$\{s \in S \mid is\_feasible\_for(p)(s)\}$. Here, $is\_feasible\_for$ is a higher-order, truth-valued function of type $P \rightarrow (S \rightarrow B)$. Thus, $F(p)$ is the set of those $s \in S$ satisfying all of the constraints of problem p.

There is a requirement that

$$\forall p \in P, \forall s \in S \cdot is\_feasible\_for(p)(s) \Rightarrow is\_feasible\_for(p_0)(s).$$

Because of this requirement it is not necessary for the feasibility predicate to be parameterised by the problem. It is therefore possible to define a further function $is\_feasible : S \rightarrow B$

$$is\_feasible \; \underline{\Delta} \; is\_feasible\_for(p_0).$$

Thus

$$is\_feasible(s) \equiv s \in F.$$

$is\_feasible(s)$ is true means that the solution, s, is a feasible solution to the initial problem. A solution feasible for a child problem must also be feasible for its parent and therefore also for the initial problem.

With the knapsack problem the feasibility test must check to see if the knapsack has been overfilled. Thus if $s = x_1 \ldots x_n$ then

$$is\_feasible(s) = \sum_{i=1}^{n} v_i x_i \leq C.$$

An *optimal solution* for a particular problem is a solution which is feasible for the problem and which is at least as good as all other feasible solutions of that problem. We note that although the optimal value is unique, more than one feasible solution may give rise to this value. Consequently, different implementations of a Branch-and-Bound algorithm may generate different optimal solutions.

An optimal solution for a constrained optimisation problem can be defined as follows:

if $F \neq \emptyset$, a feasible solution must exist, therefore an optimal solution is some $s^* \in F$ such that $\forall s \in F \cdot f(s^*) \geqslant f(s)$;

if $F = \emptyset$, there are no feasible solutions, so an optimal solution is ω, where $\omega \notin S$ may be interpreted as 'undefined'.

There are cases where an exact optimal solution is not required but a solution whose value is within a specified tolerance of the optimal value is sufficient. In such cases, an *approximate Branch-and-Bound algorithm*, guaranteeing to find an acceptable solution but not necessarily an optimal solution, may involve considerably less effort than an exact Branch-and-Bound algorithm. It has been shown (Wah and Ma 1984) that for some problems a linear reduction in accuracy can lead to an exponential reduction in computational overheads.

At any stage in a Branch-and-Bound algorithm the best solution found so far is known as the *incumbent*. The incumbent is defined to be ω until some feasible solution is known. From this point onwards, the incumbent is the element of F giving rise to the best solution found so far. On termination of the algorithm, the incumbent is the optimal solution for the initial problem. If the program terminates with an incumbent of ω, then the initial problem has no feasible solution. In a parallel implementation the incumbent must be made globally available so that it can be used by every processor. This is an important issue when considering parallel implementations of Branch-and-Bound algorithms.

In the following text we will only consider problems where a single optimal solution is required. If all optimal solutions are required then these can be found using Branch-and-Bound techniques very similar to those described below.

The root node in a Branch-and-Bound search tree corresponds to the initial given problem while the other nodes correspond to problems derived from this. The path from the root node to any other node in the tree corresponds to a sequence of elementary operations which transforms the initial problem into the DCOP corresponding to that node.

We define Σ to be a set whose elements represent the possible elementary operations which may be performed on a problem-state. Σ^* then denotes the set of all possible strings over Σ. Σ^* also contains the empty string, which we denote by ε.

In our knapsack example, the elementary operations are compound variables consisting of an item together with a tag to indicate whether that item is included or excluded from the knapsack. Σ is therefore defined as the set

$$\Sigma = \{(1, Inc), (1, Excl), (2, Inc), (2, Excl), \ldots, (n, Inc), (n, Excl)\}.$$

Σ^* is thus the set containing all possible permutations of zero or more of the items being included or excluded.

For each $a \in \Sigma$ there is a partial function $a : P \to P$ which applies operation a on a particular problem to generate a further problem. Corresponding to each string of operations, $\sigma \in \Sigma^*$, is a partial function, $\sigma : P \to P$. If $\sigma = \varepsilon$, the empty string, then $\sigma(p) = p$, otherwise the string of operations is considered as a single operation followed by the remaining operations. This may be described formally:

$$\text{if } \sigma = a\tau \text{ for } a \in \Sigma \text{ and } \tau \in \Sigma^*, \text{ then } \sigma(p) = a(\tau(p)).$$

If the initial problem is denoted by p_0 then a further problem, p, is in P if and only if it is possible to generate p from p_0 by application of a string of operations in Σ^*, i.e.

$$p \in P \Leftrightarrow \exists \sigma \in \Sigma^* \text{ such that } \sigma(p_0) = p.$$

In our example the initial problem corresponds to an empty knapsack. A further problem, p, can only be in the problem space if it is possible to generate

p by applying operations to the initial problem, i.e. by including or excluding items from the knapsack. The partial function, $a(p)$, is defined as follows for $a = (i, Inc)$ or $a = (i, Excl)$.

If x_i has already been assigned a value then $a(p) = p$, otherwise $a(p) = p\prime$ where $p\prime$ is derived from p by assigning a value to x_i, according to whether item i is included or excluded from the knapsack.

Nodes in the search tree represent a set of *problem-states* given by:

$$X = \{(p, \sigma) \in P \times \Sigma^* \mid \sigma(p_0) = p\}.$$

The initial problem-state, x_0, is represented by (p_0, ε), the initial problem and an empty string of operations.

In a Branch-and-Bound problem each problem-state, $x \in X$, is labelled with a value from $S \cup \{\omega\}$. We define the function

$$label : X \to S \cup \{\omega\}.$$

to be the function assigning labels to problem-states in the search tree. Either $label(x)$ will be a feasible solution for p_0 or it will be ω, such that:

$$label(x) \in F(p) \quad \Rightarrow \quad label(x) \text{ is an optimal solution for p}$$
$$label(x) = \omega \quad \Rightarrow \quad F(p) = \emptyset.$$

A problem-state, $x \in X$, becomes *bounded* when it is known that branching from x would not lead to a better incumbent. A bounded problem-state will not be expanded by the Branch-and-Bound algorithm. We define a dynamic function, $bound : X \to W$ such that while x is unbounded the value of an optimal solution for p is no better than $bound(x)$.

One possible value for the bound function is the 'cost', $f(label(x))$, of the current problem-state. This bound can often be improved, however, by including some estimate of the additional cost that must be incurred when deriving a feasible solution from x. If $g(x)$ gives such an estimate and $c(x) = f(label(x))$, then we could define $bound(x)$ to be $c(x) + g(x)$.

Better estimates $g(x)$ are likely to lead to tighter bound values and may lead to greater pruning of the search tree. There is, however, a trade-off between the potential for greater pruning of the search tree and the extra effort required in generating an accurate function, $g(x)$.

In the 0/1 Knapsack problem, the objective function is to maximise the profit over all possible assignments of items to the knapsack. The bound on any particular problem-state must therefore be at least as great as any possible solution which may be found in this section of the search space. We define $c(x)$ to be the profit obtained from the items already included in the knapsack and we calculate $g(x)$ to be an (over) estimate of the profit that can be achieved using the remaining (unassigned) items and any remaining space in the knapsack. One effective method for estimating the value of $g(x)$ is to relax the integer constraint and to find the maximum possible profit attainable if we allow fractions of objects. This value gives a fairly accurate estimate of the potential profit and is guaranteed to be an upper bound since a solution involving only whole items can never be better than one in which fractions of items are permitted. This value is easily computed by a greedy algorithm that includes items in order of greatest profit/volume ratio.

For each problem-state, x, which is expanded by the Branch-and-Bound algorithm there is a set of child problem-states derived by applying to p all those elementary operations in Σ that are defined on p. We define a function, $child : X \rightarrow 2^X$, where $child(x)$ is a finite (and possibly empty) subset of the set of children problem-states of x comprising only those children of x that are to be considered in finding an optimal solution to the initial problem. To guarantee that Branch-and-Bound returns an optimal solution for the initial problem, we impose the requirement that there is some $x\prime = (p\prime, \sigma\prime) \in child(x)$ such that the value of an optimal solution for $p\prime$ is equal to that for p

If $\lambda = label((p, \sigma))$ is feasible for p and hence is also feasible for p_0, then λ is considered as a possible optimal solution for p_0. Otherwise, x is *branched* from by generating $child(x)$. These children are then added to the active pool of problem-states. Initially, the active pool is the singleton set containing the initial problem-state, (p_0, ε).

Our knapsack example implements the child function by choosing an unassigned item and generating two child problems. One of these problems includes the item in the knapsack while the other excludes it. If there is insufficient capacity in the knapsack to hold the selected item, then only the child excluding the item will be generated.

$x \in Active$ becomes bounded (i.e. $bound(x)$ is set to $\bot$, the bottom element of W), if either of the following tests hold:

1. $F(p)$ is known to be $\emptyset$;

2. $f(incumbent) \geqslant bound(x)$.   **[pruning by incumbent]**

These two tests are present in all Branch-and-Bound algorithms and any kernel which implements these rules is referred to by McKeown *et al.* (1991b) as an implementation of *Basic Branch-and-Bound*. We shall later discuss further rules for pruning problem states. Kernels which implement these extra rules will be referred to as an implementation of *Extended Branch-and-Bound*.

Problem-states are *selected* from the active pool in order of precedence. Only problem-states that have not been bounded are selected. The precedence ordering is established by a priority function, defined on a set of values, V, where V is a partially ordered set with ordering $\supseteq$; thus, for $v_1, v_2 \in V$

$$v_1 \supset v_2 \Leftrightarrow (v_1 \supseteq v_2) \wedge (v_1 \neq v_2).$$

Then, $priority : X \rightarrow V$ is a total function such that x has precedence over y if and only if $priority(x) \supset priority(y)$.

Common priority schemes include breadth-first, depth-first and best-first searches as well as numerous heuristic search strategies. The basic search strategies are described as follows.

- A *breadth-first* algorithm expands the search space on a level by level basis. A priority function that always chooses to expand the problem with the smallest depth value will lead to a breadth-first expansion of the search space.

  An advantage of a breadth-first search is that it guarantees to find the solution of minimum depth in the tree. In a problem where solutions may be found at different depths this may be one of the optimisation criteria.

- A *depth-first* search attempts to generate a solution by performing 'stabs' down the search tree. After expanding a given problem, the algorithm attempts to expand one of its newly generated children. One way of implementing this is to always select for expansion the problem-state of greatest depth in the tree.

  An advantage of a depth-first search is that it often has the smallest storage requirements of any search strategy. This strategy also attempts to generate an initial solution quickly so that this can be used for pruning sections of the tree.

- The *best-first* strategy orders the problems for expansion according to their bound values. Problems with the best bound values are chosen to be expanded ahead of all others. An advantage of the best-first strategy is that, on a single processor, provided the bound values are unique, it will never expand a problem-state which is not expanded by all other strategies.

In many applications with sequential implementations, the following principle may be used. If the bound function generates either very tight or very loose bounds then depth-first and best-first search strategies expand similar numbers of nodes but the depth-first strategy usually has smaller memory requirements. For moderate bounds, however, the best-first strategy will usually perform fewer expansions.

For the knapsack example, we shall use a best-first priority ordering. V is therefore the set of integers, the same type as W. As we wish to maximise the possible profit, the problem-state with the largest bound value is likely to lead to a good solution. We therefore define the ordering on V as follows: $(v_1 \sqsupset v_2) \Leftrightarrow (v_1 > v_2)$.

5.5   Extensions to Basic Branch-and-Bound

As demonstrated in McKeown *et al.* (1991b) and Turpin (1991), some Branch-and-Bound algorithms make use of two further rules for pruning the search tree, referred to as pruning by *Dominance* and pruning by *Isomorphism*. We refer to a Branch-and-Bound algorithm that includes these additional rules as an *Extended Branch-and-Bound* algorithm.

$x \in Active$ becomes bounded if either of the following tests hold:

3. **[pruning by dominance]**
   $\exists x\prime \in Active \cup Dead \cdot x\prime \neq x \land (x\prime, x) \in Dom.$

4. **[pruning by isomorphism]**
   $\exists x\prime \in Active \cup Dead \cdot x\prime \neq x \land x\prime = isomorph(x).$

We now explain what is meant by these two tests.

Pruning by dominance uses a partial ordering, Dom, defined on $X \times X$ such that, $(x, y) \in Dom \land x \neq y$ implies that y may be bounded if x has been generated. We specify a truth-valued function $is\_Dom : X \times X \to B$ to satisfy:

$$is\_Dom(x, y) \underline{\Delta} (x, y) \in Dom \land x \neq y.$$

Pruning by isomorphism is possible if an equivalence relation, Iso, is apparent such that, $(x, y) \in Iso \land x \neq y$ implies that one of x and y may be bounded if both are generated. Let $[x]$ denote the equivalence class to which x belongs. The function $isomorph : X \to X$ is a dynamic function used to provide an algorithm for determining a representative problem-state of $[x]$ selected from $\{x\} \cup Active \cup Dead$.

Conceptually we view dominance as a method of pruning a tree and isomorphism as a method of pruning a graph. Note, although dominance and isomorphism can sometimes lead to a great reduction in the number of problem-states expanded there is a large class of Branch-and-Bound algorithms which do not incorporate either of these rules but rely on the rules of basic Branch-and-Bound for reducing the size of the search tree. Further details of extended Branch-and-Bound can be found in McKeown *et al.* (1991b).

Our example algorithm, for solving a 0/1 Knapsack problem, is not really suited to the extra rules of extended Branch-and-Bound but for illustrative purposes we will describe how these additional rules could be used.

The purpose of the dominance relation is to exclude problems from the search if they cannot possibly generate a solution which is better than another problem which has already been generated.

In the 0/1 Knapsack problem we may have the following situation. Assume we have three items, I_1, I_2 and I_3, with the following properties:

$$p_1 > p_2 + p_3$$
$$v_1 \leq v_2 + v_3.$$

We create a dominance relation as follows. Any sub-problem which contains item I_1 must dominate any similar sub-problem which specifically excludes item I_1 but includes both I_2 and I_3. By storing problem-states in a dead pool after they have been expanded and searching for such instances before a problem-state is expanded, it is possible to exclude large sections of the search tree.

The purpose of the isomorphism relation is to identify cases where two problem-states are essentially identical. In this case it is only necessary to expand one of the problem-states as the solution generated for one problem is equivalent to the solution generated from the other.

Suppose we have two items, I_1 and I_2, with the following properties

$$p_1 = p_2$$
$$v_1 = v_2,$$

then we can implement an isomorphism relation by checking for problem-states which include one of the items and exclude the other. Any problem-state which specifically includes item I_1 and excludes I_2 is essentially equivalent to another problem-state which includes item I_2 and excludes I_1. If one of these problems has already been expanded then the other can safely be discarded. Obviously, the discarded problem could never lead to a solution with a higher profit than the one which has already been generated. By maintaining a pool of dead nodes it is possible to identify such occurrences and prune large amounts of work.

Although isomorphism is much easier to implement than dominance, both are expensive operations due to the costs of storing and searching a large pool of dead nodes; these overheads in storage and calculation often make the implementation of these extra rules impractical. In a parallel machine the dead pool would need to be made globally available if dominance and isomorphism were to be fully implemented. This may be possible on a shared memory machine but is likely to lead to contention problems if several processors have to search the dead pool simultaneously. On a message passing machine, the dead pool may be stored centrally or may be distributed over the processors. Either way, accessing the dead pool is likely to lead to considerable communication overheads.

A possible way of reducing this overhead may be to use a local form of dominance where each processor keeps a local dead pool which includes only the problem-states expanded on that processor. Another way may be to use a tiered approach where local dead pools are stored but a central pool is also kept. This allows a processor to check its own local dead pool and to only access the global dead pool if the local one fails to find a dominating problem-state.

Many problems can however make use of a limited form of dominance that can be implemented with very small overheads, even on a parallel machine. In particular, many Branch-and-Bound algorithms involve the generation of a 'worst bound' on a solution as well as the current 'best bound'. This type of dominance occurs so frequently that Horowitz and Sahni (1978) use this form of bound test for pruning problem-states rather than the test against the current incumbent value. Note, any problem-state which is found to generate a single feasible solution may trivially generate both an upper and a lower bound with the same value. Thus this form of test against a worst bound is virtually equivalent to the test against the incumbent.

The *bound* function described earlier generates a 'best possible' bound on any solution which may be generated from this problem-state. For a maximisation problem this will be an upper bound on any possible solution. Note, this does not imply that a solution exists; it simply gives an upper bound on the value of any solution that could be generated.

We now define a second bound function *lbound* : $X \rightarrow W \cup \{\top\}$ such that, if $lbound(x) \neq \top$ then a feasible solution can be generated from this problem and it will have a bound at least as good as $lbound(x)$. Here, $\top$ denotes a value which is 'worse' than any element of W.

By using $lbound(x)$ it is possible to prune any problem-states which have a *bound* which is worse than the best *lbound* so far found. Since this new bound is a bound on the worst possible value of the incumbent, we refer to it as the *incumbent-bound*.

$x \in Active$ may now be bounded if:

5. $f(incumbent\_bound) \gg bound(x).$ **[pruning by lower bound]**

Any Branch-and-Bound algorithm which does not generate both an upper and a lower bound can have a dummy *lbound* function which returns the value $\top$. This form of dominance can be implemented very cheaply even on a message passing parallel computer because only a single value needs to be stored rather than the whole dead pool. The use of the incumbent-bound is very similar to that of the incumbent and indeed might play the part of the incumbent in practice. It may however still be necessary to keep the actual incumbent value, particularly in approximate Branch-and-Bound, where problem-states with a bound within a certain tolerance of the genuine incumbent can be pruned. Pruning of this type should not be made against the incumbent bound as a solution has not actually been found with this value.

In our knapsack example we can implement the lower bound function easily by using the current profit from a problem, $c(x)$, from the definition of *bound*. This is a lower bound on any solution which can be generated from this problem-state. Any problem-state which does not exceed the capacity of the knapsack can obviously generate a feasible solution with this value as all of the items not currently in the knapsack can trivially be excluded from it.

5.6   Parallel Branch-and-Bound

We begin by discussing the attractions of executing Branch-and-Bound algorithms in parallel. We then discuss possible sources of parallelism in the Branch-and-Bound paradigm and the implications of running such algorithms in parallel. Next, we discuss various parallel implementations and give some indications of the strengths and weaknesses of each. We finish by giving performance results for some of the parallel Branch-and-Bound kernels which have been developed and which illustrate some of the important aspects of parallel Branch-and-Bound.

5.6.1 Reasons for Parallel Branch-and-Bound

Branch-and-Bound algorithms are often used for solving NP-hard problems with very large search spaces. The time to solve such problems is normally proportional to the number of problem-states in the search tree. Problems with large search spaces therefore often place large demands on resources. A parallel implementation could allow these programs to be run over a number of machines and therefore allow them to complete more quickly.

There has been much research in the past on the possibility of using approximate Branch-and-Bound to reduce the execution time of this form of algorithm. By using a parallel machine it may be possible to generate an optimal solution in the same time as the sequential machine takes to generate an approximate solution. Even if it is not possible to generate an exact solution in reasonable time, the parallel machine should, hopefully, be able to generate a much closer approximation to the true optimal solution.

The use of parallel processing may allow larger problems to be solved than would be possible on a sequential machine. As many Branch-and-Bound problems are NP-hard the size of problem which can be solved in 'reasonable time' is most unlikely to grow linearly with the number of processors but some increase in problem size should be possible.

5.6.2 Implications of Parallel Branch-and-Bound

We begin by making a slight change to the classical, sequential, view of Branch-and-Bound. The classical definition of Branch-and-Bound (see Horowitz and Sahni 1978; Turpin 1991) requires that the state space search generates all of the children of a problem before any other active node can become the selected node. This is obviously a requirement imposed on any sequential implementation and it is inappropriate for a parallel one. Indeed, there is no reason, in a parallel system, why a child problem cannot be selected for expansion on one processor while its parent is still being expanded on another processor.

Another point to consider is that a parallel Branch-and-Bound algorithm might take longer to execute than the equivalent sequential algorithm. This result has been known for some time and is known as a 'detrimental anomaly'. Similarly, it is possible for a parallel algorithm to have super-linear speedup over the sequential one, this being referred to as a 'speedup anomaly'. Details of these anomalies together with some examples can be found in Burton *et al.* (1982) and Lai and Sahni (1984) but we now show some simple examples.

Figure 5.2 shows an example of a detrimental anomaly, where two processors take longer than a single processor. We assume that processors in the parallel version are given work to do whenever they become idle and, in this particular case, one such piece of work generates a large amount of work which is not performed by the sequential version.

In the figure shown, the sequential version expands 3 nodes (*a*, *c* and *d*)

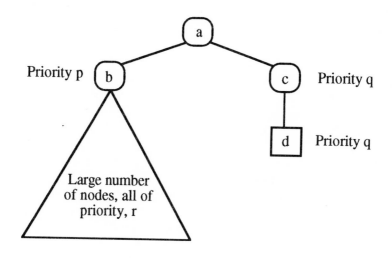

$r \supset q \supset p$, bound(d) >> bound(b)

Figure 5.2: An example of a detrimental anomaly

while the parallel one also expands the whole of the left sub-tree. This can obviously lead to a great deal of unnecessary work and may result in the parallel version taking longer to terminate than the sequential one.

Figure 5.3 shows an example of an acceleration anomaly. In this case the parallel version was able to find a solution quickly and could thus prune more problem-states. This allows the 2 processor system to have a speedup of more than 2 over the single processor one.

In the figure shown, the sequential version expands the whole of the search tree shown while the parallel one only expands 3 nodes (*a* followed by *b* and *c* in parallel). The solution node, *c*, generates a new incumbent which allows the whole of the left sub-tree to be pruned.

McKeown *et al.* have recently identified a further type of anomaly in parallel Branch-and-Bound, which they refer to as 'efficiency anomalies' (McKeown *et al.* 1991a; Rayward-Smith *et al.* 1991). These anomalies are due to the fact that a parallel system usually has more memory than a sequential one. This extra storage often allows the processors to reduce the amount of work necessary to perform problem expansions.

The expansion of a problem-state in a Branch-and-Bound algorithm might involve the generation and manipulation of a complicated data structure such as a cost matrix. It is, however, often much easier to generate this structure using the data from an ancestor problem if such data can be made available. If the parallel machine has more memory than the sequential one then it may be possible to store more data and therefore reduce the amount of effort spent

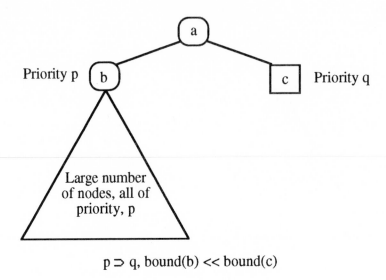

$p \supset q$, bound(b) << bound(c)

Figure 5.3: An example of an acceleration anomaly

recreating the data from scratch. As many parallel machines are created from standard sequential parts, it is often the case that a machine with m processors will have m times as much memory as one with a single processor and then efficiency anomalies can occur. Because of these anomalies it is possible for an m processor parallel machine to perform more problem expansions than a single processor one, but to still get greater than m fold speedup.

Figure 5.4 shows an example of an efficiency anomaly. Each processor has enough memory to store a single set of data from a parent problem and the time to expand a problem using this data is much shorter than the time taken if the data must be recalculated from scratch. In this example, each processor saves the data from the current problem expansion in the hope that one of its children will be expanded in the next iteration and can make use of this data. The single processor expands nodes $a, b, c, d, \ldots$ and therefore requires data to be recalculated for everything except node b. The 2 processor version expands node a and then one processor expands nodes b, d, f, h while the other expands c, e, g, i. This allows the processors to use previous data in all but 1 expansion (processor 2 must recalculate the data for node c unless it can be sent this data when it is sent the work to do). If the time to expand a problem-state is highly dependent on the availability of this data then the parallel machine can achieve super linear speedup whilst still expanding as many problem-states as the sequential one.

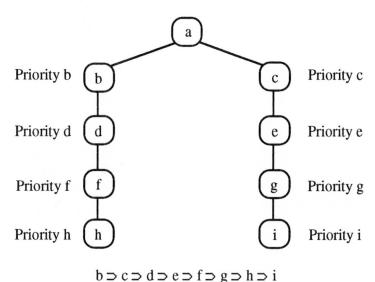

$$b \supset c \supset d \supset e \supset f \supset g \supset h \supset i$$

Figure 5.4: An example of an efficiency anomaly

5.6.3 Parallelisation of Branch-and-Bound Algorithms

There are a number of ways in which a Branch-and-Bound algorithm may be made to run on a parallel machine. We now present a number of possible sources of parallelism and discuss their advantages and disadvantages. For this discussion, we assume that an m processor parallel machine is available for solving the problem.

- *Parallel expansion of the search tree with different initial bound values.* If an upper and lower bound, U_b and L_b, are known for the problem then the processors can begin with different initial bound values in this range. Each processor is assigned an initial upper and lower bound, $(L_b, U_{b,i})$. Suitable initial values for processor i might be:

$$(L_b, L_b + i * (U_b - L_b)/m).$$

Processors that have their initial bound set too low will finish quickly without finding a solution and could then be restarted with a new value in a higher range. When a feasible solution is found its value is passed to all processors and any processor whose work becomes bounded chooses a new range of values. The whole process finishes when one processor terminates with an optimal solution.

This method of introducing parallelism should have very low communication costs but has the obvious problem that much of the work is performed many times. This is a fundamental problem with the algorithm as the initial problem-states will be expanded by every single processor. This method has the additional problem that it is not applicable for algorithms using the best-first search strategy. Using this strategy, no processor would terminate with an optimal solution before the processor running the algorithm with bounds $[L_b, U_b]$ and any processor obtaining an optimal solution would have performed exactly the same set of expansions.

Although this method does not look particularly promising it may prove effective in a small number of cases when the lower bound happens (by chance) to be close to the optimal value. In this case, the processor assigned this initial range will prune large sections of the search tree and will quickly find an optimal solution.

Similar results to those described could be obtained by using an approximate Branch-and-Bound algorithm and assigning different tolerances to the different processors. A processor searching with a large tolerance would finish relatively quickly and could start again with a lower tolerance value. Its first attempt would, however, provide reasonably tight bounds on the value of the optimal solution and these could be used by all of the processors to prune problem-states. The process terminates when a processor finishes with a tolerance of 0%. Similar principles using $\alpha - \beta$ search are described in Baudet (1978).

- *Parallel search using different algorithms.* Many optimisation problems can be solved in a number of different ways using different algorithms with different bound and priority functions. By running different algorithms on different processors it may be possible to complete the search more quickly. A possible implementation could have one processor performing a depth-first search while another used best-first search.

  Once again, the same tree is being expanded by more than one processor so some nodes will be expanded unnecessarily. As different search strategies are used, however, the overlap in the areas searched should not be too great. Although the speedup of this system may not be very great, it may have some beneficial effects on memory usage. The generation of the bounds by the processor using the depth-first strategy may allow another processor using a best-first search to prune large sections of the search tree. The use of the best-first search strategy may otherwise have been impossible due to its large memory requirements.

- *Perform the expansion of a single node in parallel.* The amount of effort required to expand a single node may be very significant and it may often be possible to split this work between a number of processors. For example, consider a Branch-and-Bound algorithm where each problem-state generates m children. Each of these children requires a bound value and a priority value to be computed. In many such algorithms these values could be generated in parallel on m different processors. If the generation of the bound and priority values forms a considerable part of the execution time then this could lead to a reasonable speedup.

This method may prove effective for a number of problems and has the advantage that it avoids the problems of detrimental speedup anomalies. Since the same search tree will be expanded, irrespective of the number of processors, such anomalies will never occur. Unfortunately, the ratio of communication to calculation is likely to be fairly high using this technique as data must be sent to processors every time a new child is generated.

An example where this could be used is in the solution of Integer Programming problems. One method of solving these problems requires an LP relaxation to be performed in the expansion of every problem. The solution of an LP relaxation can involve considerable effort and it may be possible to do this in parallel. There has been much interest in interior point methods (Karmarkar 1984) for solving the LP component of these problems. These methods have the potential for parallelism (Mitra *et al.* 1988) and may prove a particularly effective method for solving this type of problem.

The main disadvantage of this form of parallelisation is that it is likely to be very problem dependent and cannot be done in a general way. This approach may, however, prove effective for certain classes of Branch-and-Bound algorithms and the fact that its performance is predictable may make it suitable for applications where predictable performance is more important than average absolute performance.

- *Parallel evaluation of subproblems.* As the search tree is expanded there is a pool of active problem-states which could be expanded in parallel. As problem-states are independent of each other they can be expanded in parallel on different processors. This appears to be the obvious source of parallelism with a shared memory multi-processor but with a message passing machine it may be difficult to follow the priority function accurately without large communication overheads.

There are many ways in which this type of parallelism could be exploited, different approaches involving different tradeoffs between the amount of communication necessary and the average number of problem-states expanded. Possible methods of parallelisation include:

1. *Static distribution of the search tree.* One strategy which would lead to very low communication costs would be to split the search tree statically between the available processors. Incumbent values could be broadcast to allow pruning of the search tree and termination of the algorithm occurs when all processors have completed the search of their section of the tree.

The disadvantage of this method is that the amount of work required to expand a section of the search tree is not normally known in advance so it is not possible to divide the work evenly between the processors. In general, if the amount of work required to expand a section of the tree is known in advance then it is likely that the work need not be done at all!

This method may prove effective when a complete expansion of the search tree is required and where no bounding takes place. A possible use of this technique is therefore to verify that a known solution is optimal. In this case the value of the proposed solution is already known and the task is to expand the whole of the remaining tree to verify that no better solution exists. Although the actual shape of the final tree may not be known, it may be possible to estimate its shape and to use this estimate when dividing the work between the processors.

2. *Dynamic distribution with farming of available work.* Dynamic distribution of work allows work to be distributed more evenly between the available processors. Unfortunately, this also means higher communication costs. Although communication costs can be high with a message passing computer, many of the current parallel Branch-and-Bound kernels use an approach based on this principle (Roucairol 1987; Quinn 1990; McKeown *et al.* 1991a). On a shared memory machine this approach seems particularly suitable though memory contention for accessing the pool of work can prove problematical.

3. *Dynamic distribution with farming of large tasks.* Rost and Maehle (1989) have noted the large communication overheads present in the farming technique and have come up with a similar scheme where larger tasks are farmed between the processors.

The initial problem is first split into a number of sub-problems by running the algorithm sequentially on one processor using a breadth-first search. This processor continues until M problem-states have been generated where $M \gg m$. The problem-states in the active pool now represent large sections of the search space.

A farming technique is then used to spread the work over the available processors. Each processor is given one of these problems to expand and does not communicate until it generates a new incumbent or it has expanded the whole of the search tree represented by the problem-state. When the whole sub-tree has been expanded a message is sent to the master processor requesting another piece of work.

Two main disadvantages of this method have been noted. First, the initial breadth-first search is done on a single processor while all other processors are idle. Rost & Maehle do not consider this to be a major problem as the initial breadth-first search is assumed to be a very small part of the total search.

The second potential disadvantage is that, when the pool of problem-states on the master processor is exhausted, processors finishing the work allocated to them have to wait until all of the other processors have finished before the program can terminate. As the sub-trees are specifically chosen to involve a large computation component (in order to reduce communication overheads) this could lead to a considerable time when only some of the processors are busy.

4. *Dynamic distribution with local priority scheme.* In order to reduce the communication costs inherent in the farming method it is possible to store the newly generated child problems in the memory of the processor that generated them and to only follow the priority scheme locally. This implies that there is no global pool of active problem-states so care must be taken to ensure that all processors have work to do and to ensure that the priority function is adhered to. These factors can prove difficult to overcome but much work has been done in this area and many of the currently available kernels use techniques based on this principle (Mohan 1983; Pardalos and Rodgers 1990; McKeown *et al.* 1991a).

5. *Randomised algorithms.* The two methods described above perform extra work to ensure that the same problem-state is not expanded on two different processors. Randomised algorithms accept that some work may be done more than once but use a random choice of which problem to expand to minimise the chance of two processors choosing the same problem. As well as allowing work to be done several times, this method also tends to ignore the priority function and may therefore expand work which has very little chance of generating a new incumbent. Nevertheless, experiments with this approach have shown that it can be reasonably successful for some classes of algorithms (Janakiram *et al.* 1988).

5.7 Architectures for Branch-and-Bound

Although there are many ways of introducing parallelism into Branch-and-Bound algorithms, the main way in which this can be achieved is by dynamically dividing the search space between the available processors and expanding multiple problem-states in parallel. The main decision is therefore whether to have a global control, where the problem-states of globally highest priority are chosen for expansion, or to have a local control, where only a subset of the problem-states are available.

The use of a global control ensures that the priority function is closely adhered to while a local control may lead to the unnecessary expansion of a number of nodes. The use of a global control, however, requires that all processors have access to a global pool of active nodes. In a message passing machine this will lead to large communication overheads, while in a shared memory machine this may lead to memory contention in certain memory blocks. The shared memory machine may also require the use of semaphores to ensure that processors do not attempt to update the contents of the same memory location simultaneously.

In all cases the current value of the incumbent must be made available to all processors so that it can be used to prune sections of the search tree. This value can either be stored in a section of shared memory or can be broadcast to every processor each time a new incumbent is generated. This should not be an important consideration, however, as the value of the incumbent should not usually change very frequently.

Implementation of extended Branch-and-Bound requires access to the dead pool for pruning problems; this is most easily done using shared memory to store the problems after they have been expanded. Implementation on message passing machines is likely to lead to large communication overheads for accessing the dead pool.

The work required to expand a particular problem-state, and therefore the time required, is often dependent on the actual data in the problem being expanded. For example, it is often much quicker to determine that a problem is immediately feasible than it is to generate all of its children. Similarly, the expansion of problem-states at different levels in the search tree may require differing amounts of work. This tends to suggest that enforced synchronisation may cause some loss in performance as some processors sit idle, waiting for others to finish. A MIMD machine is therefore likely to be more effective at solving Branch-and-Bound problems than a similar SIMD one.

Yu and Wah (1983) and Wah and Ma (1984) have suggested a parallel architecture and operating system specifically designed to execute Branch-and-Bound algorithms. This architecture, referred to as MANIP, makes use of special hardware to improve the parallel performance of this type of algorithm.

5.7.1   The MANIP Architecture

The MANIP architecture is designed to perform a best-first expansion of the search tree and uses multiple processors to evaluate multiple sub-problems in parallel. A global data register is used for storing the current incumbent, and is implemented using sequential associative memory to prevent simultaneous updates. Other important features of the MANIP architecture are the sub-problem memory controllers which store the active pool of problem-states and the selection and redistribution network which allows the various memory controllers to communicate with each other. The m processors are split into n groups, where each group uses a single memory controller. The n controllers are then connected using a ring network which allows problem-states to be distributed between them.

MANIP is a synchronised architecture so an m processor machine should expand m problem-states in parallel but must then have a synchronisation step so that the m problem-states of highest priority can be identified and distributed to the processors. The designers of MANIP have noted that the selection phase need only identify the m problem-states of highest priority but does not require these to be sorted in any particular way. They believe that the identification of these problem-states can be performed in parallel using a ring network to connect the memory controllers. Hardware comparators are used to identify the highest priority nodes in each memory controller and these are then sent to neighbours in the ring network. The process of identifying high priority work and passing it to neighbours continues until the m nodes of highest priority are distributed evenly between the memory controllers.

The MANIP architecture has been simulated on a DEC VAX 11/780 and encouraging results have been generated showing good speedup for the vertex covering problem (Wah and Ma 1984).

5.7.2 MIMD Approaches

The most common method for performing Branch-and-Bound algorithms in parallel is to expand multiple problem-states in parallel by expanding one problem-state on each processor. There are four main approaches for performing this, depending on whether a local or global active pool is used and whether the system is synchronised or unsynchronised. These approaches are briefly mentioned above but we now explain them in greater detail. We assume a MIMD message passing machine for this discussion and we assume that a heap-tree is used for storing and sorting the active pool.

Synchronised with a Central Active Pool (SCAP)

This strategy uses a global control to ensure that only the problem-states of the very highest priority are expanded at any stage. In every iteration, each of the m processors is allocated one of the m active problem-states of highest priority to expand. Clearly, a functional requirement of this strategy is that these m can be identified and thus a master/slave approach is usually adopted. The master processor is responsible for identifying the m problem-states of highest priority at the start of every iteration and one is sent to each of the m slave processes. Note, several implementations allow a master process and a slave process to be run in parallel on a single processor.

This task can obviously be performed using a farming technique where all active problem-states are stored on the master processor. At the start of each iteration the master processor identifies the problem-states of highest priority and sends one to each of the slave processors.

During the expansion phase, each slave processor expands the problem-state it has been allocated. As described earlier, expanding a problem-state involves either generating a new feasible solution or generating the children of the problem-state. In the former case, if a new incumbent value has been generated, it is sent to the root while in the latter case the newly generated child problem-states are sent.

Termination of the expansion phase can be determined when a message is received from each processor. Termination of the algorithm is easily detected when there are no active problem-states to distribute during the selection phase.

A disadvantage of this approach is that it can lead to large communication overheads as communication time is usually proportional to message size and the amount of data that must be sent between processors is likely to be very large. The size of a problem-state depends on the definition of the problem, but it will typically require at least $\Theta(n)$ bytes and possibly $\Theta(n^2)$ or $\Theta(n^3)$ bytes to describe a single state in a problem of size n. The master processor can thus become a communication bottleneck as problem-states are sent to the master processor and then back to the slave processors for expansion.

The use of a synchronised system can lead to further overheads as some processors may be forced to remain idle while other processors complete the work allocated to them.

Asynchronous with a Central Active Pool (ACAP)

The implementation of this strategy is similar to that of SCP but does not involve a synchronisation step. Thus, this strategy requires the master processor to allocate an active problem-state (if one exists) to a processor as soon as that processor finishes its expansion phase.

If a processor becomes idle when the active heap on the root is empty, the master processor can increment a count. When new nodes are generated, this count can be decremented and the idle processors can be sent work. Termination of the algorithm occurs when the idle processor count is equal to m, the number of processors. This strategy should perform more efficiently than the SCP strategy as processors will not have to wait for the synchronisation step. It will, however, have the same large communication overheads as the earlier strategy and the root processor can become a communication bottleneck.

Synchronised with a Local Active Pool (SLAP)

This strategy uses a local form of control to reduce the communication overheads of the strategy mentioned above. In order to do this, each processor only considers the work available in its own local heap. The processor removes the highest priority problem-state from its heap, expands it, and adds any newly generated children problem-states to the heap. When a processor empties its own local heap, it does not terminate but instead sends a request to its neighbours for more work.

When a new incumbent is generated, its value can be broadcast to each processor in the network so that it can be used to prune unexpanded problem-states. Rather than communicating after each expansion, the processors now need only issue a message if they have run out of work or have generated a new incumbent.

If this system were to be synchronised, it would be necessary for each processor to synchronise after every problem expansion and this would be likely to negate the advantages of using a local control strategy. We therefore do not discuss this any further, only discussing the implications of the asynchronous version.

Asynchronous with a Local Active Pool (ALAP)

This is similar to the strategy just described, but does not have a synchronisation step so processors continually remove problems from their local heap, expand them and add the newly generated children problems to the heap.

Two potential disadvantages of this method are, firstly, that it is now much more difficult to detect when the algorithm has terminated and, secondly, that processors have no knowledge of the active problem-states stored on other processors.

The first of these difficulties means that it is necessary to use some form of distributed termination detection algorithm while the second means that the m problem-states being expanded at any moment in time may not be the m highest priority problem-states in the system. In a case where a large proportion of the search tree must be examined this approach is likely to be very efficient as communication costs are minimised, but in cases where only a small part of the search tree is examined it is likely to lead to a large number of unnecessary expansions.

Further important considerations concern how requests for work are handled when a processor empties its local heap of work and how the work is redistributed after this occurs.

As already mentioned, this strategy is unlikely to follow the priority function very accurately. This may result in more nodes being expanded than is necessary in implementations which follow the priority function more closely.

In order to spread the high priority work more evenly between the available processors it may be necessary to use some strategy for sending work to other processors in the network after certain intervals. It is likely that the most effective strategy will involve sending high priority work to other processors, but the optimal interval between each such message is likely to depend on the particular algorithm and the hardware being used. Sending work frequently will help to distribute the work more evenly but will obviously also increase the communication overheads as well as the overheads for adding and removing problems from the queue of work.

5.8 Parallel Implementations

In this section we discuss a number of implementations of parallel Branch-and-Bound algorithms using different parallel machines. We begin by discussing two implementations based on the iPSC hypercube, then one on the NCUBE and finally a number of kernels on networks of INMOS Transputers.

5.8.1   Pardalos and Rodgers

Pardalos and Rodgers (1990) have used an iPSC hypercube for parallel processing of a number of quadratic zero-one problems. They use parallel processors to expand multiple problem-states simultaneously but use a depth-first search strategy to reduce memory demands. While MANIP uses specially designed hardware to ensure that only the best nodes are chosen for expansion, Pardalos & Rodgers have opted for a much more local form of control.

This system is similar to the ALAP strategy described earlier and must solve the problem of how to request work when the local heap of work becomes empty.

Obviously it is not possible for processors to predict when a neighbouring processor will empty its heap of work so they must be prepared to receive messages at any time. Unfortunately, on the hypercube, the operations to send and receive messages are very expensive compared to other operations. It is therefore important not to issue communication commands more often than is necessary.

Ideally, messages indicating new incumbent values or requesting more work should be received and processed as quickly as possible to allow pruning or to prevent processors sitting idle. Unfortunately, for the reasons mentioned above, it is inefficient to continually check for incoming messages. The kernel therefore uses a variable MAXV which indicates how often a processor should check for communication from its neighbours.

After every MAXV problem expansions the processor issues a 'probe' operation to check for incoming messages. If messages are waiting they are dealt with immediately, otherwise the processor begins another MAXV problem expansions. The choice of value for the parameter MAXV has been found to be fairly critical and is also likely to be highly problem dependent. Too large a value indicates that processors may sit idle for long periods of time while too small a value indicates that processors spend a large proportion of their time communicating. Empirical results from the quadratic zero-one problems being investigated suggest that MAXV should be set to a value of approximately 1000.

Note that the first few iterations of the algorithm are treated as a special case to ensure that all processors have work, and it is only after this point that the MAXV parameter is used.

5.8.2   Clausen and Träff

Clausen and Träff have used a similar method to that described above to run experiments on the graph partitioning problem (Clausen and Traff 1991), also using an iPSC hypercube. The Clausen/Träff kernel uses a similar scheme to the one mentioned above, where messages may only be received after STEPS iterations of the expansion process. They refer to this kernel as the 'on demand' strategy where messages are only sent in response to a request for work.

One problem with this strategy is that it is possible for a processor to exhaust its supply of work and then to be idle for a considerable time as its neighbours complete their STEPS node expansions. A second kernel was therefore developed which uses an 'on overload' strategy. This allows problem-states to be sent to neighbouring processors if the queue of work on the current processor gets too large.

After completing STEPS expansions the processors check the size of their queue of work; if the queue is larger than some value QUEUEMAX then work is sent to the neighbouring processors. This will hopefully spread the work more evenly between the processors and it should reduce the chance that a processor will run out of work. Processors should therefore spend a greater proportion of their time actually expanding problem-states.

The designers of this system have come up with a number of rules for altering the value of the QUEUEMAX parameter as the algorithm proceeds. These rules are used to allow for the changing amount of work in the system as the algorithm proceeds and seeks to reduce the number of unnecessary messages.

Results presented for these kernels show that, for this particular problem, the use of the 'on overload' strategy increases the processor usage dramatically, but often does so at the cost of extra node expansions. This implies that the processors are kept busy but not necessarily doing useful work. This should, however, prove advantageous when a large portion of the search tree must be examined in order to identify the optimal solution.

5.8.3   Quinn

Quinn has performed a number of experiments on an NCUBE hypercube computer (Hayes *et al.* 1986) using the Travelling Salesman Problem as an example (Quinn 1990). The experiments compared 5 different strategies for the implementation of parallel Branch-and-Bound, one semi-synchronous algorithm and 4 asynchronous ones. The implementation of the Travelling Salesman Problem used by Quinn generates a binary search tree where at any stage a particular edge is chosen and may be included or excluded from the tour.

1. A semi-synchronous algorithm with a global pool of active problem-states. Each time a processor completes a node expansion it sends a request to a master processor asking for more work. This is similar to the AGAP strategy mentioned earlier and has problems of high communication overheads as the master processor is a communications bottleneck. This bottleneck actually results in a drop in performance for large networks of processors.

The asynchronous algorithms are all variations on the theme described above for ALAP. The different strategies use various methods for distributing the work between the available processors. All of these implementations require a problem-state (if available) to be sent to a neighbouring processor after every problem expansion is completed. The variants are:

2. The problem-state with the included edge is retained by the processor on which it is generated while the problem-state with the excluded edge is sent to one of the neighbours. This achieves good processor utilisation, but poor speedup as a large number of unnecessary expansions are performed. Only a few processors (those next to the root processor) perform useful work, all other processors performing low priority work which is pruned in the sequential version.

3. A similar algorithm to the one just described, but where the newly created problem-state with the smaller bound value is retained and the one with the higher bound value is sent to a neighbour. Again, a large number of unnecessary expansions are performed so poor speedup is reported.

4. Both newly generated problem-states are added to the queue of work and the problem of second highest priority is removed and sent to the neighbour. The problem of highest priority is thus kept for expansion on the current processor. This is more effective at spreading the high priority work between the available processors but processors very far from the root processor still perform little useful work.

5. Both newly generated problem-states are added to the queue of work and the problem of highest priority is sent to a neighbour. This strategy spreads the work more evenly between the available processors and all processors perform a reasonable amount of useful work. This strategy therefore achieves good speedup as well as reasonable processor utilisation.

The results from these strategies show the importance of distributing the high priority work between all of the processors. If this is not done effectively some processors may only perform low priority work and poor speedup will result. Adding and removing problems from the priority queue will lead to some overheads but these are likely to be insignificant compared to the benefits of ensuring that all processors are performing useful work.

5.8.4   McKeown *et al.*

McKeown *et al.* have developed a number of kernels for running Branch-and-Bound algorithms on a network of Transputers (McKeown *et al.* 1991a; Rayward-Smith *et al.* 1991). These kernels are based on a higher-order definition of the Branch-and-Bound paradigm and therefore allow an algorithm to be used interchangeably in any of the available kernels. The kernels, which we now describe, are referred to as 'Select Highest Overall', 'Select Highest available', 'Select Highest Locally' and a hybrid kernel known as 'Select

Highest Hybrid'. Although these kernels have been developed for a network of Transputers it is hoped that they could be implemented on other architectures and there is a current project to implement similar kernels on a shared memory machine (Boffey and Saeidi 1991).

Select Highest Overall (SHO)

SHO is a synchronised strategy which uses a global control to ensure that only the problem-states of very highest priority are expanded at any stage. It is thus similar to SGAP and suffers from the communication bottleneck mentioned earlier.

The method used for reducing the communication overheads is for each processor to store the problem-states it generates in its own local memory but to send a small token describing the problem-state to the master processor. This still allows the master processor to identify the problem-states of highest priority but reduces the amount of communication necessary. The tokens sent by the slave processors contain the priority and bound values of the newly generated problem-state together with the number of the processor which is currently storing it. The priority value is used for sorting the tokens while the bound value is used to prune problem-states that cannot lead to a better incumbent. While the master processor identifies the m nodes of highest priority, each slave sorts its active problem-states into a heap ordered by priority values, and considers pruning the heap if memory overflow is imminent.

If the master finds that a processor has more than one of the selected problem-states it sends a message indicating how the problems should be distributed. Processors are also sent copies of the incumbent if a new value has been generated in the previous iteration.

After expanding a node the slave processor compares the newly generated child problems to the current incumbent and sends a token to the master processor for each surviving child problem. Since this is a synchronised system, there is no advantage in sending the representation of each child individually so all of the new children are sent in a single packet.

Select Highest Available (SHA)

The implementation of the SHA strategy is based on that of SHO but does not involve a synchronisation step. When the master allocates a problem-state to a processor it must also send the priority and bound values of the node in order to identify it. This is necessary because the information stored by the master processor could be slightly out of date with the information stored on the slave processors. When a processor sends a problem-state to another processor it must therefore search its heap for a problem-state with the correct bound and priority values (it is likely to be near to the top of the heap) to ensure that the information in the root processor is kept up-to-date.

A strict implementation of the SHA strategy requires that the highest priority work known to the root is sent in response to a request, but a possible relaxation would be to check for work within a certain tolerance of the highest that is already on the requesting processor.

Select Highest Locally (SHL)

The SHL strategy is similar to the ALAP idea described earlier. The termination detection algorithm used in this kernel is based on one suggested in Topor (1984) which uses a spanning tree topology and sends differently coloured tokens between processors depending on the processor status.

When a processor exhausts its own local supply of work it requests work from neighbouring processors. Work is allowed to spread quickly around the network, however, as several pieces of work are sent in response to a single request.

When a processor receives a request for work it checks its status; if it has spare work in its heap (in which case it must currently be expanding a problem-state) then it sends some of the work to the requesting processor using a Fibonacci series $(2,3,5,8,13,21...)$ to decide how much work to send. It sends the nodes of 2nd, 3rd, 5th ... highest priority until the bottom of the heap is reached. The advantage of this particular approach is that the amount of work sent in response to a request is directly related to the amount of work available on the sending processor and it is also highly biased towards sending work of high priority.

If the processor is currently idle it immediately sends a message indicating this. If the processor is currently expanding a problem-state but has no spare work to send, it will save the request and wait until it either has spare work available or becomes idle. These rules are used to prevent messages swamping a processor which is busy working, but allow messages to be sent freely between idle processors.

In order to spread the high priority work more evenly between the available processors they are allowed to send high priority problem-states to their neighbours at regular intervals. In order to do this an iteration count is implemented indicating how many nodes need to be expanded before work may be sent to neighbours. Thus, when the count is equal to infinity it is a pure implementation of SHL, where work is only ever sent in response to a specific request. Otherwise it is a slight variant on SHL where unsolicited work is occasionally sent to neighbouring processors.

As an example, in SHL(5) each processor sends work to a neighbouring processor (cycling between each of its neighbours) after every 5 expansions without receiving any request. The work sent in these circumstances is the highest priority work known to the sending processor.

Select Highest Hybrid (SHH)

As mentioned previously, the SHA kernel has the potential disadvantage of large communication costs while the SHL kernel risks performing large numbers of low priority expansions. The SHH kernel attempts to overcome the worst of these problems by dividing the high priority work between the processors and then allowing them to do the work on their own.

The SHH strategy begins by running the SHA process to generate a number of problem-states and to ensure that the high priority states are distributed between the processors. The kernel then switches to the SHL strategy so that those states of high priority can be expanded without the communication overheads present in SHA.

The current version of the kernel never attempts to change back to the SHA strategy. Although this is possible it would require a considerable overhead and at present there does not appear to be any need for this to be done.

The optimal time to perform the switch between the SHA and SHL strategies is likely to be problem dependent but one choice would be to change when the active pool reaches a size of approximately $16 * m$. By using a suitable priority function for this first phase of the algorithm it should be possible to ensure that every processor has at least 3 or 4 nodes of high priority to expand whilst still only forming a small part of the search tree.

The SHH kernel appears to have many of the benefits of the system described by Rost and Maehle (1989) (see Section 3) where large packets of work are farmed between the processors, but manages to avoid most of the problems inherent in their system.

The initial search to generate a number of problems is now carried out in parallel using the SHA strategy and then continues with minimum communication using the SHL strategy. When processors exhaust their supply of work they need not sit idle, but can request work from neighbouring processors.

Experiments using a network of Transputers have shown that the amount of communication with this strategy is much lower than with the SHA strategy but the number of nodes expanded is much closer to SHA than to SHL.

Results

The Transputer kernels have been used for the implementation of a wide range of Branch-and-Bound problems and we present here some of the findings.

The first important result is that the SHO kernel generally performs poorly when compared to SHA. Although SHO performs well for a number of algorithms, its synchronisation step hinders performance in cases where the time to expand a problem-state is not constant but depends on the data in the node being expanded. The synchronisation step in SHO prevents processors which finished quickly from continuing until the other processors are ready.

Using the SHA and SHL kernels we have encountered many cases of acceleration anomalies but most cases tend to require slightly more problem expansions when more processors are used. This can partly be explained by the fact that processors are always given work to do if they are idle, even though the work may be of low priority.

The variants of SHL which send unsolicited work to neighbouring processors demonstrate the importance of following the priority function and spreading the high priority work between the processors. Generally, the kernels which distribute the work frequently have slightly higher communication costs but require considerably less node expansions.

Experiments with SHH show that this kernel has much lower communication costs than SHA or SHL(1) but expands many fewer problems than SHL(∞). In our experiments with the Transputer kernels, however, we have found that the actual communication costs for SHA are fairly small for most algorithms and the SHH strategy has proved unnecessary. Machines with different communication characteristics (such as the iPSC hypercube) may, however, find that the reduced communication in SHH allows it to outperform the other kernels.

The 0/1 Knapsack Problem

This algorithm is based on the one described earlier in this chapter and is for an instance of the problem with 750 items, each with a volume chosen randomly from the range $1 \ldots 800$ and with a profit based on the size of the item ($profit = size + 0 \ldots 50$). The capacity of the knapsack was

$$\sum_{i=1}^{n} \frac{v_i}{4}.$$

As stated earlier, the 0/1 Knapsack problem does not lead to a particularly interesting example of a Branch-and-Bound algorithm and when analysing these results it is important to consider how the algorithm is likely to proceed when expanding the search tree.

The algorithm begins by generating an initial incumbent which may be used for pruning the search tree. This is necessary to reduce the overheads of storing problem-states. If an initial incumbent is not provided then it will be necessary to store every problem-state generated until an incumbent is generated. With an initial incumbent, it is often possible to prune many of these problems immediately. This incumbent is generated by greedily including items in decreasing order of profit/volume ratio and generates a fairly tight initial bound.

The algorithm then expands the search tree as explained earlier by including items or excluding items from the knapsack. The bounding function is, however, quite tight and the search tree generated is very deep but very thin. Such a tree does not, in general, provide much opportunity for exploiting parallelism and thus we would not expect the algorithm to have very good speedup.

The shape of the search tree is due to the very tight bounds which are generated. In the knapsack problem, feasible solutions always appear at the

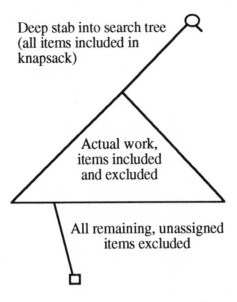

Figure 5.5: Sample search tree for 0/1 Knapsack problem

same depth in the tree and in this example they appear at a depth of 750 (every item must be included or excluded). As the expansion rule chooses items to include in the knapsack in the same order as the bound function the search begins with a deep stab into the tree. Each newly generated problem-state will have the same bound value as its parent and will therefore be selected for expansion in the following iteration.

When the knapsack is eventually filled there will be a number of objects which have been included but only marginally ahead of other similar objects. The search tree will therefore grow out from this point as similar objects are included and excluded from the knapsack to find the optimal assignment of items.

When a good assignment is found the algorithm will again stab down towards the leaf as all of the remaining items are excluded. Once again, the children problem-states will probably have the same bound as the parent (e.g. if the knapsack is completely full) and will therefore be chosen in the next iteration.

Figure 5.5 shows the probable shape of the Branch-and-Bound search tree for an example problem.

Despite these large sections of the search where there is little scope for parallelism, the results do show some interesting features and show why it is so difficult to get meaningful speed-up measures from problems of this type.

Unfortunately, in Branch-and-Bound, the amount of processing required is

Advances in Parallel Algorithms

Table 5.1: 0/1 Knapsack, SHA

Number of Processors	Nodes Expanded	Bounds Calculated	Time (seconds)	Speedup
1	4707	5088	190.07	1.00
2	1541	2175	32.65	5.82
4	3116	4388	35.69	5.33
8	6145	8856	40.27	4.72
16	14745	21969	48.83	3.89

Table 5.2: 0/1 Knapsack, SHL_∞

Number of Processors	Nodes Expanded	Bounds Calculated	Time (seconds)	Speedup
1	4704	5088	183.87	1.00
2	9491	10267	185.07	0.99
4	11278	13004	106.64	1.72
8	17950	21590	86.49	2.13
16	27479	35049	67.61	2.72

not known in advance so it is very difficult to perform load balancing. Even more importantly, it is not necessary for all of the nodes in the search tree to be expanded to generate an optimal solution to the problem (indeed the whole point of Branch-and-Bound is to reduce the number of nodes that actually need to be expanded) so processors cannot simply be given work at random. These results illustrate these difficulties and show examples of both speedup and detrimental anomalies.

Results for the experiments with the 0/1 Knapsack problem are shown in Tables 5.1—5.4. Looking at the sequential version, we see that only a very tiny section of the search tree is generated (just 4704 problem-states out of a total of nearly $6 * 10^{225}$) and that most of these only generate a single child problem (4704 expansions required only 5088 bound calculations). These results tend to support our statements about the probable shape of the search tree and show that much of the search is inherently sequential.

Table 5.1 shows an example of an acceleration anomaly where the two processor system performs far fewer expansions than the sequential one and achieves a speedup of greater than 2. This anomaly is due to the two processor system identifying a good solution more quickly than the sequential one and therefore being able to prune more of the search tree.

As the 2 processor system is almost optimal (750 expansions must be per-

Table 5.3: 0/1 Knapsack, SHL$_5$

Number of Processors	Nodes Expanded	Bounds Calculated	Time (seconds)	Speedup
1	4704	5088	183.87	1.00
2	1675	2337	33.86	5.43
4	3154	4437	32.46	5.66
8	6308	8977	32.85	5.60
16	12644	17968	33.17	5.54

Table 5.4: 0/1 Knapsack, SHL$_1$

Number of Processors	Nodes Expanded	Bounds Calculated	Time (seconds)	Speedup
1	4704	5088	183.87	1.00
2	3665	4332	77.17	2.38
4	5665	7082	62.86	2.93
8	6138	8731	36.96	4.97
16	12269	17375	37.97	4.84

formed sequentially so with 2 processors 1500 expansions should be expected) using more processors does not improve the speedup but simply causes more problem-states to be expanded.

Table 5.2 shows an example of a detrimental anomaly where the two processor system actually takes longer than the sequential one. This is due to the second processor not performing any useful work at all and possibly even providing extra unnecessary work for the first processor. As processors will always be sent work if they become idle, the search tree is split at the first possible moment (probably after the expansion of the initial problem) and the two processors independently search for an optimal solution.

Increasing the number of processors does give some improvement in speedup, but much of the work being done is unnecessary due to the algorithm following the priority function very poorly.

The difference in times between the sequential version of SHA and that of SHL shows the slight overhead associated with storing the heap of tokens as well as the heap of problem-states.

Tables 5.3 and 5.4 show how these variations on SHL perform on this particular problem. Little can be gained from these results however due to the particular characteristics of this problem which, as we have already demonstrated, can have both severe acceleration and deceleration anomalies.

Table 5.5: Travelling Salesman Problem, SHA

Number of Processors	Nodes Expanded	Time (seconds)	Speedup	Pseudo Efficiency
1	1526	203.7	1.0	100.0
2	1457	97.3	2.1	99.9
4	1456	48.8	4.2	99.6
8	1459	24.9	8.2	97.8
16	1746	15.0	13.6	97.1

Table 5.6: Travelling Salesman Problem, SHL_∞

Number of Processors	Nodes Expanded	Time (seconds)	Speedup	Pseudo Efficiency
1	1526	201.9	1.0	100.0
2	1713	104.8	1.9	108.1
4	2078	67.9	3.0	101.2
8	2211	36.6	5.5	99.9
16	3310	29.3	6.9	93.4

One interesting point that can be seen from these results, however, is the communication costs of sending problem-states too frequently. The 16 processor version of SHL_5 performs more expansions than the equivalent SHL_1 one but does so more quickly due to smaller communication overheads.

The Travelling Salesman Problem

This algorithm is based on one by Little *et al.* (1963) and uses a reduced cost matrix to generate a search tree. The tree formed is a binary tree where edges between cities are either included or excluded from the solution. The choice of edge to branch on is made on the basis of trying to maximise the cost difference between the two branches of the tree. This will, hopefully, exclude the largest possible part of the tree in one go.

The search tree generated by this algorithm is very different from the one for the Knapsack problem. Solutions may occur at different depths in the search tree (there must be 30 included edges but the number of excluded edges may vary) which will generally be fairly bushy. We would therefore expect this algorithm to get an increase in speedup as more processors are added.

The results refer to an instance of the travelling salesman problem consisting of 30 cities with inter-city costs being randomly chosen in the range $0 \ldots 99$ and using a best-first search strategy.

Table 5.7: Travelling Salesman Problem, SHL_{10}

Number of Processors	Nodes Expanded	Time (seconds)	Speedup	Pseudo Efficiency
1	1526	201.9	1.0	100.0
2	1510	94.3	2.1	105.9
4	1512	45.6	4.4	109.6
8	1766	27.7	7.3	105.4
16	1965	16.3	12.4	99.8

Table 5.8: Travelling Salesman Problem, SHL_1

Number of Processors	Nodes Expanded	Time (seconds)	Speedup	Pseudo Efficiency
1	1526	201.9	1.0	100.0
2	1462	96.5	2.1	100.2
4	1499	49.4	4.1	100.3
8	1565	25.4	8.0	101.9
16	1565	12.9	15.7	100.6

Table 5.5 shows that the SHA strategy performs consistently well and gets some slight speedup anomalies in some cases. The number of problem-states expanded is reasonably predictable and the performance is therefore also fairly predictable.

Table 5.6 however shows how important it is to ensure that high priority work is distributed evenly between the processors. Although this strategy is very efficient at expanding work, much of this work is unnecessary and speedup is therefore not as good as may be expected.

Tables 5.7 and 5.8 show that by sending unsolicited work between the processors the high priority work is spread more evenly and the number of nodes expanded is considerably reduced. A small communication overhead therefore leads to a great saving in unnecessary work, as can be seen by the 16 processor version of SHL_1 which is still achieving very near linear speedup.

We use the term 'pseudo efficiency' to give an indication of the rate at which nodes are expanded in the parallel system as compared to the sequential one. This measure is defined as follows.

$$e \quad = \quad \frac{|T_m| s_m}{|T_1| m}$$

where

$T_m \quad = \quad$ Size of tree generated by m processors

$s_m \quad = \quad$ Speedup for m processors.

Several of these results include data with a pseudo efficiency of greater than 100%, this being due to the efficiency anomalies mentioned earlier. In these cases, the use of extra memory available to the multiple processors allows them to reduce the amount of work necessary to expand a node. In SHL_{10} this allows the 4 processor system to be more than twice as fast as the 2 processor one whilst still expanding a larger number of nodes.

5.9   Summary

We have shown how the Branch-and-Bound paradigm may be expressed formally and have shown several ways in which Branch-and-Bound algorithms can be implemented on a parallel machine. We have discussed anomalies in both speedup and efficiency and have shown that both of these do occur in real problems. Results from further experiments have shown the overheads of synchronisation and the dangers of not following the priority function closely.

References

Baudet G. M. (1978) The Design and Analysis of Algorithms for Asynchronous Multiprocessors. *Technical report*, Dept. of Computer Science, Carnegie-Mellon University.

Boffey T. B. and Saeidi P. (1991) Parallel Branch-and-Bound Using Shared Memory. *Technical report*, SCM Dept. University of Liverpool.

Burton F. W., McKeown G. P., Rayward-Smith V. J. and Sleep M. R. (1982) Parallel Processing and Combinatorial Optimisation. In *Proc. of the Combinatorial Optimisation III Conference* (Ed. by Wilson L. B., Edwards C. S. and Rayward-Smith V. J.), pp. 19–36, Stirling.

Clausen J. and Traff J. L. (1991) Implementation of Parallel Branch-and-Bound Algorithms—Experiences with the Graph Partitioning Problem. *Annals of Operations Research*.

Hayes J. P., Mudge T., Stout Q. F., Colley S. and Palmer J. (1986) A Microprocessor-based Hypercube Supercomputer. *IEEE Micro*, pp. 6–17.

Horowitz E. and Sahni S. (1978) *Fundamentals of Computer Algorithms*. Computer Software Engineering Series. Computer Science Press, Maryland.

Ibaraki T. (1978) Branch-and-Bound Procedure and State Space Representation of Combinatorial Optimisation Problems. *Information and Control*, **36**, pp. 1–27.

Ibaraki T. (1988) Enumerative Approaches to Combinatorial Optimisation. *Annals of Operations Research*, **11**(1–4).

Janakiram V. K., Gehringer E. F., Agrawal D. P. and Mehrotra R. (1988) A Randomised Parallel Branch-and-Bound Algorithm. *Journal of the ACM*, pp. 277–301.

Karmarkar N. (1984) A New Polynomial-Time algorithm for Linear Programming. *Combinatorica*, **4**(4), pp. 373–95.

Kohler W. H. and Steiglitz K. (1974) Characterization and Theoretical Comparison of Branch-and-Bound Algorithms for Permutation Problems. *Journal of the ACM*, **21**(1), pp. 140–56.

Kumar V. and Kanal L. N. (1983) A General Branch-and-Bound Formulation for Understanding and Synthesizing And/Or Tree Search Procedures. *Artificial Intelligence*, **21**, pp. 179–98.

Lai T. and Sahni S. (1984) Anomalies in Parallel Branch-and-Bound Algorithms. *Communications of the ACM*, **27**(6), pp. 594–602.

Lawler E. L. and Wood D. E. (1966) Branch-and-Bound Methods: A Survey. *Operations Research*, **14**, pp. 699–719.

Little J. D. C., Murty K. G., Sweeney D. W. and Karel C. (1963) An Algorithm for the Travelling Salesman Problem. *Operations Research*, **11**, pp. 972–89.

McKeown G. P., Rayward-Smith V. J., Rush S. A. and Turpin H. J. (1991a) Using a Transputer Network to Solve Branch-and-Bound Problems. In *Proc. of the TRANSPUTING '91 Conference*, California. IOS Press.

McKeown G. P., Rayward-Smith V. J. and Turpin H. J. (1991b) Branch-and-Bound as a Higher Order Function. *Annals of Operations Research*.

Mitra G., Tamiz M. and Yadegar J. (1988) Experimental Investigation of an Interior Search Method Within a Simplex Framework. *Communications of the ACM*, **31**(12), pp. 1474–82.

Mitten L. G. (1970) Branch-and-Bound Methods: General Formulation and Properties. *Operations Research*, **18**(1), pp. 24–34.

Mohan J. (1983) Experience with Two Parallel Programs Solving the Travelling Salesman Problem. In *Proc. of the 1983 International Conference on Parallel Processing*, pp. 191–3.

Pardalos P. M. and Rodgers G. P. (1990) Parallel Branch-and-Bound Algorithms for Quadratic Zero-One Programs on the Hypercube Architecture. *Annals of Operations Research*, **22**, pp. 271–92.

Quinn M. J. (1990) Analysis and Implementation of Branch-and-Bound Algorithms on a Hypercube Multiprocessor. *IEEE Transactions on Computers*, **39**(3), pp. 384–7.

Rayward-Smith V. J., Rush S. A. and McKeown G. P. (1991) Efficiency Considerations in the Implementation of Parallel Branch-and-Bound. Internal report, University of East Anglia, Norwich.

Rost J. and Maehle E. (1989) Implementation of a Parallel Branch-and-Bound Algorithm for the Travelling Salesman Problem. In *Proc. of the 1988 CONPAR Conference* (Ed. by Jesshope and Reinartz), pp. 121–8. Cambridge University Press.

Roucairol C. (1987) A Parallel Branch and Bound Algorithm for the Quadratic Assignment Problem. *Discrete Applied Mathematics*, **18**, pp. 211–25.

Topor R. W. (1984) Termination Detection for Distributed Computations. *Information Processing Letters*, **18**, pp. 33–6.

Turpin H. J. (1991) *The Branch-and-Bound Paradigm*. PhD thesis, School of Information Systems, Norwich.

Wah B. W. and Ma Y. W. E. (1984) MANIP—A Multicomputer Architecture for Solving Combinatorial Extremum-Search Problems. *IEEE Transactions on Computers*, **C–33**(5), pp. 377–90.

Yu C. F. and Wah B. W. (1983) Virtual Memory Support for Branch-and-Bound Algorithms. In *Proc. of the 1983 COMPSAC conference*, pp. 618–26.

CHAPTER 6

Algebraic Transformation Approach for Parallelism

E. V. Krishnamurthy
Australian National University, Australia

6.1 Introduction

A major strategy used in the design of a parallel algorithm is the algebraic transformation approach. In this approach, a given problem belonging to one domain is transformed into another domain and solved; the resulting solution is then inversely transformed to obtain the solution in the original domain.

Obviously, the most important conditions for using such a strategy are:

1. There exists a structure preserving map from one domain to another. That is, the algebraic structure is preserved under the chosen transformation and there exists a constructive homomorphism or isomorphism between the two structures. From the computability point of view this implies that there is an exact solution for recursive equations in the transformed domain, and such a solution can be expressed in a canonical form.

2. The transformed domain is simpler so as to solve the problem more efficiently. From the computational complexity point of view this means that the transformation leads to several simpler mutually independent homomorphic image problems each of which can be solved exactly, independently and in parallel using several processors, thereby reducing the communication and computation time.

This article provides a basic introduction to a class of parallel symbolic iterative algorithms that use the above strategy for exact inversion of rational (real and complex) matrices, and rational polynomial matrices. These algorithms use modular or p-adic representation, and arithmetic for integers, rationals (real and complex) and rational polynomials, in conjunction with the Hensel's lemma to exploit the full power of very large data-level parallel machines—such as the systolic arrays or a connection machine—to provide solution in linear time. Thus it is possible to solve algebraic problems exactly and in parallel achieving a very high speed. Finally, we also indicate the failures that can arise in the use of these mapping techniques.

6.2 Similar Algebras and Morphism

One of the most elegant and useful concepts of universal algebra is the concept of morphism which is a structure preserving map from one algebraic system to another. We will briefly explain in this section how this concept is useful in parallel computation (see Collins *et al.* 1984; Lipson 1981; Loos 1987; MacLane and Birkhoff 1967).

Two algebras (A, P) and (A', P') where A and A' are elements and P and P' are algebraic operations are defined to be similar, if there is a bijection from P to P' with the property that the corresponding operations $p \in P$ and $p' \in P'$ have identical arity (e.g. nullary, unary, binary...).

A function $\phi : A \to A'$ is said to be a morphism (or homomorphism) from (A, P) to (A', P'), if for any $p \in P$ and corresponding $p' \in P'$ and for any $a, b, c, .. \in A$

$$\phi(p(a, b, c, \ldots)) = p'(\phi(a), \phi(b), \phi(c) \ldots). \tag{6.1}$$

In other words, the morphism ϕ preserves corresponding operations of A and A'. That is, the same result is obtained whether:

1. an operation is performed in A and the result is mapped into A' through ϕ using the left hand side of equation 6.1 or

2. the arguments from A are first mapped into A' (through ϕ) and then the corresponding operation of A' is performed as in the right hand side of equation 6.1.

6.2.1 Constructive Isomorphism

A and A' are called isomorphic if ϕ is bijective (one-to-one and onto). The following classical example illustrates the practical utilization of the concept of isomorphism. Consider the map:

$$\phi : x \to \log(x)$$

from $(R^+, .)$ the set of positive reals under multiplication, to $(R, +)$ the set of all reals under addition. Then

$$\phi(xy) = \log xy = \log x + \log y = \phi(x) + \phi(y)$$

so that ϕ is a morphism; also ϕ is bijective with the inverse given by the exponential function $x \to \exp x$. Thus ϕ is an isomorphism. This mapping enables us to carry out the multiplication operation of two positive reals by the three step procedure:

1. forward mapping: taking logarithms of the two operands;

2. calculus: adding the logs;

3. an inverse mapping: exponentiating or taking antilog.

Although two isomorphic algebras can have identical algebraic properties, their computability properties and computational complexities may be quite different. In the above example, for instance, the complexity of the operations in the two domains differs. Further, it is possible that in an arbitrary isomorphism $A \leftrightarrow A'$, there may be no algorithm for finding the element $a \in A$ given $a' \in A'$ and conversely. Hence it is necessary that there is an algorithm to find an element in A given an element a' from A'. Such an isomorphism, for which an algorithm exists to determine the elements in A from those in A', is called 'explicit or constructive isomorphism' (see Shepperdson and Frolich 1956).

6.2.2 Constructive Epimorphism

If $\phi : A \rightarrow A'$ is surjective (or onto), we call it an epimorphism. A' is then called a homomorphic image of A. Then equation 6.1 indicates that we can obtain at least partial information about the value of $p(a, b, c, \ldots)$ in A by computing the corresponding value $p'(\phi(a), \phi(b), \phi(c), \ldots)$ in A'. However, due to the many to one property of ϕ, we may not be able to determine p uniquely, given p'. In this sense, p' is a projection of p, and to compute p uniquely we may need many such projections. If we have an algorithm to obtain p uniquely then we say there exists a 'constructive epimorphism' between A and A'. Therefore, when p is not unique, failures can arise in these techniques.

The widely used 'evaluation and interpolation' technique is an illustration of the concept of constructive epimorphism; this is explained below using two examples: polynomial multiplication and functional matrix inversion.

Example 6.2.1 Consider the multiplication of two single variable polynomials $a(x)$ and $b(x)$ each of degree n. This can be achieved directly by multiplication of the coefficients and obtaining the $(2n + 1)$ coefficients of the resulting polynomial $c(x)$ of degree $2n$. Alternatively one can evaluate $a(x_i)$ and $b(x_i)$ at $(2n + 1)$ points x_i, then perform pointwise multiplication of the values, and finally interpolate the resulting values to obtain the degree 2n polynomial. If the number of points chosen are distinct and greater than or equal to $(2n+1)$ we have an isomorphism; otherwise, we have only an epimorphism. In particular, if the number of distinct points is less than $(2n + 1)$ we can only obtain a polynomial of a lower degree, resulting in a loss of information.

Example 6.2.2 The evaluation-interpolation technique can be readily extended to invert matrices whose elements are polynomials or other functions over any field, by a proper choice of the number of evaluation / interpolation points, so that the elements of the inverse matrix are represented to a desired accuracy. Such a technique uses evaluation, inversion and rational interpolation to obtain the elements of the inverse matrix as a continued fraction expansion at the points of evaluation. We illustrate this technique with a simple example. Consider the polynomial matrix:

$$A(x) = \begin{bmatrix} 3+x & 1+x \\ 2+x & 2+x \end{bmatrix}.$$

An estimate of the maximal degree D of both the numerator and denominator of the elements of the inverse matrix is first obtained; this turns out to be roughly the product of the size of the matrix and the maximal degree of any element in A. We then evaluate the matrix at $(2D + 1)$ distinct points and find the inverse (or Moore-Penrose inverse since matrices may be singular) of all these $(2D + 1)$ matrices numerically. Then using the corresponding elements in these numerical inverse matrices as values, we use rational interpolation to reconstruct the inverse of $A(x)$. This produces the elements of the inverse of $A(x)$ in continued fraction form with numerator and denominator degrees equal to D. For the above example if we assume $D = 1$, we evaluate $A(x)$ at 3 points $x = 1, 2, 3$. The resulting matrices are then inverted and we obtain:

$$A^{-1}(1) = \begin{bmatrix} 1/2 & -1/3 \\ -1/2 & 2/3 \end{bmatrix};$$

$$A^{-1}(2) = \begin{bmatrix} 1/2 & -3/8 \\ -1/2 & 5/8 \end{bmatrix};$$

$$A^{-1}(3) = \begin{bmatrix} 1/2 & -2/5 \\ -1/2 & 3/5 \end{bmatrix}.$$

The rational interpolation of the corresponding elements yields, (Murthy *et al.* 1992):

$$A^{-1}(x) = \begin{bmatrix} p & q \\ r & s \end{bmatrix}$$

where

$$
\begin{aligned}
p &= 1/2; \\
r &= -1/2; \\
q &= -1/3 + (x - 1)/(-24 + (x - 2)/ - (1/6)) \text{ and} \\
s &= 2/3 + (x - 1)/(-24 + (x - 2)/(-1/6)).
\end{aligned}
$$

That is,

$$A^{-1}(x) = \begin{bmatrix} 1/2 & -(1+x)/(4+2x) \\ -1/2 & (3+x)/(4+2x) \end{bmatrix}$$

(see also Section 6.5.2 for an algebraic method illustrated by the same example).

The above technique has been implemented efficiently using a very high degree of data parallellism in the connection machine (Krishnamurthy and Pin 1992).

A clever way to achieve speed-up in evaluation and interpolation is by a judicious choice of the points at which evaluation is done. The well-known fast Fourier transform (FFT), proposed by Cooley and Tukey in 1961, is based on the evaluation-interpolation or epimorphism strategy; here, the evaluation points are chosen to be the nth roots of unity (see Aho *et al.* 1974; Kronsjö 1979).

REMARKS: In computer science theory (e.g. Lambda calculus) evaluation corresponds to application of a function to an argument; and interpolation is a very elegant and simple form of abstraction, where the functional form is a polynomial of restricted degree over a field. In particular, the Newton interpolation algorithm provides a simple recursive abstraction scheme based on divided differences to compute the coefficients of the polynomial from the functional values at a set of points.

The constructive epimorphism strategy also turns out to be computationally advantageous, if the algebraic objects are Euclidean domains, such as integers Z (or polynomials $F[x]$ over a field F), and also for the field of quotients constructed from Z (or $F[x]$) namely, the field of rationals Q (or the field of rational polynomials $Q(x)$). In these cases, we can construct several different A' such that the operations in A' are simpler than those in A. As a result, we can execute several operations simultaneously in many processors and achieve speed through vector parallelism.

In the following section we will describe the two important epimorphism techniques that are widely used for computation in Euclidean domains and in the field of quotients constructed from them. We then briefly explain how these techniques can be combined to exactly invert rational (real and complex) and rational polynomial matrices.

6.3 Multiple and Single Homomorphic Image Techniques

6.3.1 Euclidean Domains

The two different constructive epimorphism techniques for the Euclidean domains are:

Multiple homomorphic image technique The forward mapping technique creates many independent homomorphic images from integers in Z (or polynomials $Z[x]$ over integers) to simpler domains such as Z_p (or to the polynomials $Z_p[x]$ over Z_p) by taking residue modulo a prime p; here Z_p denotes the field of residue class of integers modulo p, namely, $\{0, 1, \ldots, p-1\}$. Then the computation is performed in each domain. The inverse mapping algorithm is based on the Chinese remainder theorem and is described in Section 6.4. This algorithm, called the Chinese remaindering algorithm (CRA) combines the results from several domains to obtain the solution in the original domain Z (or $Z[x]$) (see Szabo and Tanaka 1967).

Single homomorphic image technique This technique creates a single image in an elementary domain Z_p, (or $Z_p[x]$) and finds the solution to the problem in this elementary domain. If required, one can then lift or build up this solution iteratively to a larger domain Z_p^n, where n is a positive integer (or $Z_p^n[x]$), by using Hensel's lemma, until the larger domain Z_p^n (or $Z_p^n[x]$) contains the given original domain Z (or $Z[x]$). This technique is known as 'Hensel's Lifting Algorithm' (HLA).

The HLA can be used to find the multiplicative inverses of elements in Z_p^n, as well as the formal power series expansion of the inverse of an element in $Z_p^n[x]$.

The single and multiple homomorphic image techniques can be combined and extended to perform computations in the field of quotients constructed from Z and $Z[x]$, namely, the field of rational numbers Q, and the field of rational functions $F(x)$ respectively, by using another constructive isomorphism, as will be explained in the next two subsections.

6.3.2   Rationals

The basic idea is to map the rational operands from Q into the set of integers Z_p and carry out the arithmetic operations in $(Z_p, +, .)$ and then map the integer results back onto the appropriate rational numbers. It would be ideal if the relationship between $(Q, +, .)$ and $(Z_p, +, .)$ were an isomorphism. Unfortunately, no such mapping exists.

What we can get is only a mapping from the subset Q^* of Q, defined by: $Q^* = \{a/b : gcd(b, p) = 1\}$ onto Z_p through ab^{-1} mod p where b^{-1} is the multiplicative inverse of b in the field Z_p. But now each element k in Z_p is the image of the infinite subset Q_k of rational numbers. The question therefore is whether or not we can identify a unique element in each generalised residue class Q_k with which we can establish a one-to-one mapping onto the integers in Z_p. If this could be done, we can have a one-to-one and onto mapping between these unique elements in the generalised residue classes $\{Q_0, Q_1, \ldots, Q_{p-1}\}$ and their images in Z_p and this mapping would have an inverse. Unfortunately, however, we can only identify a unique element in some of the generalised residue classes Q_k, but not all of them. As a consequence, we must settle for a one-to-one and onto mapping between those unique elements which can be found and their images in Z_p. This is done as below.

Let F_N be the finite subset of Q^* given by $F_N = \{a/b \in Q^* : gcd(a, b) = 1$ and $0 \leq a \leq N, 0 \leq b \leq N\}$ where $N > 0$ is an integer. We call this set F_N the set of order-N Farey fractions. It can be proved that if p is chosen such that $2N^2 + 1 \leq p$ and Q_k contains the order N Farey fraction $x = a/b$ in Q_k, then it is the only order N Farey fraction in Q_k (see Gregory and Krishnamurthy 1984). Hence the mapping from F_N to Z_p is bijective and has an inverse. If p is chosen properly then there is a constructive isomorphism between $(Q^*, +, .)$ and $(Z_p, +, .)$. For carrying out arithmetic in F_N, we can therefore use the three step procedure:

1. Forward map the operands from F_N to Z_p by finding ab^{-1} mod p; b^{-1}

can be obtained using the Extended Euclidean Algorithm (EEA) to be described in Section 6.4. This gives the Hensel code of ab^{-1} denoted by $H(p, 1, a/b)$.

2. Carry out arithmetic in $(Z_p, +, .)$

3. Inverse map the integer results back into F_N, using the Extended Euclidean algorithm (EEA) to be described in Section 6.4.

CAVEAT: Note that if some of the results are not order N Farey fractions, we will not obtain the correct result! So any computational result should be:

1. defined, that is, relatively prime to p; otherwise, the result is not defined and we need to choose a new prime, and

2. within the allowed range, that is, $N \leq \sqrt{p-1}/2$. For this purpose, we have several alternative choices. We may choose:

   (i) a single large p satisfying $p \geq 2N^2 + 1$; then carry out arithmetic in $(Z_p, +, .)$ in Step 2; finally, convert the integer result back to F_N using EEA; this choice does not have a significant computational advantage; or

   (ii) use a positive power (say r) of p and carry out the arithmetic in $(Z_p r, +, .)$ in Step 2; this method is called the Hensel code method (also called the p-adic method); then convert the integer result back to F_N using EEA; or

   (iii) use the multiple homomorphic image technique for forward mapping of F_N by choosing many primes pi. Then in Step 2 we can carry out the arithmetic in several $(Z_p i, +, .)$. Finally, we can combine the several results using CRA, and obtain the integer result; this integer is mapped back into F_N using the EEA (see Krishnamurthy and Murthy 1987; Murthy 1988b).

We now consider the arithmetic of rational polynomials.

6.3.3   Rational Polynomials

We define the following sets:

(a) $Z[x] = \{a(x) = \sum_{i=0}^{m} a_i x^i : a_i \in Z, m$ a nonnegative integer$\}$
(b) $Z_p[x] = \{a(x) = \sum_{i=0}^{m} a_i x^i : a_i \in Z_p, m$ a nonnegative integer$\}$
(c) $Z_p[[x]] = \{a(x) = \sum_{i=0}^{\infty} a_i x^i : a_i \in Z_p\}$
(d) $F(x) = \{a(x)/b(x) : a(x), b(x) \in Z[x], b(x) \neq 0\}$
(e) $F_p(x) = \{a(x)/b(x) : a(x), b(x) \in Z_p[x], b(x) \neq 0\}$
(f) $F_p^*(x) = \{a(x)/b(x) : a(x), b(x) \in Z_p[x], gcd(b(x), x) = 1, b(x) \neq 0\}$
(g) $P(L/M, N, x) = \{a(x)/b(x) \in f(x) : gcd(a(x), b(x)) = 1,$
$0 \leq |a_i| \leq |b_i| \leq N, 0 \leq \deg a(x) \leq L, 0 \leq \deg b(x) \leq M\}$.

We call $P(L/M, N, x)$, order-N, degree (L/M), Pade' rational polynomial. As seen from (g), it is a rational polynomial whose numerator and denominator have integer coefficients with absolute values not exceeding N and degrees at most L and M respectively.

$$(h) \quad P(L/M, F_p(x)) \;=\; \{a(x)/b(x) \in: gcd(b(x), x) = 1,$$
$$0 \leq \deg a(x) \leq L, 0 \leq \deg b(x) \leq M\}$$

The Pade' rational polynomial $P(L/M, N, x)$ coincides with the order-N Farey fraction for $L = M = 0$.

For computation with Pade' rationals the forward mapping is a succession of two mappings as below:

$$P((R-1)/(R-1), N, x) \rightarrow P((R-1)/(R-1), F_p^*(x)).$$

Here we map the rational polynomial (with integer coefficients) whose numerator and denominator degrees are at most $(R-1)$ and maximum coefficient size N to the rational polynomial $F_p^*(x)$ using the following rule, after factoring out purely powers of x, if any, in $a(x)$ and b(x): choose $p \geq 2RN^2 + 1$ and set $a_i := a_i$ if $a_i > 0$; otherwise, set $a_i := p + a_i$; similarly for b_i.

$$P((R-1)/(R-1), F_p^*(x)) Z_p[[x]] Z_p[x]$$

Here we first map the Pade' rational over a finite field to its formal power series in $Z_p[[x]]$ by finding $[a(x)b^{-1}(x)]$. Note that $b^{-1}(x)$ exists in the formal power series field, by definition, and can be computed either by EEA or HLA to be described in Section 6.4. The infinite series $ab^{-1}(x)$ is then truncated and mapped to a subset $Z_p^*[x]$ of $Z_p[x]$ by the rule:

$$Z_p^*[x] \;=\; \{[a(x)b^{-1}(x)] \bmod x(2R-1) :$$
$$a(x)/b(x)(P(R-1)/(R-1), F_p^*(x))\}.$$

In other words $Z_p^*[x]$ denotes the set of images of the Pade' rationals $\alpha(x) = P((R-1)/(R-1), N, x)$. This is called the Hensel code of the Pade' rational and denoted by $H(p, 2R-1, \alpha(x))$. It has been proved by Krishnamurthy (1985), that the above rules ensure that the mappings are one to one and onto and hence invertible. Thus for arithmetic of Pade' rationals we can use the following three step procedure:

1. Forward map the operands from $P((R-1)/(R-1), N, x)$ to $Z_p^*[x]$

2. Carry out arithmetic of Hensel code in $Z_p^*[x]$

3. Inverse map the result using the succession of the mappings:

$$M1: H(p, 2R-1, \alpha(x)) \rightarrow P((R-1)/(R-1), F_p^*(x));$$

this mapping converts the power series to a rational function over Z_p. This is carried out by using EEA (to be described later) over polynomials.

M2: $P((R-1)/(R-1), F_p^*(x)) \rightarrow P((R-1)/(R-1), N, x)$

where $N \leq \sqrt{(p-1)/2R}$; this mapping converts the rational polynomial over Z_p to a rational polynomial over Z. This is carried out by using EEA over integers (described in Section 6.4).

From the above discussion it is clear that the three basic techniques required for computation in the field of quotients are:

Extended Euclidean algorithm (EEA) for computing the multiplicative inverse of an elment over a field and for inverse mapping of Hensel codes to rationals or rational polynomials;

Chinese remaindering algorithm (CRA) for reconstructing an integer (or a polynomial) from the residues (or evaluated values);

Hensel's lifting algorithm (HLA) that finds the formal power series expansion of the inverse of an element in Z_p^n or $Z_p^n[x]$.

We will describe these techniques in Section 6.4.

In the next subsection we will briefly explain how the above techniques can be put together to invert a rational or rational polynomial matrix using a symbolic iterative method.

6.3.4 Inverting Rational and Rational Polynomial Matrices

Two basic considerations are needed in applying the algebraic transformation approach to invert matrices whose elements are integers or polynomials:

Since the inverse matrix may contain rational or rational polynomial entries, we must either employ the single or multiple homorphic image techniques applicable to the field of quotients.

Since we use Hensel's lifting algorithm that mimics Newton's non-linear iterative method over the real field, we must prove that the computation terminates and produces results that are contained within the domain of computation; otherwise the algorithm fails.

6.4 Algorithms: CRA, EEA and HLA

6.4.1 Chinese Remainder Algorithm(CRA)

One of the oldest known algorithms for computing the solutions to a simultaneous system of congruences over the integers Z is based on a classical theorem from the theory of numbers called 'Chinese remainder theorem' (CRT); hence we call this algorithm 'Chinese remaindering algorithm' (CRA). From an algebraic point of view, it turns out that the use of Chinese remainder theorem is equivalent to the interpolation of a polynomial over a field.

Given a set of remainders (residues) $\{r_0, r_1, r_2, \ldots, r_n\}$ with respect to a set of moduli $\{p_0, p_1, p_2, \ldots, p_n\}$ which are pairwise relatively prime, both problems reconstruct an element r such that $r_i = r \bmod p_i$ for $i = 0, 1, 2, \ldots, n$ in respective domains. r is uniquely determined, if

$$\text{size } r < \text{size} \prod_{i=0}^{n} p_i = M$$

where size is defined suitably thus: for integers the 'size' is the magnitude and for polynomials the 'size' is the degree. Element r is defined as

$$r = \sum_{i=0}^{n} (M/p_i) r_i T_i \bmod M$$

where T_i is the solution of $(M/p_i)T_i = 1 \bmod p_i$. The equivalence between the CRT and the Lagrange polynomial interpolation is readily seen from the following correspondence (Krishnamurthy 1985):

$$r(x) = n\text{th degree polynomial in a field } F$$

$$r(x_i) = r_i, \qquad i = 0, 1, 2, \ldots, n$$

$$p_i(x) = (x - x_i)$$

$$r(x) \bmod p_i(x) = r_i$$

$$M = \prod_{i=0}^{n} p_i(x).$$

The Lagrange interpolant is obtained from

$$Ln(x) = \sum_{i=0}^{n} r_i T_i \prod_{k=0}^{n} (x - x_k), \ k \neq i,$$

where

$$T_i = 1/\prod_{k=0}^{n}(x_i - x_k) = (M/p_i(x))^{-1} \bmod (x - x_i), \ k \neq i.$$

Since in polynomial interpolation we restrict ourselves to linear polynomials $(x - x_i)$, $i = 0, 1, 2, \ldots, n$, the remainder or residue computation for a polynomial $r(x)$ is equivalent to $r(x) \bmod (x - x_i) = r_i$, by remainder theorem.

Thus the reconstruction of an integer r in the range $0 \leq r \leq M - 1$ from a set of residues $\{r_0, r_1, r_2, \ldots, r_n\}$ with respect to a set of primes

$\{p_0, p_1, p_2, \ldots, p_n\}$ and the reconstruction of a polynomial $r(x)$ of degree at most n from a set of residues $\{r_0, r_1, r_2, \ldots, r_n\}$ with respect to a set of distinct linear polynomials $\{p_0(x), p_1(x), p_2(x), \ldots, p_n(x)\}$ can be achieved by identical algorithms for suitably defined data domains. In other words, the residue based techniques as well as the interpolation-evaluation techniques are particular cases of the more general class of epimorphism strategy.

A parallel version of the above algorithm that uses $(n + 1)$ processors is given below.

Algorithm

This algorithm takes as inputs the residues $\{r_0, r_1, \ldots, r_n\}$ and moduli $\{p_0, p_1, \ldots, p_n\}$. For integers r is reconstructed such that $r_i = r \bmod p_i$ where

$$0 \le r \le M - 1 \text{ and } M = \prod_{i=0}^{n} p_i$$

For polynomials $r(x)$ is constructed such that $r(x) \bmod p_i = r_i$, where $p_i(x) = x - x_i$; here $r(x)$ is a polynomial of degree n over a field.

We assume that there are i processors $(i = 0, 1, \ldots, n)$ each with five registers R_i, D_i, S_i, P_i and M_i. The operations subtraction, multiplication and inversion are carried out in the appropriate field for each processor.

for $i := 0$ **to** n **do**
begin
         $R_i := r_i$;
         $P_i := p_i$;
end;
for $j := 0$ **to** n **do**
begin
         $D_j := R_j$;
         for $s := j + 1$ **to** n **do in parallel**
         begin
         $S_s := R_j$;
         $R_s := R_s - S_s$;
         $M_s := P_{j-1} \bmod P_s$;
         $R_s := R_s M_s \bmod P_s$;
         end;
end;

$$r := D_0 + \sum_{j=1}^{n} D_j \prod_{s=0}^{j-1} p_s$$

REMARK: In the case of polynomials, we have $P_j = (x - x_j); P_s = (x - x_s)$

Table 6.1: CRA Example

	$P_0 \bmod 5$	$P_1 \bmod 7$	$P_2 \bmod 11$	$P_3 \bmod 13$	d_i	P
r_i	1	5	9	11	$d_0 = 1$	1
$S_s := R_j$		1	1	1		
subtract		4	8	10		
$M_s = 5^{-1}$		3	9	8		
$M_s R_s$		5	6	2	$d_1 = 5$	$+5 \times 5$
$S_s := R_j$			5	5		
subtract			1	10		
$M_s = 7^{-1}$			8	2		
$M_s R_s$			8	7	$d_2 = 8$	$+8 \times 5 \times 7$
$S_s := R_j$				8		
subtract				12		
$M_s = 11^{-1}$				6		
$M_s R_s$				7	$d_3 = 7$	$+7 \times 5 \times 7 \times 11$
						$= 3001$

The last three assignment statements in the above algorithm are:

$$M_s := (x_s - x_j)^{-1}; \; R_s := R_s M_s \text{ and } r := D_0 + \sum_{j=1}^{n} D_j \prod_{s=0}^{j-1} (x - x_s).$$

Example

Let $p_i = \{5, 7, 11, 13\}$; $r_i = \{1, 5, 9, 11\}$. Table 6.1 illustrates CRA.

6.4.2   Extended Euclidean Algorithm (EEA)

This algorithm finds the multiplicative inverse of an element over a field or truncated formal power series of the reciprocal polynomial in $Z_p[x]$; also, as will be explained below in remarks, by changing the inputs to this algorithm, it can be used for encoding a Farey rational or a Pade' rational into Hensel code or decoding Hensel code into Farey or Pade' rational. Thus this algorithm plays a very central role in the mapping strategy.

Algorithm

Given p and an element x we compute x^{-1} such that $x x^{-1} \bmod p = 1$.

begin
$a := p$; $a_0 := x$;
$b := 0$; $b_0 := 1$;

while $a_0 \neq 0$ **do**

    begin

            $q := quotient(a, a_0);$

            $ra := a - q * a_0;$

            $rb := b - q * b_0;$

            $a := a_0; a_0 := ra;$

            $b := b_0; b_0 := rb;$

    end {while};

$x^{-1} := b \bmod p;$

end;

Examples

By assigning different values to a, a_0, b, b_0 and properly choosing the termination conditions, we can compute several of the required results as mentioned below:

1. If we set $a := p^r$; $a_0 := x$ and use the above algorithm we obtain as the result $x^{-1} \bmod p^r$.

   Example: choose $a := 5^4$, $a_0 := 3$; then we get $3^{-1} = 417$.

2. If we set $a := p^r, a_0 := g$ and $b := 0, b_0 := f$, where $f/g \in F_N$ and we terminate when $ra = 1$, then $rb \bmod p^r$ gives the Hensel code $H(p, r, f/g)$ for a desired p and r.

   Example: choose $a := 5^4, a_0 := 11$ and $b := 0, b_0 := 7$; then we get $rb \bmod 625 = 512$; this is the Hensel code $H(5, 4, 7/11)$.

3. If we set $a := x^r$; $a_0 := f(x)$ and use polynomial arithmetic over $Z_p[x]$, we obtain the truncated $(r-1)$th degree power series reciprocal of $f(x)$.

   Example: choose $a = x^5$; $a_0 = (3 + x + x^2)$ over Z_{601}; then $(3 + x + x^2)^{-1} = 401 + 267x + 178x^2 + 52x^3 + 324x^4$

4. If we set $a := p^r$ and $a_0 := H(p, r, \alpha)$, the Hensel code of a rational number a, and we terminate the algorithm when ra and rb are both less than or equal to $N(= \sqrt{(p-1)/2})$ we get the equivalent Farey rational ra/rb (see Gregory and Krishnamurthy 1984).

   Example: choose $a := 5^4$ and $a_0 := 289$; then we get $ra = 7, rb = 13$. Thus $H(5, 4, 7/13) = 289$.

5. If we set $a := x^{2R-1}$ and $a_0 = H(p, 2R - 1, \alpha(x))$, the Hensel code of a rational polynomial $\alpha(x)$, and we terminate the algorithm when the degree of both ra and rb are less than or equal to $(R - 1)$, we get $P((R - 1)/(R - 1), F_p^*(x))$; this is the mapping M1 of Section 6.3.3.

   Example: choose $a := x^5$ and $a_0 := H(601, 5, 1+600x+2x^2+595x^3+24x^4)$; we get the result $P(2/2, F_{601}^*(x)) = (2x^2+5x+1)/(6x^2+6x+1)$.

6. If we set $a := p$ and $a_0 :=$ coefficient in $F_p^*(x)$, and terminate the algorithm when ra and rb are both less than or equal to $N(= \sqrt{(p-1)/2R})$ we convert the coefficients in Z_p coefficients over the rational; this is the mapping M2 of Section 6.3.3.

    Example: choose $a := 601, a_0 := 403, R = 3, N = 10$; then the rational corresponding to 403 is 7/3.

7. If we set $a := x^r, a_0 := g(x)$ and $b := 0, b_0 := f(x)$, where $f(x)/g(x)$ belongs to $P((R-1)/(R-1), F_p^*(x))$ and we terminate when $ra = c$, a constant in Z_p, then $c^{-1}rb$ realises the forward mapping from $P((R-1)/(R-1), F_p^*(x))$ to its Hensel code $H(p, 2R-1, f(x)/g(x))$.

    Example: choose $a := x^3, a_0 := (1+x)$, and $b := 0, b_0 := (2+x)$, where $(2+x)/(1+x)$ belongs to $P(1/1, F_3^*(x))$; then we get $H(3, 3, (2+x)/(1+x)) = 2 + 2x + x^2$.

6.4.3   Hensel's Lifting Algorithm(HLA)

Let p be a prime in Z and $f(x)$ be a given polynomial in $Z[x]$. Let $G1(x)$ and $H1(x)$ be two relatively prime polynomials in $Z_p[x]$ such that $f(x) = G1(x).H1(x) \bmod p$. Then for any integer $i > 1$ there exist polynomials $Gi(x)$ and $Hi(x)$ in $Z_q[x]$ where $q = p^i$, such that $f(x) = Gi(x)Hi(x) \bmod q$ with $Gi = G1 \bmod p; Hi = H1 \bmod p$; this statement is called the Hensel's lemma.

Zassenhaus extended the Hensel's lemma for $q = p^{2^i}$ where the modulus increases quadratically; this variant is useful for faster convergence. The Hensel's lemma and its Zassenhaus extension provide a method for lifting factors of a polynomial over a finite field Z_p step by step towards a larger subdomain $Z_p^n[x]$ and eventually leads towards the desired result in $Z[x]$. Thus, we can factor a polynomial over $Z[x]$. This algorithm known as Hensel's Lifting Algorithm (HLA) can be employed in a manner analogous to the Newton's method to invert Hensel codes of rationals and rational polynomials, as explained below.

In Section 6.5 we explain how this method can be extended to invert rational polynomial matrices.

Reciprocal of Hensel Code of a Rational

Recall that Newton's method computes a root of $f(x) = 0$, in the real field, by generating a sequence of approximations $x[0], x[1], \ldots$ using the update rule:

$$x[i+1] = x[i] - f(x[i])/f'(x[i]), i = 0, 1, 2, \ldots$$

where $x[0]$ is suitably chosen to ensure convergence. Newton's method can be used to find the reciprocal of a number A by choosing $f(x) = (1/x) - A$; then the Newton's update rule is given by:

$$x[i+1] = x[i](2 - Ax[i]), i = 0, 1, 2, \ldots$$

This sequence converges to the reciprocal of A quadratically. This update rule can be used to reciprocate the Hensel code:

$$H(p, r, \alpha) = \sum_{i=0}^{r-1} a_i p^i, a_0 \neq 0.$$

For this purpose we use the rule:

$$
\begin{aligned}
H(p, 2^k, 1/\alpha) &= H(p, 2^{k-1}, 1/\alpha) \\
&\times [H(p, 2^k, 2) - H(p, 2^k, \alpha)H(p, 2^{k-1}, 1/\alpha)]
\end{aligned}
$$

for $k = 1, 2, \ldots$ with $H(p, 2^0, 1/\alpha) = a_0^{-1} \bmod p$.

Example

Given that $H(5, 4, 3/11) = 3 + 4 \times 5 + 0 \times 52 + 3 \times 53$, we can find $H(5, 4, 11/3)$ thus: since $3^{-1} \bmod 5 = 2$, we get $H(5, 1, 11/3) = 2$; using the update rule, we then get $H(5, 2, 11/3) = 2 + 2 \times 5$, $H(5, 4, 11/3) = 2 + 2 \times 5 + 3 \times 5^2 + 1 \times 5^3$, and so on.

Inverting Hensel Code of a Rational Polynomial

Let

$$H(p, r, \alpha(x)) = A(x) = \sum_{i=0}^{r-1} a_i p^i, a_0 \neq 0;$$

and let its inverse be denoted by

$$H(p, r, \alpha^{-1}(x)) = A^{-1}(x) = \sum_{i=0}^{m} c_i x^i.$$

Then starting with $c_0 = a_0^{-1} \bmod p = H(p, 1, \alpha^{-1}(x))$, we successively obtain:

$$H(p, 2, \alpha^{-1}(x)), H(p, 4, \alpha^{-1}(x)), \ldots$$

using the Newton update rule.

Example

Consider the reciprocation of $H(3, 2, (1 + x))$; we start with

$$H(3, 1, (1 + x)^{-1}) = 1^{-1} \bmod 3 = 1;$$

$$H(3, 2, (1 + x)^{-1}) = 1 + 2x;$$

$$H(3, 4, (1 + x)^{-1}) = 1 + 2x + x^2 + 2x^3 \text{ and so on.}$$

6.5   Parallel Rational Matrix Inversion

6.5.1   Matrices with Rational Entries

In order to apply HLA to matrix inversion we assume that the given matrix containing rational entries is rescaled to obtain an integer matrix. This step is necessary for estimating the size of the Farey rationals F_N of the elements of the inverse matrix and hence the choice of p, and the number of iterations needed for obtaining the correct inverse mapping, as will be explained below.

Let A be an $n \times n$ matrix with integer elements and let A be non-singular modulo p where p is a prime. (Note that the non-singularity of A mod p is vital). The first step of the algorithm is to find A^{-1} (mod p) by using any standard direct method and then use this as an approximation $H(p, 1, B_1)$ to find A^{-1} (mod p^r) by the Newton update rule:

$$
\begin{aligned}
H(p, 2^k, B_{2^k}) \;=\; & H(p, 2^{k-1}, B_{2^{k-1}})[H(p, 2^k, 2I) \\
& - H(p, 2^k, A)H(p, 2^{k-1}, B_{2^{k-1}})] \text{ mod } p^{2^k}
\end{aligned}
$$

for $k \geq 1$ until $(2N^2 + 1) \leq p^{2^k}$; that is the lifted domain contains the rational elements of the inverse matrix. For this purpose we choose p such that

$$
p^{2^k} > 2 \prod_{j=1}^{n} \|c_j\|_2 \text{ for } \|c_j\| \neq 0
$$

where $\|c_j\|_2$ is the Euclidean norm of the j th column of A. For this choice the rational elements of the inverse matrix will lie in F_N where N is the largest integer satisfying the inequality

$$
(2N^2 + 1) \leq p^{2^k}.
$$

We then inverse map the elements of $H(p, 2^k, B_{2^k})$ using EEA to the rational form.

Example

Consider the matrix

$$
A = \begin{bmatrix} 1 & -1 & 2 \\ 3 & 2 & 4 \\ 0 & 1 & -2 \end{bmatrix}
$$

by choosing $p = 3$, we get

$$
A \text{ mod } 3 = \begin{bmatrix} 1 & 2 & 2 \\ 0 & 2 & 1 \\ 0 & 1 & 1 \end{bmatrix}.
$$

We then compute the inverse of A mod 3:

$$A^{-1} \bmod 3 = B_1 = \begin{bmatrix} 1 & 2 & 2 \\ 0 & 2 & 1 \\ 0 & 1 & 1 \end{bmatrix}$$

then using the Newton update rule we get:

$$B_2 = A^{-1} \bmod 9 = \begin{bmatrix} 1 & 0 & 1 \\ 6 & 7 & 2 \\ 3 & 8 & 5 \end{bmatrix} \quad B_4 = A^{-1} \bmod 81 = \begin{bmatrix} 1 & 0 & 1 \\ 60 & 61 & 20 \\ 30 & 71 & 50 \end{bmatrix}$$

$$B_8 = A^{-1} \bmod 6561 = \begin{bmatrix} 1 & 0 & 1 \\ 4920 & 4921 & 1640 \\ 2460 & 5741 & 410 \end{bmatrix}.$$

Since the Euclidean norm of A is 1440, it is adequate to stop the iteration at this stage. Using the EEA we can inverse map B_8 and obtain:

$$A^{-1} = \begin{bmatrix} 1 & 0 & 1 \\ -3/4 & 1/4 & -1/4 \\ -3/8 & 1/8 & -5/8 \end{bmatrix}.$$

REMARK: The above method can be modified to obtain the generalised inverse of a matrix (see Gregory and Krishnamurthy 1984).

6.5.2 Polynomial Matrices

Let us first assume that $A = (a_{ij}(x))$ is a nonsingular $n \times n$ matrix whose elements are over $Z_p[x]$. In order that the elements of $A^{-1}(x)$ belong to $P((R-1)/(R-1), F_p(x))$ we need to perform k iterations such that $2^k \geq (2R-1)$. We then inverse map the elements using EEA to obtain the rational polynomials. The HLA is described below. The first step of the algorithm is to find A^{-1} (mod x) by using any standard direct method and then use this as an approximation $H(p, 1, B_1)$ to find A^{-1} (mod x^{2^k}) by the Newton update rule:

$$\begin{aligned} H(p, 2^k, B_{2^k}) &= H(p, 2^{k-1}, B_{2^{k-1}}) \times [H(p, 2^k, 2I) \\ &\quad - H(p, 2^k, A)H(p, 2^{k-1}, B_{2^{k-1}})] \bmod x^{2^k} \end{aligned}$$

for $k \geq 1$, until $2^k \geq (2R-1)$.

Example

Let

$$A = \begin{bmatrix} x^2 & x^2 + 2x + 2 \\ x + 2 & x + 2 \end{bmatrix} \quad \text{in } F_3(x).$$

The maximal degree of the determinant decides the degree of the denominator polynomial. This is 3. Hence we iterate $k = 3$ times so that $2^3 \geq (2R-1) = 7$. We have

$$A \bmod x = \begin{bmatrix} 0 & 2 \\ 2 & 2 \end{bmatrix} = B_1$$

$$B_2 = \begin{bmatrix} 1+2x & 2+2x \\ 2+x & 0 \end{bmatrix} \quad B_4 = \begin{bmatrix} 1+2x+x^2+2x^3 & 2+2x+2x^3 \\ 2+x+2x^2+x^3 & 2x^2 \end{bmatrix}$$

$$B_8 = \begin{bmatrix} p & q \\ r & s \end{bmatrix}$$

where $p = 1+2x+x^2+2x^3+x^4+2x^5+x^6+2x^7$, $q = 2+2x+2x^3+2x^5+2x^7$, $r = 2+x+2x^2+x^3+2x^4+x^5+2x^6+x^7$, and $s = 2x^2+2x^4+2x^6$.

Inverse mapping of B_8 using EEA gives the Pade' rationals, and we obtain:

$$A^{-1} = \begin{bmatrix} 1/1+x & 2+2x+x^2/2x^2+1 \\ 2/x+1 & 2x^2/2x^2+1 \end{bmatrix}.$$

Inversion of Matrices in Q(x)

In order to use the above method for inverting matrices whose entries belong to $Q(x)$, we need to choose a prime p or a product of primes

$$M = \prod_{i=1}^{s} p_i,$$

such that p (or $M) \geq 2RN^2 + 1$ where $R - 1$ is the maximum degree of the numerator and denominator polynomials in the inverse matrix and N is the maximal coefficient size in the numerator or denominator polynomials. Thus estimates of N and R are needed to determine the number of iterations. If the $n \times n$ matrix A has unit denominators to begin with, and $d - 1$ is the maximal degree among its elements, then $(R-1) \leq n(d-1)$. Also if m is the maximal coefficient size in any entry of A, the maximal coefficient size in the entry of the inverse matrix is $N = n! m^n d^{n-1}$. Naturally this estimate grows exponentially and is one of the essential difficulties that arise in polynomial matrix computations.

Example

Consider

$$A = \begin{bmatrix} 3+x & 1+x \\ 2+x & 2+x \end{bmatrix}.$$

Let us assume $N \leq 4$ and $(R-1) = 1$ or $R = 2$. Hence we choose $p = 101$. We need to iterate k times such that

$$2^k \geq (2R-1) = 3 \text{ or } k = 2.$$

The iterations are as follows:

$$B_1 = A^{-1} \bmod x = \begin{bmatrix} 51 & 25 \\ 50 & 26 \end{bmatrix}$$

$$B_2 = \begin{bmatrix} 51 & 25 + 63x \\ 50 & 26 + 63x \end{bmatrix} \quad B_4 = \begin{bmatrix} 51 & 25 + 63x19x^2 + 41x^3 \\ 50 & 26 + 63x + 19x^2 + 41x^3 \end{bmatrix}$$

this on conversion to P(1/1,F 101(x)) becomes:

$$A^{-1} = \begin{bmatrix} 1/2 & -(1 + x)/(4 + 2x) \\ -1/2 & (3 + x)/(4 + 2x) \end{bmatrix} \in P(1/1, 4, x).$$

REMARKS: When N is very large we can choose many distinct primes and perform multiple homomorphic image technique. We can extend the above method to compute the generalised inverse of matrices over Q(x).

6.6 Extension to Complex Field

The algebraic mapping strategy can further be extended to complex field by using Gaussian primes which are primes of the form $p = \alpha^2 + \beta^2$ (α, β integers); these can be factored into: $(\alpha + i\beta)(\alpha - i\beta); i = \sqrt{-1}$. It was shown by Euler that primes of the form $4k + 1$, k an integer, have this property. Using a Gaussian prime we can construct a code for a complex rational, analogous to the Hensel code for a rational (see Despain *et al.* 1985; Murthy 1988a). We call this code 'Gauss-Hensel code' (GHC). The GHC essentially splits a complex rational(or a matrix) into a pair of mutually independent Hensel codes on which elementwise (convolution-like) arithmetic operations can be performed without developing cross products. Also this code permits the use of Hensel's lifting technique for the inversion of matrices.

6.6.1 Encoding a Complex Farey Rational to GHC

Consider the set of complex rationals in the canonical form $z = a/b + ic/d$ where $gcd(a, b) = 1, gcd(c, d) = 1$ (gcd = greatest common divisor) and

$$0 \leq |a| \leq N, 0 \leq |b| \leq N, 0 \leq |c| \leq N, 0 \leq |d| \leq N,$$

where N is a positive integer denoting the size of computational result. We call this set 'Farey complex rationals of order N', denoted by F_C. Let $z = a/b + i\,c/d$ and $p = (\alpha + i\beta)(\alpha - i\beta)$; then to establish a one-to-one and onto mapping between F_C and an ordered pair of integers modulo p^r we use the following three step algorithm.

Let p denote a Gaussian prime such that

$$p = (\alpha + i\beta)(\alpha - i\beta) = \alpha^2 + \beta^2.$$

Then to establish a one-to-one and onto mapping (isomorphism) between the elements of F_C and an ordered pair of integers modulo p^r such that $p^r \geq 2N^2 + 1$, we use the following three step algorithm:

1. Map $i = \sqrt{-1}$ to $\sqrt{p^r - 1} \bmod p^r = w_r$ (Note: $w_1 = \sqrt{p - 1} \bmod p = ba^{-1} \bmod p$, since $\beta^2 + \alpha^2 = p$ implies $\beta^2/\alpha^2 = -1 \bmod p$).

   The choice of r for a given p is made to suit the requirement that $p^r \geq 2N^2 + 1$

2. Construct $z \bmod p^r = (ab^{-1} + icd^{-1}) \bmod p^r$ where b^{-1}, d^{-1} are multiplicative inverses modulo p^r (which are computed using EEA). This results in Hensel codes for the real and imaginary parts. Let $ab^{-1} \bmod p^r = e$; $cd^{-1} \bmod p^r = f$

3. We then define $GHC(p, r, z)$ or simply GHC(z) as an ordered pair (x,y) thus:

$$x = (e + w_r f) \bmod p^r; \ y = (e - w_r f) \bmod p^r.$$

Finding w_r

Note that w_r is to be chosen as one of the square roots of $\sqrt{p^r - 1} \bmod p^r$. It can be proved that w_r exists, if w_1 exists; this is a consequence of Hensel's lemma (Krishnamurthy 1985). Since w_1 exists for all the Gaussian primes (which are primes of the form $4k + 1$, k an integer) w_r also exists. Note that $w_r^2 = -1 \bmod p^r$. Thus,

$$w_r^{-1} = -w_r \bmod p^r = (p^r - w_r) \bmod p^r.$$

In order to compute w_r, we use the square rooting algorithm based on Hensel-Newton method (Krishnamurthy 1985). For this purpose, we define:

$$w_1 = \sqrt{p - 1} \bmod p = \beta \alpha^{-1} \bmod p = a_0.$$

We then obtain the (i + 1) digits a_i of the square root of $(p^{i+1} - 1) \bmod p^{i+1}$ defined by

$$a_i = \sum_{j=0}^{i} b_j p^j$$

using the following recurrence:

$$\begin{aligned} b_0 &= a_0, \\ b_i p^i &= (p^{i+1} - a_{i-1}^2 - 1)/2a_{i-1} \bmod p^{i+1}. \end{aligned}$$

If we start with $a_0 = -\alpha^{-1}\beta$ we can compute the other square root whose value is $-w_r = p^r - w_r = w_r^{-1}$.

   Example: Evaluate w_4 for $p = 5$. We have

$$5 = 1^2 + 2^2 = \alpha^2 + \beta^2;$$

therefore $b_0 = w_1 = 2 = a_0$; hence $b_1 5 = [(5^2 - 4 - 1)/(2 \times 2)] \bmod 5^2 = 5$; thus $b_1 = 1$.

Hence, $a_1 = b_0 + b_1 5 = 2 + 1 \times 5 = 7 = w_2$ and $a_1^2 = 49$. Similarly, $b_2 \times 5^2 = [(5^3 - 49 - 1)/14] \bmod 5^3 = 50$ or $b_2 = 2$.

Thus, $a_2 = 2 + 1 \times 5 + 2 \times 5^2 = 57 = w_3$ and $a_2^2 = 3249$ Similarly, $b_3 \times 5^3 = (5^4 - 3249 - 1)/6498 \bmod 5^4 = 125$ or $b_3 = 1$ and $a_3 = 2 + 1 \times 5 + 2 \times 5^2 + 1 \times 5^3 = 182$. Thus, $w_4 = \sqrt{5^4 - 1} \bmod 625 = \sqrt{624} \bmod 625 = 182$ and $-w_4 = 443 = w_4^{-1}$. We will use these results in the following sections.

REMARK: This algorithm for computing w_r is based on Hensel's lemma and is linearly convergent, i.e. it produces the digits of w_r one by one. However, it is possible to use Zassenhauss-Hensel lemma (Krishnamurthy 1985) to devise an algorithm that has quadratic or higher order convergence.

The following recurrence gives the nth order convergence

$$b_{f(i)} p^{f(i)} = ((p^{f(i+1)} - 1 - w_{f(i-1)}^2)/2w_{f(i-1)}) \bmod p^{f(i+1)} \text{ where } f(i) = n^i.$$

Accordingly, if w_r is needed with a large number of digits this would produce faster results. For $n = 2$, we get quadratic convergence.

6.6.2 Example for GHC Encoding

Let $z = 1/2 + 1/3\,i$; find $GHC(5, 4, z)$. Here we have chosen $p = 5, r = 4$; therefore $w_4 = \sqrt{624} \bmod 625 = 182$. We have then

$$e = 1 \times 2^{-1} \bmod 625 = 313$$
$$f = 1 \times 3^{-1} \bmod 625 = 417.$$

Thus

$$x = (313 + 182 \times 417) \bmod 625 = 582$$
$$y = (313 - 182 \times 417) \bmod 625 = 44$$

and $GHC(5, 4, z) = GHC(1/2 + 1/3\,i) = (582, 44)$.

REMARK: In future, where the context is clear and p and r are specified, we simply denote $GHC(p, r, z)$ by $GHC(z)$.

6.6.3 Uniqueness of GHC and Decoding

The decoding of GHC to its equivalent complex Farey rational is possible only if it is defined and unique. Using arguments similar to that in Theorem 5.14, page 27 (Gregory and Krishnamurthy 1984), it is easy to prove that the GHC is unique provided that for each pair of order N-Farey rationals a/b, c/d have the condition

$$2N^2 + 1 \leq M = p^r.$$

The decoding of GHC proceeds in two steps:

Step 1: Finding e and f from x and y. For a given GHC, we compute

$$
\begin{aligned}
e &= 2^{-1}(x + y) \bmod p^r \\
f &= 2^{-1}w_r^{-1}(x - y) \bmod p^r.
\end{aligned}
$$

Note that $w_r^{-1} = -w_r$ and it exists for a Gaussian prime p.

Step 2: Finding an element in F_C corresponding to e and f. For this purpose, we use the EEA.

Example: Let $GHC(5, 4, z) = (582, 44)$; find z.

Step 1: Finding e and f.

$$
\begin{aligned}
e &= 2^{-1}(582 + 44) \bmod 625 = 313 \\
f &= 2^{-1}(-182)(582 - 44) \bmod 625 = 417
\end{aligned}
$$

Step 2: Finding F_N using EEA; this gives $a/b = 1/2$; $c/d = 1/3$.

6.6.4   GHC Arithmetic

We assume that p and r are specified. Then, the arithmetic algorithms addition (+), subtraction (-), multiplication (*) and division (/) are carried out in parallel for the pair (x,y), using the same principles as for Hensel (real) codes.

Definitions

Complex Conjugate From the construction of GHC, it is clear that if $GPC(z) = (x, y)$ then $GPC(z^*) = (y, x)$ where $z^* =$ complex conjugate.

Zero The additive null element $(0 + i\,0)$ is denoted by $(0, 0)$.

Unity The multiplicative unit element $(1 + i\,0)$ has the GHC representation $(1, 1)$.

Imaginary unity The imaginary unit $(0 + i\,1)$ has the representation $(w_r, -w_r)$.

Algorithms

Let $GHC(z_1) = (x_1, y_1)$ where $z_1 = e_1 + i\,f_1 \bmod p^r$ and $GHC(z_2) = (x_2, y_2)$ where $z_2 = e_2 + i\,f_2 \bmod p^r$

Addition (+), subtraction (-)

$$
GPC(z_1 \pm z_2) = (x_1 \pm x_2 \bmod p^r, y_1 \pm y_2 \bmod p^r)
$$

Multiplication (*)

$$
GPC(z_1 * z_2) = ((x_1 * x_2) \bmod p^r, (y_1 * y_2) \bmod p^r)
$$

Division (/)

$$GPC(z_1/z_2) = ((x_1 * x_2^{-1}) \bmod p^r, (y_1 * y_2^{-1}) \bmod p^r).$$

We omit the proof of these results.

Example

Let $z = (4 + i)$; find z^{-1}. To compute the reciprocal of a rational complex number, we need to choose an appropriate size of r to get the result in the required range of the Farey rational, F_N. In this case, $N = 17$ and so we need

$$p^r \geq 2N^2 + 1 = 578.$$

If we choose $p = 5$, we need $r \geq 4$ to represent z^{-1} uniquely. Thus, $w_r = 182$ and $w_r^{-1} = -182$. We have,

$$
\begin{aligned}
x &= 4 + 182 \times 1 = 186 \bmod 625 \\
y &= 4 - 182 \times 1 = -178 \bmod 625 = 447 \bmod 625.
\end{aligned}
$$

Thus,

$$
\begin{aligned}
GHC(z) &= (186, 447) \\
GHC(z^{-1}) &= (186^{-1}, 447^{-1}).
\end{aligned}
$$

Using EEA we can compute inverses and obtain:

$$GHC(z^{-1}) = (-84, 158) \bmod 625 = (541, 158) \bmod 625.$$

Decoding $GHC(z^{-1})$, we obtain $e = (158 - 84)/2 = 37$ and $f = (-182)(-84 - 158)/2 = 147$ Decoding of e and f using EEA results in $F_C = 4/17 - 1/17 i$.

6.6.5 Matrix Computations

Let M be an $m \times n$ matrix with complex rational entries or

$$M = ((m_{ij})) = A + iB = ((a_{ij})) + i((b_{ij}))$$

where $1 \leq i \leq m$ and $1 \leq j \leq n$. We then write, for a specified p and r,

$$GPC(M) = (X, Y)$$

where $X = A + w_r B$ and $Y = A - w_r B$ or in elementwise form we have,

$$((x_{ij})) = ((a_{ij})) + w_r((b_{ij})) \text{ and } ((y_{ij})) = ((a_{ij})) - w_r((b_{ij}))$$

Thus,

$$(X + Y)/2 = A; \ (X - Y)/2 \, w_r = B.$$

(i) Addition/Subtraction

If $GPC(M_1) = (X_1; Y_1)$, where $M_1 = A_1 + iB_1$ and $GPC(M_2) = (X_2; Y_2)$ where $M_2 = A_2 + iB_2$ and M_1 and M_2 are conformable, and a_{ij} and b_{ij} respectively the real and imaginary parts of the elements of M satisfy

$$|a_{ij}|, |b_{ij}| \leq (p^r - 1)/2,$$

then $GPC(M1 \pm M2) = (X_1 \pm X_2; Y_1 \pm Y_2)$.

(ii) Multiplication

Let M_1 and M_2 be two conformable matrices for multiplication. Also, let the elements of product matrix $M = M_1 M_2$ be m_{ij}, such that the elements a_{ij} and b_{ij} respectively the real and imaginary parts of the elements of M, satisfy

$$|a_{ij}|, |b_{ij}| \leq \sqrt{p^r - 1}/2$$

then $GPC(M_1 M_2) = (X_1 X_2, Y_1 Y_2)$.

(iii) Transposition

If $GPC(M) = (X, Y)$, then it is easily shown that M^T (transpose of M) has

$$GPC(M^T) = (X^T, Y^T).$$

(iv) Conjugate Transposition

If $GPC(M) = (X, Y)$ then the conjugate transpose M^* satisfies

$$GPC(M^*) = (X^*, Y^*) = (Y^T, X^T).$$

(v) Inversion

Let M be a non singular square matrix modulo p^r. Let $GPC(M) = (X, Y)$. Then,

$$GHC(M^{-1}) = (X^{-1}, Y^{-1})$$

provided the elements c_{ij} of C and d_{ij} of D, the real and imaginary parts respectively, of the elements of

$$M^{-1}(= C + iD) \text{ satisfy } |c_{ij}|, |d_{ij}| \leq \sqrt{p^r - 1}/2.$$

Thus, the inversion of a complex rational matrix is achieved by inverting the individual components in the pair and then decoding the result. In fact, we have a stronger result for the generalised inversion of matrices.

(vi) Generalised Inversion

If M is an $m \times n$ matrix, then we say G is a generalised inverse if $MGM = M$. If, however, G satisfies the additional conditions:

$$GMG = G$$
$$(GM)^* = GM$$
$$(MG)^* = MG$$

then, G is called the Moore Penrose generalised inverse (Gregory and Krishnamurthy 1984; Krishnamurthy 1985) of M and denoted by M^+. The Moore-Penrose inverse has many applications in filtering, signal processing and estimation. It is easy to show that, if $GHC(M) = (X, Y)$ then $GHC(M^+) = (X^+, Y^+)$ provided that the elements c_{ij} and d_{ij} respectively the elements of the real and imaginary elements of M^+ satisfy

$$|c_{ij}|, |d_{ij}| = \sqrt{p^r - 1}/2.$$

6.6.6 Example

We will illustrate the complex matrix inversion by a very simple example. The crucial fact to remember is that the elements of the inverse matrix

$$\sqrt{p^r - 1}/2.$$

Therefore, at the very first step, we need to estimate the size of the elements of the inverse to make a choice of p and r so that the final decoding results in the correct answer without any overflow. Let

$$M = \begin{bmatrix} 2+i & 1+i \\ 1-i & 2-i \end{bmatrix} = A + iB$$

with

$$A = \begin{bmatrix} 2 & 1 \\ 1 & 2 \end{bmatrix}, \quad B = \begin{bmatrix} 1 & 1 \\ -1 & -1 \end{bmatrix}.$$

Since the elements of M^{-1} can be estimated to be of size $N = 3$, we choose $p = 5, r = 2$ so that $N \leq \sqrt{5^2 - 1}/2$. We then find $w_2 = \sqrt{5^2 - 1}$ mod 5^2. We have $w_2 = 7$. Hence, $GHC(5, 2, M) = (X, Y)$ where

$$X = (A + 7B) \bmod 25 = \begin{bmatrix} 9 & 8 \\ 19 & 20 \end{bmatrix}$$

$$Y = (A - 7B) \bmod 25 = \begin{bmatrix} 20 & 19 \\ 8 & 9 \end{bmatrix}.$$

We then find X^{-1}, Y^{-1} modulo 25 using HLA. We obtain,

$$X^{-1} \bmod 25 = \begin{bmatrix} 15 & 14 \\ 2 & 3 \end{bmatrix}$$

and

$$Y^{-1} \bmod 25 = \begin{bmatrix} 3 & 2 \\ 14 & 15 \end{bmatrix}.$$

Hence,

$$M^{-1} = [X^{-1} + Y^{-1}]/2 + i[X^{-1} - Y^{-1}]/2w_r \bmod 25 = \begin{bmatrix} 9 + 8i & 8 + 8i \\ 8 - 8i & 9 - 8i \end{bmatrix}$$

which on decoding using EEA gives

$$\begin{bmatrix} 2/3 - 1/3i & -1/3 - 1/3i \\ -1/3 + 1/3i & 2/3 + 1/3i \end{bmatrix}$$

as the final result.

6.6.7 Remarks

We now make the following remarks concerning the GHC and their applications.

Complexity The transformations described to construct GPC(M) of a matrix M and the final conversions are O(1) operations if $2n^2$ processors are used in parallel. The matrix inversion procedure based on HLA is essentially based on matrix multiplication which is O(n). Thus the complex matrix can be inverted exactly in almost linear time using data level parallelism machines (such as the connection machine or systolic processors).

Use of several primes The methods described can be extended to several powers of primes $(p_i^{r_i})$; however, in such a case, the results have to be combined first using the Chinese remaindering algorithm; then the Euclidean algorithm is to be applied to convert the elements to the rationals.

Use of Fermat-Gauss-primes The primes $p = 2^{2^n} + 1$ are known as Fermat primes for $n = 1, 2, 3, 4$ for which $p = 5, 17, 257, 65537$. Note that these are also Gaussian primes. Using these primes the matrix multiplication process can possibly be speeded up by Fourier transform methods.

Hensel's Lifting Techniques Using the GHC, we can devise iterative techniques for finding the reciprocal of complex numbers and generalised inversion of complex matrices similar to those discussed for real p-adic systems in Section 6.5. These techniques can now be extended to invert rational and complex matrices.

6.7 Limitations on Using Mapping Techniques

It is necessary at this point to indicate some of the failures associated with the use of mapping techniques and indicate some precautionary measures to avoid them. The failures of residue methods in matrix computations essentially arise due to the fact that there need not be a one-to-one correspondence between the rank of a matrix over a real or complex field and its mapped version over the finite or the p-adic field. As a result, the intermediate steps may breakdown, especially when the chosen primes are not relatively prime to the determinant of the given matrix. Hence there is no unique solution and the epimorphism technique fails. Gregory and Krishnamurthy (1984) suggest some precautionary measures based on the comparison of ranks for different choices of the prime. These measures consist of aborting and restarting the computations, with a new set of primes. Such measures may result in a serious loss in speed and hence would undermine the purpose of using these techniques. On the other hand, if the algebraic mapping techniques are not used, one has to use an infinite precision rational system, which lacks parallelism (see Roch 1989). Thus experimentation is needed to find clever strategies to avoid failures in the mapping techniques. Until recently the mapping methods could not be subjected to a large scale experimentation since the conventional sequential machines have been too slow to handle arbitrary precision objects that are numbers or polynomials, or matrices containing them. The availability of data parallel machines, as well as new programming languages, now permit us to carry out such large scale experiments with these algebraic objects very effectively (see Neun and Melenk 1989; Norman and Mitchell 1989). Villard (1989), has conducted some preliminary experiments with the HLA method of exact matrix inversion. It is hoped that further new experiments would be conducted to design failure—free techniques that can produce exact results in algebraic computations to achieve the desired parallelism and speed.

References

Aho A. V., Hopcroft J. E. and Ullman J. D. (1974) *The Design and Analysis of Computer Algorithms*. Addison-Wesley.

Collins G. E., Mignotte M. and Winkler F. (1984) Arithmetic in Basic Algebraic Domains. In *Computer Algebra* (Ed. by B. Buchberger, *et al.*), Springer Verlag, pp. 189–210.

Despain A. M., Peterson A. M. and Rothans O. S. (1985) Fast Fourier Transform Processors using Gaussian Residue Arithmetic, *J. Parallel and Distributed Computing* **2**, pp. 219–37.

Gregory R. T. and Krishnamurthy E.V (1984) *Methods and Applications of Error-free Computation*. Springer Verlag, New York.

Krishnamurthy. E. V. (1985) *Error-free Polynomial Matrix Computations*. Springer Verlag, New York, 1985.

Krishnamurthy E. V. and Murthy V. K. (1987) Error-free Parallel Rational Arithmetic for Optical and VLSI Computing. *Applied Optics*, **26**, pp. 4819–22.

Krishnamurthy E. V., and Pin C. (1992) Data parallel evaluation interpolation algorithm for polynomial matrix inversion *Parallel Computing*, in press.

Kronsjö L.I (1979) *Algorithms: their complexity and efficiency*. John Wiley, London.

Lipson J. D. (1981) *Elements of Algebra and Algebraic computing*. Addison-Wesley, Reading, Mass.

Loos R. (1987) Computing in Algebraic Extensions. In *Computer Algebra* (Ed. by Buchberger B. *et al.*), Springer-Verlag, pp. 172–87.

MacLane S. and Birkhoff G. (1967) *Algebra*. MacMillan, New York.

Murthy V. K. (1988a) Exact Parallel Matrix Inversion Using Para Hensel Codes with Systolic Processors. *Appl. Opt.* **27**, pp. 2022–4.

Murthy V. K. (1988b) Parallel complex rational and matrix arithmetic using Gaussian prime codes. Unpublished Report.

Murthy V. K., E. V. Krishnamurthy and Pin C. (1992) Systolic algorithm for rational interpolation and Pade′ approximation. *Parallel Computing*, in press.

Neun W. and Melenk H. (1989) Implementation of the LISP-Arbitrary precision Arithmetic for a vector processor. In *Computer Algebra and Parallelism* (Ed. by Della Dora J. and Fitch J.), Academic Press, London, pp. 76–89.

Norman A. J. and Mitchell J. (1989) Factorization in Functional Language. In *Computer Algebra and Parallelism* (Ed. by Della Dora, J. and Fitch J.), Academic Press, London, pp. 133–41.

Roch J.-L.(1989) Towards a Parallel Computer algebra Co-processor. In *Computer Algebra and Parallelism* (Ed. by Della Dora J. and Fitch J.), Academic Press, London, pp. 3–50.

Shepperdson J.C and Frolich A. (1956) Effective Procedures in Field theory. *Phil. Tran. Royal Society of London*, **248 A**, pp. 407–32.

Szabo N. Z. and Tanaka R. I. (1967) *Residue Arithmetic and its Applications to Computer Technology*. McGraw-Hill, New York.

Villard G. (1989) Exact Parallel Solution of Linear Systems. In *Computer Algebra and Parallelism* (Ed. by Della Dora J. and Fitch J.), Academic Press, London, pp. 198–205.

CHAPTER 7

Discrete Event Simulation in Parallel

A. J. Wing

University of East Anglia, UK

7.1 Introduction

Simulation has seen increasing use in a wide range of applications, including such diverse areas as VLSI design, computer and communication networks, aerospace and manufacturing systems. The large scale nature of many simulations and their high computational requirements provides strong motivation for a move from sequential to parallel execution in order to decrease very long run-times. In this chapter we are concerned with the application of parallelism to discrete simulations.

For stochastic simulations, replicated trials have been suggested in which separate sequential simulations are run independently on different processors (see Rajagopal and Comfort 1989). There is no theoretical need for processors within different simulations to communicate (except to collect final results) so that communication overheads are also minimised. Clearly this approach is not applicable to primarily deterministic simulations. Another way of introducing parallelism involves using functional units dedicated to carrying out particular simulation sub-tasks (Krishnamurthi *et al.* 1985). This approach can offer only limited speed-up as there is no attempt to exploit the natural parallelism within the model. *Concurrent Simulation* has been suggested by Jones (1986) and implemented on a shared memory machine (Jones *et al.* 1989). In this approach the simulation of an event is considered to consist of two phases. A state variable phase involves the inspection and then modification of state variables required by the event. A second, event-scheduling phase, adds any further events generated to a pending event set. The state variable phase must be executed sequentially but some concurrency is available in event scheduling. This approach can only be applied on shared memory machines and available parallelism is limited.

Another possible way in which parallelism may be introduced is to use several processors to simulate concurrently active parts of the system. This approach has wide-spread applicability as it exploits the concurrency inherent in those real-world systems with component parts or sub-systems that function simultaneously. A central issue in any simulation carried out in parallel is the mechanism by which simulation time is advanced. In traditional discrete simulation (see Mitrani 1982), time is commonly advanced from event to event (discrete event) or in fixed increments (time-stepped). A similar dichotomy exists with regard to simulations carried out in parallel. Either different parts of the simulation advance together in time steps or independently (subject to causality constraints) from event to event. The latter provides greatest potential for parallel speed-up in systems with irregular and unpredictable event-times. It is upon this that we therefore concentrate.

The conventional algorithm for discrete-event simulation relies upon a single event list from which events are removed and simulated in their correct time order. This is not readily executed in parallel because of the bottle-neck inherent in access to the event list. Consequently a radically new approach must be adopted. Early work in this field was carried out by a number of authors working largely independently (Bryant 1977; Chandy and Misra 1979; Chandy and Misra 1981; Peacock *et al.* 1979b; Peacock *et al.* 1979a). These authors used the term *distributed simulation*, as simulations were primarily intended to be implemented on a network of loosely coupled computers. Later work sometimes uses the term parallel simulation and we do so in this document.

We first consider the principles underlying parallel simulation and discuss some of the inherent problems that may limit the potential for speed-up (Section 7.2). One of the major issues in this area is that of synchronisation between parts of the parallel system at different simulation times. We discuss the two major types of synchronisation, conservative (Section 7.3) and optimistic (Section 7.4). There are a number of issues which relate to parallel simulation regardless of the synchronisation method used and these are examined in Section 7.5. A comparison of conservative and optimistic synchronisation is then made in Section 7.6. We give the arguments for and against using either method and discuss whether these are supported by empirical and analytical studies. The chapter is completed with some concluding remarks on the future prospects for parallel simulation (Section 7.7).

7.2   Principles of Parallel Simulation

The central idea of parallel simulation is to decrease simulation run-time using several processors, thus exploiting the natural concurrency inherent in many systems. To this end a system should not be considered as an indivisible entity but modelled as a set of separate physical processes interacting to produce the overall system behaviour. A simulation can then be constructed in which every physical process (PP) would be modelled by a corresponding logical process (LP). In this way the computation involved in a large simulation can be effectively partitioned and executed on several processors. Any communication between physical processes should be matched within the simulation by communication of logical processes. Some authors use the concept of a directed link or channel between communicating logical processes. Misra (1986) refers to source and sink processes. The former can only generate, whilst the latter may only receive, messages.

A number of general points must be made about this approach. There is no need for data to be known globally throughout the simulation system and all communication may be carried out using message passing. This is an advantage as it means that parallel simulation can be carried out with both shared memory and message passing MIMD machines. However, an obvious disadvantage is the extra communication overheads as compared to an equivalent sequential simulation. These will tend to limit performance improvement in systems with high process interaction. Another disadvantage is that the overall degree of parallelism available is limited by the number of physical and thus logical processes. This natural parallelism may not be easily exploitable by the simulation if there is a poor mapping of processes to the available physical hardware. Dynamic process allocation may rectify this problem but is likely to carry an additional performance penalty.

There are two other problems that must be dealt with in parallel simulation. Firstly, termination of the computation as a whole must be detected. This is a non-trivial task in distributed systems and the general case has been considered by several authors (see, for example, Torpor 1984). Secondly, different processors needing to communicate must do so at the correct points in their respective calculations. This brings us onto the important subject of causality.

7.2.1 Causality

In a conventional discrete simulation, time may be advanced in fixed increments or via a varying amount from event to event. The simulation clock-time is identical to the time associated with the currently executing time-step or event. We may represent an event and its associated time as a tuple (E_n, t_n) and therefore the operation of a sequential simulation from start to some arbitrary time t_{n+1} as a sequence of tuples,

$$(E_0, t_0), (E_1, t_1), \ldots (E_n, t_n), (E_{n+1}, t_{n+1}).$$

In a simulation, events must always be executed in increasing time order, i.e. t_{n+1} must be greater than t_n. A *causality error* would occur if this condition did not hold. Anomalous behaviour might then result as an event simulated earlier (in real-time) affected state variables used by subsequent events. In the physical world this would represent a nonsensical situation in which future events could influence the present.

For a simulator composed of separate LPs the situation is somewhat more complex. If we assume the simulation progresses in a synchronous, rigidly time-stepped fashion, then all processes taken together will again represent a single clock-time analogous to that in a sequential simulation. However, in a parallel simulation where time advances in an event driven fashion, each LP will have an individual clock-time set by the last event simulated in that process. In this document we will generally refer to this as the process's *logical* or *simulation* time. The lack of an overall global time in the simulation has profound consequences for the preservation of causality.

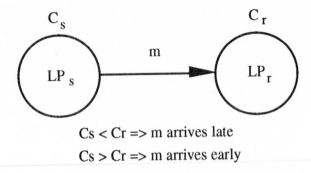

$$Cs < Cr => m \text{ arrives late}$$
$$Cs > Cr => m \text{ arrives early}$$

Figure 7.1: Possible causality errors

7.2.2 Synchronisation

Each logical process carries out all known events within that LP in their correct time sequence and thus maintains a local event list. This is a necessary, but not sufficient condition, to ensure causality. A message sent between logical processes may still violate causality in two possible ways. Assume a logical process LP_s at simulation time C_s sends a message to another logical process LP_r with simulation time C_r (see Figure 7.1).

If C_s is greater than C_r, the message arrived *physically early* and LP_r may update state variables as a result of events which, relative to its own logical clock, have not yet occurred. On the other hand, if C_s was less than C_r, the message arrived *physically late* and LP_r may have already modified the simulation state from what it would have been at time C_s. A further problem is that without additional information on the logical time at which the message was generated there is no possible way for LP_r to know that an error has occurred.

A partial solution is to use *time stamps*. Messages sent between logical processes are associated with a time value representing the clock time of the sending process. Any logical process can temporarily ignore received messages with a time stamp greater than the local logical time until this has advanced sufficiently. In practice physically early messages may be placed in a buffer or possibly returned to the sending process. This would remove the problem of early messages but not necessarily that of late messages. These must be prevented by some form of synchronisation between processes.

Synchronisation has received a great deal of attention in literature dealing with parallel simulation. A number of very different approaches have been suggested, but many authors try to classify all of these into either *conservative* or *optimistic* methods. Conservative (pessimistic) methods will generally involve the complete prevention of causality errors. In contrast optimistic approaches may allow causality errors but these are then detected and any anomalous behaviour is corrected. For convenience we generally adopt this dichotomy in discussing synchronisation methods.

However, in some circumstances it may be worthwhile to use a more sophisticated classification. Reynolds (1988) suggests nine design variables by which we may classify synchronisation in a simulation. There may be *partitioning*, in which sets of LPs may use non-identical synchronisation strategies. The degree of *aggressiveness* indicates the optimism with which an LP processes events. Thus systems showing aggressiveness process events on the basis of conditional knowledge and may produce incorrect computation and messages. The *risk* exhibited by a process is the extent that any erroneous messages may be propagated. We will later discuss synchronisation methods which are aggressive, but do not have risk. *Accuracy* is not incompatible with aggressiveness but implies that all events will ultimately be processed in the correct order. The *synchrony* is a measure of how closely coupled are the logical times of LPs. For instance, a tightly coupled simulation will have near-identical simulation times for all logical processes. Synchronisation may also involve a varying degree of knowledge *embedding, dissemination* and *acquisition*. Knowledge embedding involves using application specific information to improve synchronisation efficiency. Knowledge dissemination implies that LPs send unsolicited messages (which do not have any physical equivalent) to enable other processes to progress. In contrast knowledge acquisition implies that processes actively demand or ask for information. If any of the above factors can alter dynamically then *adaptability* is shown.

7.3   Conservative Methods

Conservative synchronisation methods have been suggested by a number of authors and a useful survey in this area is presented by Misra (1986). We may define a conservative method as any in which logical processes are prevented from receiving messages with a time stamp value less than their logical clock. Most commonly, an LP is prevented from proceeding (blocked) if this would advance the logical clock beyond that of another LP that could communicate with it. Physically early messages are buffered. In general the literature in this area assumes a known and unchanging pattern of inter-process communication.

There are a variety of ways in which we may characterise conservative synchronisation methods. Many algorithms can be considered as either synchronous or asynchronous in nature. In the former case some overall co-operation between LPs is required and global knowledge is used to identify those events which it is safe to process. In contrast, with asynchronous algorithms no logical process has overall system knowledge. Synchronisation is achieved with each logical process knowing only its own state and that of neighbouring LPs. Synchronisation may also be achieved by methods which lie between these two extremes. We first discuss synchronous methods (7.3.1) and then asynchronous methods (7.3.2).

In this area, Fujimoto has emphasised the importance of *lookahead*. This is an LP's ability to predict behaviour into the simulation future. We discuss the application and importance of lookahead in Section 7.3.3.

7.3.1 Synchronous Methods

Several authors have suggested synchronous conservative methods which rely on logical processes in the system co-operating at intervals to determine which events are safe to process. The simulation advances in a series of iterative steps. There are a number of ways in which the safety of events may be found. An important idea in this area is distance, defined as the minimum increase in time stamp before an event at one LP may affect the state of another process. Distance is part of the wider concept of lookahead (Fujimoto 1989a, 1990a). Using lookahead an iteration may produce a higher time value, T, up to which it is safe to process events. The period from the current logical time up to T is often termed a time window.

Lubachevsky (1989a, 1989b) suggests a *bounded lag* algorithm. In each iteration a logical process determines which events are safe to process by checking with other processes that may directly, or indirectly, send it a message. To limit the communication overhead involved an LP pre-computes the distance to other processes. In any iteration no process i attempts to interrogate the state of a process j such that the distance (or lag) $j \rightarrow i$ is greater than a *bound* time value B. Thus a bounded region of test is produced and communication costs diminished. A natural consequence of this approach is that in any iteration we cannot possibly determine the safety of processing an event in the future beyond our bound value. The algorithm therefore produces a time window which is limited in size by an upper limit of B. A difficulty in applying this method is that the determination of an effective bound value requires application specific information. Too low a value may result in not enough events being included in the window. Too high a value will impose the additional overheads inherent in the algorithm whilst producing very much the same behaviour as if the window was not present. For a well-chosen bound Lubachevsky reports a speed-up of 16 on the 25 processors of a MIMD Sequent Balance.

Another algorithm for synchronisation has been suggested in which a time window is produced without a fixed bound value. In YAWNS (Nicol *et al.* 1989) the lower edge of the window is simply the higher bound on the previous iteration. Each LP finds the lowest time at which any message will be generated for another process. All processes communicate to find the lowest of these values which then forms the higher bound of the window. The simulation of all those events in the time window proceeds in parallel. This method relies on having pre-knowledge of when messages will be generated and on the ratio of events to messages being high.

Ayani (1989) uses the idea of distance in a similar manner to Lubachevsky. A preprocessing step ensures that each LP is aware of the distance from itself to every other process. The algorithm is applied iteratively in a three phase process, each phase terminated by barriers. A (*marking*) phase determines the lowest time-stamped event in each LP. A second (*concurrency control*) phase then finds which of the selected events may be executed concurrently given the distance between the relevant LPs. The final (*evaluation*) phase, processes events and adds any events created. Ayani has implemented his approach on a Sequent Balance 8000 with 10 processors. On a queuing network simulation a speed-up of 5 was reported. A further study (Ayani and Rajaei 1990) in which a Generalised Cube Multistage Interconnection Network was simulated showed an almost linear speed-up for up to 15 processors.

Chandy and Sherman (1989a) suggest a conditional events approach. In a sequential simulation the event list will contain a single definite event which has the lowest time-stamp. All other events have higher time stamps and are conditional, since they may be removed from the event list as a consequence of previously simulated events. For distributed simulation there will also be a definite event (with the lowest associated time) and many conditional events. The lowest event time can be determined by communication between all logical processes. In addition if LPs can generally predict a lower bound on any message sent then very many more events may be assumed to be definite and not conditional. The algorithm suggested by Chandy and Sherman is applied iteratively. LPs communicate lower bounds on the next output message to other processes. This enables receiving processes to convert as many conditional events as possible to definite events The execution of these definite events in parallel then enables further events to be converted from conditional in the next iteration. In this approach communication of synchronisation messages does not have to follow the same routes as real messages. Arbitrary pairs of processes may communicate. The algorithm will always make progress since there will always be at least one definite event. This approach has been implemented by Chandy and Sherman on a Intel iPSC/1 with varying numbers of processors. They measured the performance of their algorithm using a queuing network simulation and reported a speed-up of 9 using 24 processors.

Barrier Implementation

Barriers are commonly used in shared memory machines implementing synchronous methods. A logical processes reaching some barrier point in the algorithm control flow must wait until other processes in the network have reached that point. All processes are free to proceed when the last completing LP reaches the barrier. There will be one or more barriers during the progress of a distributed calculation.

Barrier synchronisation has the disadvantage that, in effect, each process must communicate with all other processes. This will produce considerable communication overhead in systems with large numbers of processes. Another problem is that individual processes arriving at a barrier must idle until the final process arrives. There is effectively sequential execution as the last process completes. The delays caused by uneven distribution of workload and varying process execution time will therefore be maximised.

A traditional way of implementing barrier synchronisation relies on a shared counter written to by each logical process. This method is sometimes known as a two-lock implementation. Prior to a barrier the counter is initialised to zero. Each logical process reaching the barrier increments the counter but simultaneous writes by more than one process are prevented. Processes idle at the barrier until the shared counter equals the known number of processes. At this point all processes may proceed. Brooks (1984) has suggested an alternative approach. The 'Butterfly Barrier' is intended to avoid the bottleneck present in sequential access to a shared counter. In Brook's scheme a barrier actually consists of several sequential stages, the number increasing as the natural logarithm of the total number of processes. At each stage a process synchronises with a single, more-distant, neighbour.

Some work has been carried out to quantify the effects of barrier synchronisation. Axelrod (1985) compares performance of the Butterfly Barrier with two-lock barrier synchronisation. For large numbers of processors the conclusion is drawn that the Butterfly Barrier can show significantly better performance. Greenbaum (1989) carries out a theoretical analysis of several forms of synchronisation that can be implemented synchronously. She concludes that for large numbers of processors the performance of barrier synchronisation is significantly poorer than other synchronous methods.

7.3.2 Asynchronous Algorithms

In an asynchronous method no global knowledge is assumed to be available to logical processes in the system. Instead an individual logical process determines that events may be safely processed via communication with neighbouring processes. This gives rise to an emphasis on the idea of *channels*. A directional channel AB is said to exist from process A to process B if A may send a message to B (which corresponds to communication in the physical system). A clock time, corresponding to the time stamp of the most recent message sent or buffered on a channel, is associated with each link. In the general case, a process may only advance simulation time up to the lowest clock-time of any incoming channel. In fact this restriction is too harsh if a process can predict future behaviour from the current inputs and state, i.e. has lookahead. We discuss lookahead later (in Section 7.3.3).

An asynchronous simulation may *deadlock*. This occurs when the modelled physical system would proceed, but within the simulation all (or a subset) of the logical processes remain in the same state indefinitely. Each deadlocked logical process waits on another to proceed. This definition excludes normal distributed termination and any situation in which the physical system would itself have deadlocked. In essence the deadlock situation arises because of lack of global knowledge among logical processes. At any point one process must hold the lowest timestamped event or message which is safe to process, but be unaware of the fact. An example of deadlock is illustrated in Figure 7.2 which shows part of a wider LP network (with processes able to predict future

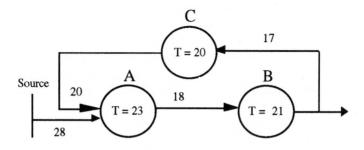

Figure 7.2: Example of deadlock

behaviour). The LPs A, B and C respectively have simulation times of 23, 21 and 20 and wait upon channels CA, AB and BC. Channel clock-times are as shown on each link. If all links are currently empty and no message is being processed in the cycle ABC, the process A could safely process the buffered message from the source channel. Instead it waits indefinitely for a message on channel CA that will never arrive.

In dealing with the problem of deadlock some authors distinguish between deadlock avoidance (DA) and deadlock recovery (DR) synchronisation. Deadlock avoidance involves preventing deadlock from ever occurring. In contrast, deadlock recovery implies that deadlock may be allowed within the simulation but is then detected and broken. Most algorithms in this area are based upon work originally carried out by Bryant (1977) and Chandy and Misra (1979, 1981). This has given rise to the term Chandy-Misra or Chandy-Misra-Bryant (CMB) synchronisation. We discuss various asynchronous methods below.

Blocking Table

The blocking table algorithm (Peacock *et al.* 1979b) can be implemented asynchronously. A waiting process LP_s sends a special type of message to all those processes with empty input links to itself. The message contains the identity of the blocked process and the next event time. Any receiving process LP_r stores this data and thus constructs a table of those processes which it is blocking and their relevant next event times. LP_r also replies to LP_s with the set of logical processes that it in turn is waiting on. If LP_s has not previously sent a request to a process in this set then it does so.

In this fashion a logical process waiting on input links can generate a blocking table consisting of all those processes which could possibly send a message with time stamp lower than the local clock time. The waiting process will remain blocked if any process in the blocking table has a next event time lower than its own. However some logical processes will find that all those logical processes with empty links to itself have next event times greater than their own. These processes may be unblocked and their events processed.

As the simulation proceeds the blocking tables of processes will be updated. Any logical process has available the table of processes which it is blocking. If the next event time increases beyond that in a blocked process then a message can be sent to update the blocking table and the process will restart. Similarly links may become empty. In this case the destination process must add the link's source process to its own blocking table.

Null Messages

The null message approach prevents deadlock from occurring. Logical processes send all messages that correspond to physical communication in the simulated system. In addition they may send null messages which have no physical equivalent and contain only a time value. A null message is transmitted by an LP to another part of the system as a guaranteed lower bound on the time stamp of any further message from that process. The time value sent will be the minimum of any input links to the LP plus any additional time margin provided by lookahead. A receiving process, blocked by the sender, will update its local clock to the time given. For an LP with several outward links null messages will be sent on each branch.

The strategy for generation of null messages may vary. One possible approach is for an LP to send null messages whenever an event increases the logical clock time. Each LP receiving the null message will therefore also generate further null messages. Another approach which generates a lower proportion of null messages is for an LP to only send a null message when it is blocked by other processes. In this scheme an LP will transmit all normal messages below the lowest time on any input link. It will then send a single null message (on each outward link) before waiting for any further message.

It can be shown that null messages will ensure correct termination of the simulation and that deadlock is prevented for acyclic networks (Chandy and Misra 1979). In cyclic networks deadlock may only occur where a cycle exists traversable by messages with zero time-stamp increment, i.e. lookahead is zero. In such cases it may be impossible to implement a simulation using null messages.

In other situations the performance of null messages is variable. An obvious disadvantage of null messages is the increase in communication overhead necessitated. Null messages will tend to proliferate at branches within the simulation network and thus acyclic networks with a lower degree of branching consistently show better performance. The amount of buffering allowed on channels can also make an impact. Reducing the buffer size below some implementation dependent level results in sending processes waiting for receiving processes over a greater period of time (Misra 1986).

Modifications to the Null Message Algorithm

A variety of modifications have been suggested to improve the performance of the basic null message scheme. The majority of these attempt to decrease null-message traffic as a proportion of the overall message population. The simplest approach is to introduce a time-out period. A logical process does not send a null message until after some set period of physical time has elapsed. No null message may be transmitted at all if messages arriving at the logical process have in the meantime rendered this unnecessary.

De Vries (1990) has suggested considering the overall network of processes as composed of sub-networks. By using detailed knowledge of the local network of which it forms a part, an LP can improve lookahead and diminish the number of null messages required. Cai and Turner (1990) also attempt to diminish null message transmission by knowledge of the simulation network. They suggest that such information can be obtained dynamically with the use of an additional type of null message. The *carrier null* may be propagated through the system and carries additional information on lookahead and the route taken.

A more radical change in approach is for null messages to be generated on a demand driven basis. After some set timeout period a waiting logical process sends out null messages to those processes blocking it. A receiving process will then return the local clock value assuming that this is greater than the clock time of the channel concerned. Otherwise null messages are sent to those logical processes with channel clock values identical to the receiving process. A null message will continue to propagate back through the simulation until a logical process with a clock value greater than the appropriate channel time is reached. The null message then starts to propagate forward, following in reverse the route on which it passed back. At each stage the returning null message updates the clock time of receiving logical processes until the original point of generation has been reached. Clearly, for this system to work, a null message propagating back through the network must store the route on which it passes so that this can be subsequently retraced.

We illustrate this in Figure 7.3, where logical process x sends out a null message to y having been waiting on the empty link (y,x). In Figure 7.3a t_y (the clock value of y) is greater than the channel clock time and this value is returned to x. Otherwise the channel value $t_{y,x}$ must be identical to t_y (which is illustrated in Figure 7.3b) and the null message must be propagated back to logical process z and then forward again.

If a null message propagating back through the network passes through the same logical process more than once then deadlock has occurred. This must be resolved, typically by identifying the earliest event in the set of deadlocked processes. The clock time of the appropriate LP is then advanced to this time and the simulation may proceed.

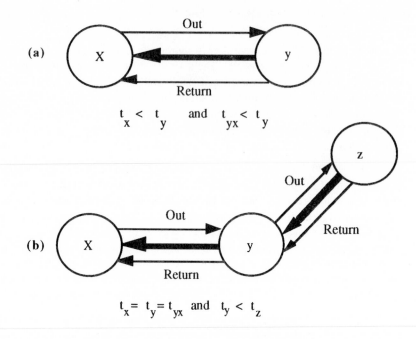

$$t_x < t_y \quad \text{and} \quad t_{yx} < t_y$$

$$t_x = t_y = t_{yx} \quad \text{and} \quad t_y < t_z$$

Figure 7.3: Demand driven null messages

SRADS

The Shared Resource Algorithm for Distributed Simulation (SRADS) is essentially conservative in nature but exhibits a limited degree of optimism. Causality errors are limited but not entirely avoided. SRADS was suggested by Reynolds (1982) for use in simulation of systems where future message times are known or largely predictable. A convenient terminology for use in this area is that of *readers* and *writers*. Processes receiving messages are readers, those sending messages are writers.

A reader knows the writer processes that could send a message and a logical time at which this is likely to occur. Given that messages have not already arrived with a time stamp greater than the predicted next arrival time, the reader must synchronise with writers by *polling*. A reader advances the local time to the point that it expects a message may arrive, simulating any known events in the interim. A poll is then initiated with the reader blocking and sending a special timestamped message to the appropriate writers. All writers return an acknowledgement once their own local logical time has advanced to the given time. Once all such replies have been reached the reader may unblock and proceed to the next predicted arrival time where this procedure is repeated.

This algorithm is conservative as LPs may be blocked because of messages that may, or may not, arrive. However, there is an element of inherent optimism.

Reader processes advance their local logical time to the next polling time whenever there are no local events to simulate in the interim. A writer process that has not advanced local time up to the polling time may send another message before an acknowledgement. This message will arrive in the logical past of the reader LP and a causality error (or *timeslip*) will be produced. In Reynolds terminology (Reynolds 1988) SRADS is aggressive, has risk and may not be accurate.

Deadlock Detection and Recovery

Chandy and Misra (1981) suggest an algorithm for parallel simulation involving deadlock detection and resolution. The simulation is allowed to proceed with conservative process blocking. Deadlock is detected via a controller process which then initiates a phase interface computation. This resolves the deadlock and enables one or more processes to proceed. During the phase interface a distributed computation occurs to determine which logical processes may be restarted. By finding the minimum timestamped message in the set of deadlocked process we may restart a single logical process. By propagating minimum next message values through the system from source to destination, several processes can proceed.

Misra (1983) suggests a circulating marker for detection and resolution of deadlock. We assume that we have (or by adding channels can make) an Eulerian network in which a circulating marker may traverse every channel during a single cycle. The particular logical process where the marker starts, and the subsequent route taken, is determined prior to run-time via a preprocessing stage. An LP will hold a marker as long as there are messages to send. After remaining idle for some particular timeout period the LP sends the marker onwards and sets a flag to ON. This flag will be reset to OFF if, subsequently, messages are received or sent to that LP. Deadlock is detected when the marker has passed through all channels in the network and finds that all the LPs that have been visited during this cycle were in the ON state.

The deadlock can be rectified relatively simply if we store on the marker the minimum event time of any LP in the ON state. On detecting deadlock the next event time in the system is instantly available and a central process broadcasts a message to every LP to update their clock time to this value. Alternatively, if the marker also stores the identity of the LP in which the event with the lowest time occurs, then deadlock may be resolved by restarting the specific LP.

Fujimoto (1989a) has examined the performance of a deadlock detection and resolution algorithm implemented on a MIMD shared memory machine. In his conclusions Fujimoto reports the existence of a *message avalanche* effect. The performance of the algorithm is poor for low message populations but improves dramatically at high message populations. The message density at which message avalanche occurs is inversely linked to the degree of lookahead exhibited by the system.

7.3.3 Lookahead

If any process can predict all future events from the current local logical clock for a period of π then that process has lookahead π. Typically lookahead enables an LP to determine a lower bound on future message time stamps. The success of conservative synchronisation is often critically dependent upon lookahead.

The Role of Lookahead

Lookahead is essential for a synchronous method. At each iteration a greater degree of lookahead will allow the identification of more events as safe to process for longer into the simulation future. The frequency of synchronisation is thus inversely related to the amount of lookahead available to logical processes. In the worst case, with no lookahead, only a single event with the lowest time stamp could be executed in any iteration.

Asynchronous conservative algorithms are generally hampered in their exploitation of available parallelism by unnecessary waiting. A process may have to idle given that the times of all available events are greater than the channel clock values on any empty links. In many cases the known events could be executed, as a message that violated causality would not subsequently arrive. The problem arises because of lack of knowledge on the part of individual logical processes as to future messages and their own behaviour. Thus lookahead is essential to prevent unnecessary waiting. For a deadlock avoidance algorithm such as null messages the link is obvious. Greater lookahead enables processes to send more null messages with higher associated times. Therefore more logical processes can work simultaneously on processing available events and there is less unnecessary waiting.

For a deadlock recovery algorithm the link between lookahead and efficient speed-up is less obvious. In a simulation with good lookahead, an LP at some particular clock time C_t may be able to predict a message with time stamp T_s, where T_s is greater than C_t. This message may then be sent immediately (assuming no further messages must be transmitted with time stamps in the interval $[C_t, T_s]$). In a system where there is poor lookahead, prediction of such a message may not be possible or transmission would be delayed. This will lower the overall percentage of links containing messages in the system and thus increase the number of deadlocks.

Generation of Lookahead

An important concept in this area is *minimum time stamp increment*. In any system with minimum time stamp increment T, the simulation of an event with time stamp t_1 cannot schedule another event or message with time stamp t_2 such that $t_2 - t_1 < T$. Allied to minimum time stamp increment is the idea of *distance* (or minimum propagation delay) between logical processes. If the distance of process a from process b is D_{ba}, then no event executed at b at

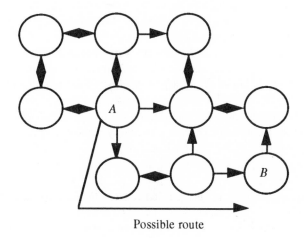

Possible route

Figure 7.4: A set of LPs

simulation time T_b may affect the execution of a at a later time T_a such that $T_a - T_b < D_{ba}$.

We illustrate this idea in Figure 7.4 which shows a set of logical processes used in a parallel simulation.

We assume that communication may only take place on the shown channels (with a double-headed arrow actually indicating two channels in opposite directions) and that there is a minimum time stamp increment of T. The minimum distance from process A to B is thus $3T$. This is not the same as in the opposite direction B to A. In fact A is unreachable from B and in this case the distance is $+\infty$.

Lubachevsky (1989a) suggests the idea of opaque periods. An LP may enter a state in which there is a guarantee that a message will not be sent on a specific link with time stamp below a minimum value. In a non-preemptive queuing network simulation the processing of a task with some particular service time on an LP will generate an opaque period for that process.

A number of authors have argued that detailed knowledge of either the simulation topology, or the application, may be used to generate lookahead. De Vries (1990) and Cai and Turner (1990) suggest modifications to the null message algorithm (see Section 7.3.2) which improve lookahead using topology specific information. Nevison (1990) argues that lookahead can be increased for the closed loop sub-systems found in Flexible Manufacturing Systems (FMS). Nicol (1988) considers stochastic simulation of a FCFS (First-Come First-Served) queuing network model (QNM). By pre-sampling the appropriate distribution, processing times at a server can be generated in advance of the actual arrival of customers. The service time of the next customer to be processed, s, is stored in a *future list* so that lookahead of the server is increased by s. Nicol's work was extended by Wagner and Lazowska (1989) to improve lookahead in other queuing models.

For the case in which a simulation has very poor natural lookahead an approach has been suggested to combine logical processes so that cyclic feedback loops are avoided (Lin *et al.* 1990). A preprocessing step identifies *strongly connected* subsets of the total number of processes (two processes that form part of such a subset must be connected by a traversable path of directed links in each direction). A single process with purely sequential operation is used to replace each subset so that any feedback loops are eliminated. The simulation may proceed with processes blocking as necessary, but without the possibility of deadlock.

Some initial work has been carried out on the automation of lookahead calculations. Cota and Sargent (1990) suggest the use of a control flow graph (CFG) to represent the actions of an LP. It should be possible for a preprocessing step using a CFG to automatically evaluate minimum delay times before an LP in one state may reach some other state. This can then be used to generate lookahead values.

7.4   Optimistic Methods

We may define optimistic synchronisation methods as those in which the logical time in particular LPs may be advanced without being limited by the possible reception from any other logical process of a message with lower time stamp. An LP processes events and advances the local clock-time on the optimistic assumption that such messages will not arrive. However, some method must be available to recover from causality errors when this assumption turns out to be false.

Below we first present a survey of different approaches, all of which contain some degree of optimism. Overall implementation and performance issues are then considered. The most well-known optimistic method is the Time Warp mechanism (Jefferson and Sowizral 1985; Jefferson 1985) which we discuss in Section 7.4.1. In the literature, a wide variety of optimisations and variants upon the basic mechanism have been suggested, as well as some closely related algorithms (Section 7.4.2). A few authors have suggested synchronisation methods which combine elements of conservative and optimistic approaches. We consider these also (Section 7.4.3).

7.4.1   The Time Warp Mechanism

The Time Warp Mechanism may be implemented asynchronously on both shared memory and message passing MIMD machines. There are some similarities to conservative methods. Each logical process maintains a form of event list and simulates events in increasing time order. Physically early messages from other logical processes must be buffered. In a similar fashion to asynchronous methods, each logical process has a local clock time which may be different from any other.

Unlike conservative methods the local clocks of different logical processes may move back as well as forward (hence the name of Time Warp). If this happens the LP is said to *rollback* and the state at this point must be restored. For this reason an LP must periodically save the state of all variables local to the process. In addition, all communication by that process is kept recorded with time stamp ordered queues for input and output messages. At any one point several states may be stored, stretching back into past simulation time. Time warp does not assume that messages are received in the same order as they are sent, nor that there are statically defined links between processes.

A message may arrive at an LP with lower time stamp than the local clock-time. Such a message is known as a *straggler* and indicates that a causality error has occurred. On receiving a straggler an LP must carry out a number of actions to restore the state of the simulation to one consistent with the new message sequence. All of these can be considered to fall into one of three phases (Jefferson and Sowizral 1985). The first, *Restoration Phase*, involves inserting the straggler message into the correct position within the sequence of received messages and rolling back the LP to an earlier state. Ideally the state would be restored to that at the simulation time exactly matching the time stamp of the straggler. In practice the state stored at the highest simulation time which is also lower than the straggler time stamp is used. The logical clock-time must be set backwards to this value. In a *Coasting Forward* phase the LP may then roll forward to the simulation time corresponding to that of the straggler. There is no difference between the actions of an LP during the Coasting Forward Phase and those of normal forward progress except that message transmission is disabled. Clearly we do not wish to resend messages which were correctly transmitted before (indeed doing so might have disastrous consequences). Both events and previously received messages are simply reprocessed.

Restoration and Coasting Forward phases ensure side-effects caused by not receiving messages in the correct order are removed. Causality errors are corrected within the affected LP which can now deal with the straggler at the correct logical time. However, a *Cancellation Phase* is necessary to deal with side-effects of the causality error created in other logical processes. In executing incorrect computation an LP may have sent messages to other processes which in turn erroneously updated their own state and logical clocks. There must therefore be some method by which a roll-backed LP can transmit a warning to those processes sent a message before rollback occurred.

The approach used involves *antimessages*. Those messages sent during the normal forward progression of an LP are considered to be positive. After rollback an LP sends out antimessages (also known as *negative* messages) to all processes erroneously sent a message during the previous phase. An antimessage can be generated for each positive message by examination of the record of positive messages sent. The meeting of message and antimessage causes the destruction of both.

On reception, a logical process treats an antimessage similarly to any positive message. The anti-message will be placed in the message queue. Assuming that both positive message and corresponding antimessage arrive before the logical clock has been advanced beyond their common time stamp then both are destroyed in the queue. There is no effect upon the state of the process. However, a rollback will be necessary if the local clock has advanced beyond the time stamp of the antimessage. In the case that the positive message has arrived then rollback is immediate. Otherwise the process may simply wait for the positive message and rollback occurs when this arrives.

An LP rolled back as the result of receiving an antimessage may well generate and send antimessages as a result. In turn other processes will do the same. This *cascade* effect occurs as an initial incorrect calculation may have sparked a chain of incorrect computation through the system. The inhibition of message transmission during Coasting Forward ensures that the worst situation that could occur is a rollback of all LPs to the state preceding the original causality error.

It has been argued that this would very rarely occur. An LP acting on incorrect information will propagate further messages with at least as high a time stamp as any messages received. Thus messages spreading an incorrect computation through the system will tend to arrive increasingly physically early as they pass from process to process. Consequently at each stage there is a longer time before an LP acts on the incorrect message and the speed at which the incorrect computation will spread decreases. From these arguments Jefferson (1985) has suggested a *temporal locality principle*, viz most messages arriving at an LP will arrive physically early or cause a relatively low degree of rollback.

There is not a requirement for logical processes to store past states and messages stretching back to the start of the simulation. Instead the Time Warp mechanism requires each LP to have at least one state stored at a simulation time predating the time stamp of any message that may be received. If all logical processes and all messages in transit have associated logical time greater than some value T, then this is a lower limit on the time stamp of any message or rollback that may occur. We term T the Global Virtual Time (GVT). One possible means of determining GVT is for the entire simulation to be halted at some point so that all processes and message buffers may be examined. This will provide an exact value for GVT. There are also algorithms for calculating GVT in a distributed fashion which run in parallel to the main system simulation. These algorithms produce a lower limit value for GVT rather than the exact, almost certainly greater, value. We discuss distributed determination of GVT in the following section.

Typically GVT calculations will be started at regular real-time intervals. The use of frequent GVT updating has a number of advantages. Each LP informed of a new value can discard data stored prior to this point. This is termed *fossil collection* with both stored states and messages being removed. The result is more efficient memory management by the Time Warp mechanism.

In addition a new GVT value may allow the simulation system to carry out input and output. These actions cannot normally be undone and must be delayed until there is no possibility of rollback. Termination of the system as a whole can be established by calculation of GVT. An LP sets the local clock to an arbitrarily large value, say $+\infty$, whenever it has no further events or messages to process. If the result of a GVT calculation is $+\infty$ then all processes must have terminated and none may restart.

In some situations an LP may find itself running out of available memory. Eventually the problem may be corrected by fossil collection with a new GVT value. In the shorter term the process may reduce the memory in use by returning some buffered messages to the original sending process. In effect there is a message rollback. On reception of a returned message a process will execute rollback to some state preceding the original send time. Rolling forward the LP will then resend the same message as before. On this occasion, we hope the destination LP will have sufficient memory to spare but, otherwise, the same sequence of actions will be repeated.

Calculation of GVT

One of the most straightforward means of determining Global Virtual Time involves a controller process which initiates and terminates the calculation. For simplicity we assume the controller runs on an otherwise unused processor, although this is not strictly necessary. To start GVT calculation the controller broadcasts a START message to every processor. These respond with an ac-knowledgement, the arrival of replies at the controller being termed a *collection*. Once all processors have responded the controller broadcasts a STOP message. On receipt of a STOP message each processor finds the minimum message time-stamp (MMT) and the minimum logical time (MLT). MMT is the mini-mum time stamp of any message sent between the arrival of START and STOP messages (or in transit when a START message arrived). MLT represents the logical time of the furthest behind LP on the processor when the STOP mes-sage arrived. We term the minimum of MMT and MLT as the least logical time (LLT) and this value is returned to the controller. The controller finds GVT by taking the minimum of all LLT received and then broadcasts this value to every LP in the system.

There is not an exact correspondence between the times at which START and STOP messages arrive at different processors. The important fact is that these time periods must all overlap; at some real time moment (RTM) all processors will have received a START message and none a STOP message. The eventual MMT value for a processor is therefore a lower bound on the messages sent up to this point. It is not the minimum time stamp of any message that could be sent in the future as some processes may have logical times lower than MMT. We take account of this by making least logical time the minimum of MMT and MLT.

A broadcast requires sending at least an integer to all N processors, so that

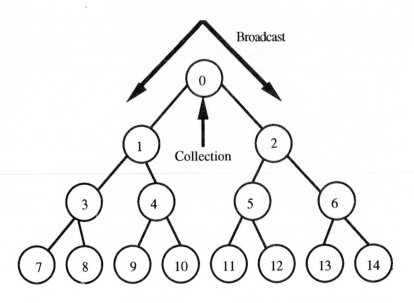

Figure 7.5: A binary tree

the order of this algorithm will be O(N). Bellenot (1990) considers a number of algorithms for GVT calculation which use different connection topologies to reduce the communication overhead. One possibility is to connect processors in a closed loop so that each node only communicates with two neighbours. START and STOP messages are sent as tokens, passing in a single direction from P_i to $P_{(i+1) mod\ N}$. Alternatively binary tree forwarding, which offers a reduction in communication overhead to O(log N), may be used. This requires processors to be connected as a binary tree with the controller at the apex (as illustrated in Figure 7.5 for 15 processors). A broadcast now involves processors forwarding messages down the binary tree so that a processor m only communicates with $(2m+1)$ and $(2m+2)$. A following collection starts immediately the message reaches the lowest leaves and passes back up the tree.

Bellenot suggests an algorithm which uses a preprocessing stage to define a message routing graph (MRG). Nodes in the graph represent processors whilst vertices indicate directed links on which communication can take place. Each processor executes code to configure itself and thus helps to define the local part of the MRG. In Figure 7.6 we illustrate the MRG produced with 14 processors. For calculation of GVT the zeroth processor sends a START message which propagates through the system, passing forward along the arcs of the MRG. Once the final node in the graph has been reached STOP messages will propagate backwards through the MRG.

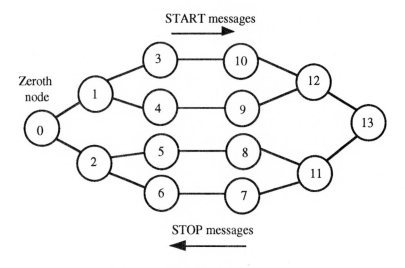

Figure 7.6: MRG for 14 processors

7.4.2 Variants of Time Warp and Related Algorithms

Since the original proposal of the Time Warp Mechanism (Jefferson and Sowizral 1985) a great many possible modifications of the basic algorithm have been suggested. Given the high performance cost of roll-back many approaches involve an attempt to make the *cancellation strategy* (the way in which erroneous simulation work is corrected) more efficient. Alternatively the amount of optimism in the simulation can be limited in an attempt to diminish the amount of incorrect computation carried out. Below we discuss some common optimisations, as well as some more elaborate modifications in this area.

Lazy Cancellation

In the original Time Warp mechanism proposed by Jefferson and Sowizral a roll-backed LP should immediately cancel all messages sent during the period of erroneous calculation by sending antimessages. This is termed *aggressive cancellation*. In some cases if the message triggering the roll-back had arrived at the correct point in the LP's logical time there might have been no change to the message sequence generated. This reflects the fact that a correct computation can still be achieved with partial or incorrect data. In such a situation the sending of antimessages and consequent roll-backs is clearly wasteful.

Lazy cancellation has been suggested to ensure that unnecessary antimessages and rollbacks are prevented. After rollback an LP does not immediately send antimessages. Instead the LP is allowed to roll forward and the messages generated are compared with those previously sent. An antimessage is not sent until the logical time proceeds beyond the previous send time without reproducing the original message. Any new messages not transmitted before rollback occurred are sent immediately.

Lazy cancellation will avoid unnecessary rollback via antimessages. However, there are disadvantages to this approach as compared to an aggressive strategy. By checking if an antimessage is necessary, a lazy strategy has greater overhead costs and may cause incorrect computation to spread further before correction. Reiher *et al.* (1990) compare the two approaches for a queuing simulation running on a MIMD shared memory machine. In general, under a wide range of conditions, the time performance of lazy cancellation was equivalent or slightly better than aggressive.

Lazy Rollback

This method (also known as jump forward or lazy re-evaluation) is similar in concept to lazy cancellation, but deals with state vectors rather than messages. After receiving a straggler, rolling back, and processing the message, there may be no change to the state of an LP. In this case the process may leap forward, restoring the state reached before the straggler arrived. Clearly exactly the same messages as before will be generated so there is no need to send antimessages. If a straggler message does alter the state then lazy rollback offers no advantage. In this case the LP must proceed forward from the rollback with aggressive or lazy cancellation of messages.

Lazy rollback entails the additional overheads of saving, and comparing, states when rollback occurs. The significance of this increases with the size of the state vectors used. Fujimoto (1990a) reports that the additional code complexity required may lower the maintainability of the Time Warp code. A further problem is that lazy rollback may not be suitable for those problem domains where messages normally result in a change of state.

Direct Cancellation

Fujimoto (1989b) suggested direct cancellation for shared memory multiprocessors. This is intended to limit the spread of incorrect simulation by quickly identifying erroneously simulated events. A causal link between two events E_0 and E_1 exists if the simulation of E_0 causes the scheduling of E_1 at some later logical time. In fact a single event E_0 may generate a set of further events $\{E_1, E_2, \ldots E_n\}$. Direct cancellation requires that such links should be recorded using pointers. This imposes an additional overhead in terms of memory storage. However, after rollback, those events generated by the initial incorrect simulation are more easily identified. Cancellation of messages and rollback of the relevant LPs is therefore quicker. Direct Cancellation has been successfully implemented by Fujimoto (1990b) and by Konas and Yew (1991).

Wolf Calls

Wolf calls (Madisetti *et al.* 1988) are intended to quickly stop the spread of erroneous computation. On reception of a straggler an LP sends special control messages to all those other processes which may have been affected by the causality error detected. Any LP receiving such a control message immediately freezes computation, but does not roll back unless this is subsequently necessary. The number of LPs affected by the incorrect computation will form a non-proper subset of those actually halted. Therefore correct simulation work may be stopped unnecessarily in one or more LPs. The number of control messages required is dependent on the speed at which incorrect computation may have spread. Application specific information is needed to determine a value for this and must be embedded in the algorithm used.

Time Warp with Phase Decomposition

In parallel simulation each LP models some part of the physical system over the overall simulation time period. Phase decomposition (Reiher *et al.* 1991) takes this a stage further, not only splitting the system into LPs but further subdividing simulation time into time phases. A phase consists of some interval of simulation time over which an LP is simulated. For the phases of a single LP there is no overlap in terms of simulation time, but when taken together the actions of the process must be defined over the entire simulation period. The final state of one phase must be transmitted as the initial state of its successor (in time order) for a particular LP. Different phases of the same LP can be executed on different processors. In some respects this is similar to dynamic process reallocation but concurrency may be allowed, i.e. in real-time the execution of phases can overlap. Rollback is allowed within phases and consequently may also occur at phase boundaries.

Time Warp with Causality Errors Penalized

This is intended to improve performance of the Time Warp mechanism by limiting optimism. Processes that have advanced local time with greater speed than others may well tend to have a higher incidence of causality errors as they receive messages from less advanced processes. In this case the incidence of several recent causality errors would be an indication that further errors were likely. We might therefore penalise the LP by slowing the progress of simulation time.

An algorithm using these ideas has been implemented for assessment purposes (Reiher *et al.* 1989) on the Time Warp Operating System (Jefferson *et al.* 1987). They use a mapping of many LPs to processors with time-sharing of a processor between different processes. Those processes with a high incidence of causality errors are given a lower execution priority. They execute less often as compared to other processes and their logical clock therefore increases at a slower rate. Similarly, Ball and Hoyt (1990) suggest *Adaptive Time Warp*

(ATW) for use in a hardware simulator. LPs with high causality error rates do not advance the local clock immediately to the next event but enter a period of inactivity.

Filter

The Filter algorithm (Prakash and Subramanian 1991) is intended to work in tandem with the normal Time Warp mechanism, reducing cascaded rollback by checking the spread of erroneous computation through the system. Unlike Wolf calls (see Section 7.4.2), the progress of correct computation should not be impeded by the algorithm.

An implementation of Filter imposes additional overheads that are not necessary in the standard Time Warp mechanism. Each LP maintains a count of messages so far sent and two lists. The *assumption list* contains data on processed messages which are assumed to have been sent in correct order. These may not necessarily have been sent to the LP maintaining the list, but all must remain valid if the process is not to roll back at a future stage (valid in this context implies that no message will be sent from the LP concerned with higher sequence number and lower time stamp). The second, *rollback list*, notes all rollbacks so far carried out in the simulation, each entry giving details of the straggler message responsible. For both lists an entry comprises a tuple $< lp_i, n, t >$, respectively the identifier of the ith LP from which the message was received, a sequence number indicating that this is the nth message from that process, and the time stamp of the message.

Messages and antimessages are sent in a similar fashion to the normal Time Warp Mechanism. The main difference is that to each positive message we add a sequence number and the assumption list of the sending process. Any LP receiving a positive message compares the attached assumption tuples against local assumption and rollback lists. Any conflict with the rollback list causes the message to be ignored as it derives from a rolled back state. Any conflict with the assumption list will require the process to roll back to a state preceding the time stamp of the message that is no longer valid. Antimessages are sent to appropriate LPs, whilst special control messages transmitted to all processes allow update of their rollback lists. If the message is consistent with both assumption and rollback lists then the new assumptions are added to those stored for the LP.

Assumption and rollback lists may grow arbitrarily large. However, there are a number of factors that will limit growth and prevent storage required from monotonically increasing. Both lists can be treated as part of the general state and pruned with the arrival of a new GVT value. Any rollback to a time T means that all assumptions with an associated time greater than T can be removed. We may arbitrarily prune assumption lists or store only the latest assumption about each LP that has sent messages.

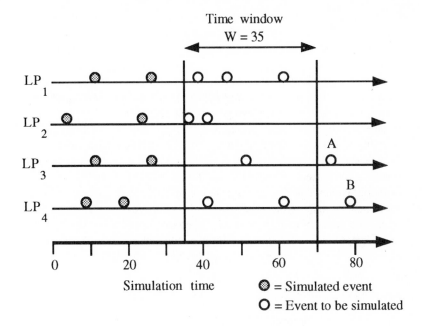

Figure 7.7: Action of a time window

Optimistic Time Windows and Moving Time Window

Some authors have suggested limiting the optimism present in the Time Warp mechanism in order to reduce the number of causality errors. Penalizing processes which have had recent causality errors is one possible approach. Alternatively we may use an optimistic windowing technique which attempts to limit how far a logical process in a simulation may advance simulation time beyond other processes. Thus known events in the future of the simulation may not be carried out in case they are subsequently proved invalid.

If the time difference between furthest behind and most advanced LP is bounded from above by T, a simulation can be said to be using a time window of width T. In many respects this approach is similar to the time windows generated by conservative synchronous algorithms (Section 7.3.1). We illustrate the action of a time window in Figure 7.7 which shows the set of events in a simulation with 4 LPs. Only 8 events can currently be simulated with a window width of 35 time units. Events A and B may never be simulated if they are proved to be invalid before inclusion in the window.

The most well-known optimistic time window approach is Moving Time Window (MTW) suggested by Sokol, Briscoe and Wieland (Sokol *et al.* 1988). In this algorithm the position of the window in simulation time is updated and broadcast to each LP whenever the number of executable events within the current window falls below some threshold value. In each update a lower bound on the time window is generated by polling processes to find the furthest behind in simulation time. The upper bound on the time window is this value increased by the set window size. MTW differs from Time Warp in a number of ways other than the use of windowing. On discovery of a causality error a number of possible correction strategies are available. Many of these require drawing a distinction between *side-effecting* (SE) and *non-side-effecting* (NSE) events. Rollback and antimessages are generally avoidable if an NSE event is generated by a message. Sokol and Stucky (1990) report that over 95% of causality errors were corrected without rollback during a battle-field simulation.

The success of any optimistic window approach will critically depend upon choosing the correct size of window. It must be wide enough so that there are sufficient events that can be simulated in parallel. However, this must be balanced against making the window so wide that there is no significant influence on the incidence of causality errors. Determination of an appropriate window width is application dependent. A time window also has the disadvantage that correct, as well as incorrect, computation may be impeded.

Space Time Approach

The space time approach (Chandy and Sherman 1989b) represents a radical departure in view-point from the Time Warp Mechanism. However, space time can be viewed as a general approach that may be instantiated to emulate different synchronisation algorithms.

Any simulation is an attempt to define the behaviour of a system over some time period. We can view ourselves as filling in a two-dimensional space time graph, the x-axis representing state variables and y-axis the passage of simulation time. For conventional parallel simulation each LP is separately simulated over the relevant time period. The x-axis is therefore partitioned into intervals representing the state vectors of each process. We show a space time graph in Figure 7.8 with three processes x, y and z. The simulation is assumed to have terminated at T.

For conservative synchronisation the space-time graph will be filled in as columns of different heights, a column of height h_i representing the state of an LP i confirmed up to time h_i. If the Time Warp mechanism is used all columns representing confirmed behaviour will be at the same height h_{tw} which is the GVT value. In addition there will be a further region of fluctuating size which represents estimated behaviour beyond GVT.

We are only really concerned with defining the space-time graph, not with how this is achieved. The space-time region can be partitioned into arbitrary, disjoint regions of irregular shape. Processes are assigned to define a set of state variables (not necessarily all from the same process) over possibly non-uniform time periods. Clearly a region in the space-time graph is influenced by other regions with which there is a common boundary. A horizontal edge indicates that the final state of one region is the initial state of the next. A vertical edge indicates message communication between LPs. Consequently there are

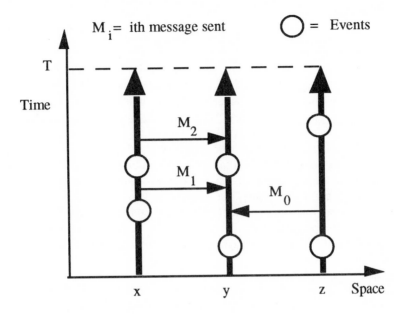

Figure 7.8: Space time graph

inflows and outflows of data from each region. The situation is complicated by cyclical dependency, the behaviour of a region a depending on region b implies that b also relies on a.

To determine each space-time region Chandy and Sherman suggest a relaxation algorithm in which processes converge to a fixed point after n iterations. In the first iteration each process may execute with an arbitrary estimate of inflow, thus producing an estimate of outflow to pass to other regions. For the following $(n - 1)$ iterations processes compare estimated inflow against received inflow in the previous iteration. Any process that correctly anticipated inflow will not recalculate the region for which it is responsible in that iteration. Other processes do so to produce new outflow values. The space-time graph is correctly filled in when no message has been sent but not received and no process is re-executing.

7.4.3 Hybrid Methods

Below we discuss a variety of hybrid approaches which combine elements of both conservative and optimistic synchronisation methods. A common feature is the use of limited optimism. In SRADS/LR and SPEEDES this takes the form of a local rollback mechanism and the prevention of the spread of erroneous calculation. Using Reynolds terminology, we say that aggressiveness is shown but there is no risk.

SRADS/LR

We have already discussed SRADS (Section 7.3.2). Dickens and Reynolds (1990) have suggested an extension of this approach which uses a local rollback mechanism. Polling in SRADS requires the logical process carrying out the poll to suspend processing. The LP may only proceed once all acknowledgement messages have been received indicating that the polled processes have reached at least the polling time. In SRADS/LR this is not necessary. Given that some processes have responded to a poll, but at least one has not yet replied, an LP may process those messages that have been received.

Messages generated are not immediately sent and the state of the process must be periodically saved. If subsequently a message arrives from a polled process (that had not previously responded) with time stamp less than the polling time then local rollback must be implemented. Stored messages generated by our optimistic processing will be sent once the polled processes respond to indicate that their local time is greater than the message time stamps.

Synchronisation in SPEEDES

The Synchronous Parallel Environment for Emulation and Discrete Event Simulation (SPEEDES) has been developed by Steinman (1991) for use in interactive parallel simulations. The SPEEDES algorithm limits optimism so that rollback need only be local in scope. This will completely prevent cascaded roll-back, but imposes additional communication overheads.

Using SPEEDES a simulation advances in a series of two stage iterations. In the first stage of an iteration all processes simulate events in parallel with state saving as the logical clock advances. Further events generated for the process are dealt with in the same iteration, but messages generated for other processes are stored without being immediately sent. The *local event horizon* (LEH) is the time stamp of the first message generated for another process. Once this point has been reached an LP broadcasts a message indicating the fact, but continues to advance simulation time. Once all processes have reached LEH they halt and a second synchronous stage of the iteration begins. All messages stored are sent. Those with time stamp below the *global simulation time* (the minimum of all LEHs) cannot cause a causality error. Other messages will result in local rollback at their destination if the simulation time has advanced beyond their time stamp.

Filtered Rollback

The bounded lag conservative approach was discussed in Section 7.3.1. A disadvantage to this approach is that the distance between processes is always taken to be the smallest possible time stamp increment between processes. If the percentage of messages which actually undergo this minimum increase is very small then performance may be degraded in order to prevent very rare causality errors.

Filtered Rollback (Lubachevsky *et al.* 1989) circumvents this problem. Identically to the Bounded Lag algorithm, the simulation proceeds in iterations. An LP communicates with any processes within a bounded region to determine the earliest time at which events at other nodes may effect execution. Unlike Bounded Lag the distance from a process j to a process i is not strictly based upon the minimum possible time stamp increase. We take into account an expected time stamp increase from j to i (ETI_{ji}). If the distance value used for Bounded Lag is D_{bl}, and that for Filtered Rollback is D_{fr} then

$$D_{bl} \leq D_{fr} \leq ETI_{ji}$$

The amount of optimism increases as D_{fr} becomes closer to the expected time stamp increase between j and i. A consequence of this optimism is that fewer processes must be checked in each iteration by i, but on occasion a causality error will occur. A mechanism must exist to correct this problem and Lubachevsky suggests *antievents*. These annihilate their corresponding event and are equivalent to antimessages in the Time Warp mechanism. In any iteration an LP may insert antievents in the local event queue or in the queues of other processes.

Lubachevsky (1990) has reported initial progress on an implementation of Filtered Rollback intended to simulate the action of a set of rigid colliding disks. Initial studies are encouraging but no performance comparison is yet available.

7.5 General Issues

In this section we deal with the use of shared memory machines, critical path performance, and load management.

7.5.1 Use of Shared Memory Machines

Parallel simulation was originally intended for use on processors which communicated solely via message passing. The overall system might well consist of a set of heterogeneous nodes, possibly computers in their own right, which are loosely coupled on a local or wide area network. More recently the transputer (Mitchell *et al.* 1990) has been developed to allow the exploitation of parallelism. This can act both as a powerful processor in its own right and as part of a wider, homogeneous MIMD machine. Technical developments have also spawned a new generation of efficient, shared memory machines such as the Sequent Balance^{tm}. In multiprocessor machines of this type processors are tightly coupled, being fully interconnected and accessing a common memory. The concept of sending a message from processor i to j thus becomes semantically identical to processor i changing state variables which j subsequently examines. A number of studies of parallel simulation have taken place on shared memory machines (for instance, Reed *et al.* 1988; Fujimoto 1989b). It therefore seems worthwhile to examine the impact of shared memory on performance.

A major limitation of shared memory multiprocessors is lack of scalability as technology limits the number of processors in a shared memory machine (typically to under twenty). There must also be some mechanism to prevent more than one processor writing concurrently to the same memory location. Thus sequential access might be enforced to commonly used data such as shared counters. However, there are also a number of very obvious advantages in using shared memory MIMD. Communication between processors is quicker and carries a lower computational cost. Full interconnection means that there is no need to route messages across intermediate processors. In addition the partitioning problem, to assign processes to balance both communication costs and processor utilisation, simplifies to a load balancing problem. The lower cost of inter-processor communication also makes dynamic reassignment of processes more practicable.

A number of authors have suggested modifying synchronisation algorithms to take advantage of shared memory structure. In an implementation of CMB deadlock recovery Reed, Malloney and McCredie (1988) detect and recover from deadlock using a central controller which forces all processes to re-synchronise at a barrier. Wagner *et al.* (1989) suggest a similar controlling process. They also introduce *lazy blocking avoidance* for use in CMB deadlock avoidance. In deciding whether known events may be simulated a process does not check the time stamp of messages received. Instead shared variables are examined to determine the next possible message that might be sent from source processes. Once blocked a process becomes idle and does not consume further processor resources. Unblocking is carried out by idle processors which periodically examine the appropriate variables to check if the process should be restarted. Wagner *et al.* examined the performance of lazy blocking avoidance on queuing models and reported performance improvements as compared to normal CMB.

For Time Warp Fujimoto (1989b) has suggested the use of Direct Cancellation (see Section 7.4.2). This involves the use of pointers to record causal links between events. Direct Cancellation can only be efficiently implemented on a shared memory multiprocessor.

7.5.2   Critical Path Performance

The concept of a critical path has been successfully used in areas such as project management (Barnetson 1969). It can generally aid analysis in situations where a large task may be split into sub-tasks, some of which must be executed sequentially. The application of critical path analysis to parallel simulation was first suggested by Berry and Jefferson (1985). By viewing the execution of a simulation as composed of event sub-tasks, an upper limit on the performance of a synchronisation algorithm may be derived.

In a physical system, events are strictly time ordered. This is reflected in a sequential simulator with events simulated in their time stamp order. For a parallel simulation there is no global system time and LPs have local logical

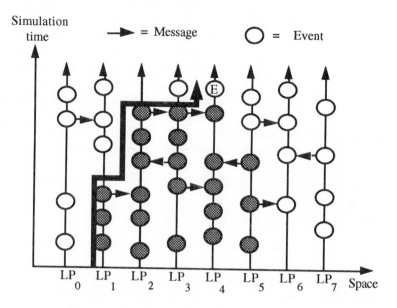

Figure 7.9: Space time diagram showing critical path

times. However, given conservative synchronisation an LP simulates an event only if no event exists with lower time stamp, or could be generated by a future message. This implies a partial ordering of events. A directed path exists from a process LP_j to a process LP_i if simulation of an event in LP_j may ultimately affect the state of LP_i. Given such a path, execution of an event with time stamp t in LP_i means that every event in LP_i and LP_j must have been simulated in correct order up to t. This is not a total ordering on events as if no path exists then LP_j may be arbitrarily ahead or behind LP_i in logical time and events may be executed concurrently in the two processes. Viewing the system behaviour in this way makes obvious the importance of process fan-out and fan-in. A high degree of interaction between LPs will fundamentally limit the potential parallelism available.

We illustrate these ideas on a space time diagram (Figure 7.9) where vertical lines represent the advance of an LPs state in simulation time and horizontal lines are messages. Event E in LP_4 may only occur after the shaded events. In general, if an event E_n is reachable from an event E_m travelling only horizontally or vertically then every event on the path $E_n \rightarrow E_m$ must be carried out sequentially. Assuming that the time to execute an event is fixed and finite the critical path corresponds to the longest path present (the heavy line on the diagram). It also represents a minimum possible time for completion of the simulation with conservative synchronisation. Lin and Lazowsky (1990) refer to simulations completing in the critical path time as conservative optimal.

Optimistic systems may exhibit better than critical path performance. We term such behaviour as *supercritical speed-up*. The Time Warp mechanism, as originally proposed by Jefferson and Sowizral (1985), will never beat the critical path. Any events simulated out of the partial order imposed by the critical path will always be rolled back before the completion of the simulation. However, Jefferson and Reiher (1991) identify possible supercritical performance in the Space-Time approach (Chandy and Sherman 1989b) and Time Warp with lazy cancellation, lazy rollback or phase decomposition. This is achieved as only a subset of the set of events cp_events $\{E_0, E_1, \ldots E_{i-1}\}$ preceding an event E_i in the critical path, actually affects execution. In those systems capable of showing supercritical speed-up members of cp_events may be executed concurrently and those shown not to affect the subsequent execution of E_i are not recalculated.

7.5.3   Load Management in Parallel Simulation

It is not our intention to address load management in the general case as ample references are available (for example, see Ni *et al.* 1985; Nicol and Saltz 1988; Lo 1988). Instead, we discuss how static and dynamic schemes may be implemented in parallel simulation and the possible effect upon performance.

We first consider static load management. In this case the application of algorithms developed for the more general case can often be applied to parallel simulation with little modification. Instead of tasks we allocate LPs to processors. Konas and Yew (1991) used several heuristics based upon the network topology but not upon knowledge of computational loads imposed by particular processes. This is relevant to situations in which computational load cannot be predicted statically. The performance of static load management alone was compared with that produced by using both static and dynamic schemes. In general the latter showed better performance by a consistently wide margin implying that knowledge of computational load is important for static schemes.

Dynamic load management in parallel simulation may take the form of creating any new processes on under-utilised processors, or the migration of existing processes. For obvious reasons dynamic process allocation is easier on shared memory hardware, although dynamic schemes have been implemented in message passing environments (Roberts *et al.* 1990). For conservative synchronisation, creating new processes may be impossible given a reliance upon a fixed network topology. A possible way of dealing with this problem is to create spare processes which may become active later in a simulation. However, the potential synchronisation cost of mostly inactive processes is likely to be prohibitively high. For the Time Warp mechanism dynamic process creation and deletion is not as correspondingly difficult.

Process migration with CMB synchronisation may be carried out in a number of ways. Reed *et al.* (1988) have idle processors obtaining work from a shared queue of unassigned processes. Processes may only be added to this queue given that there is outstanding work to be done. A previously idle process will be added by the sender of new messages. This scheme is only practicable on shared memory.

Shanker and Padman (1989) suggest a scheme which takes account of the process topology. Processors on which relocation is necessary are identified by finding the ratio of the rate of message arrival to the rate at which messages are processed. Processes are moved from processors where this ratio is high and moved to those where it is low. The *neighbours* of a process (predecessors or successors in the network) influence both the choice and destination of moved processes. Only processes with neighbours on different processors can be moved and processors have priority as a destination if they already run a neighbour process. Shanker and Padman report that on a queuing network model this scheme generally performed as good as or slightly better than when only using a static process allocation. Processor utilisation also became much more even.

A dynamic load management scheme has been especially developed for the Time Warp Mechanism (Reiher and Jefferson 1990) and implemented on TWOS (Jefferson *et al.* 1987). This approach assumes phase decomposition (Section 7.4.2). Logical processes may undergo *temporal splitting* with the overall time period over which the object is simulated divided into time phases. The phase is a basic scheduling unit so that phases of the same object may actually be executed at different processors. The aim of the load management scheme is to minimise run-time by balancing *effective utilisation* of processors. Phases on those processors carrying out a higher than average proportion of useful work are migrated to processors performing a lower than average proportion. On a battle-field combat simulation run-times with static and dynamic load management were generally comparable with those for static allocation alone.

A number of points may be made about these studies of load management. It should be noted that most dynamic schemes were actually carried out on shared memory machines so that less can be concluded about performance on a message passing MIMD machine. In general the use of the latter implies a greater role for static load balancing. Static allocations are capable of good performance given that the computational loads produced by different logical processes are known and do not alter. In none of the studies did a dynamic scheme strongly outperform a static allocation in these circumstances. However, dynamic load management (in shared memory machines at least) generally seemed to perform at least as well in most cases. Konas and Yew's work (1991) indicates that it may be essential for good performance when computational loads are difficult to predict or vary over the simulation run.

7.6   Comparison of Conservative and Optimistic Methods

Work on conservative methods is dominated by the basic deadlock detection/recovery and deadlock avoidance algorithms (Bryant 1977; Chandy and Misra 1979; Chandy and Misra 1981; Peacock *et al.* 1979b; Peacock *et al.* 1979a). For optimistic methods the Time Warp mechanism (Jefferson and

Sowizral 1985; Jefferson 1985) has achieved a similar pre-eminence. We therefore concentrate on these methods. In this section we first discuss some of the general arguments concerning the relative benefits and costs of choosing conservative or optimistic synchronisation. The literature in this area often contains conflicting views. We look at more rigorous comparisons, provided by performance and analytical studies, to provide some clarification of the issues involved.

7.6.1 General Arguments

The importance of lookahead for successful implementation of conservative methods has been emphasised in both synchronous and asynchronous algorithms. If this is not available, a parallel simulation may achieve disappointing speedup even on applications with a very high degree of parallelism. Lack of global knowledge about other parts of the simulation will block the simulation of known events even though they may actually be safe to process. In contrast, optimistic methods may exploit the natural parallelism of a system without the need for lookahead. In the Time Warp mechanism, blocking never occurs with logical processes continuing processing on the assumption that all necessary data has arrived. Thus the need for lookahead knowledge of future arrivals is dramatically curtailed. This is a strong argument for the use of optimistic methods where there is little lookahead.

To identify lookahead may be difficult and require a detailed knowledge of the simulated system. For the majority of conservative methods the lookahead available must be explicitly embedded in the synchronisation method (the exception being deadlock recovery techniques). Thus any implementor must concern himself with the details of synchronisation. This may well act as a barrier to any general user wishing to implement a parallel simulation. In contrast the user of a Time Warp implementation need not be concerned with synchronisation details.

Conservative methods are often accused of lack of robustness. An implementation using application specific information will be difficult to modify or adapt, the maximising of all possible lookahead tending to emphasise this problem. Small changes in the system to be simulated, when reflected in the simulation, may dramatically affect performance. In contrast optimistic methods, such as Time Warp, are less reliant on exploitation of lookahead and thus details of the simulated system. They therefore tend to cope better with changes or modifications.

A consequence of greater robustness is that, once developed, a Time Warp simulator such as TWOS (Jefferson *et al.* 1987) may be used in very different applications with minimum alteration. Balancing this is a greater initial development cost as compared to a conservative method. The software required is likely to be complex and difficult to design and develop. Since Time Warp systems can carry out incorrect work there needs to be more sophisticated error handling to recover from arbitrary errors created. Testing is very much complicated by issues such as cascaded rollback.

Most conservative algorithms require a static configuration of logical processes. The dynamic creation of processes and rearrangements of the logical network are difficult or impossible. Some hybrid approaches combining optimistic and conservative synchronisation (for instance SRADS) will share this problem. For Time-Warp any process may dynamically communicate with any other and there is no requirement for a static network of LPs. In some situations this will be essential.

A possible cost of Time Warp is the carrying out of incorrect computation at the expense of correct computation. In the situation that many processes are mapped to a single processor this would occur whenever an executing process was carrying out subsequently rolled back processing whilst correct computation was delayed. In the worst case the system may have many cascaded rollbacks with antimessages continually chasing after spreading incorrect computations. For feedback loops of logical processes in the system this might be especially dangerous, causing a repeating cycle of rollbacks. Lookahead might profitably be embedded in a simulation to limit the amount of incorrect computation and prevent this problem (but then it may be argued that we are losing generality).

Limitation of optimism in ATW, Filtered Rollback and Moving Time Window approaches is intended to limit the number of rollbacks. Those synchronisation methods using only local rollback mechanisms (SRADS, SPEEDES) are extreme examples of this idea with cascaded rollback entirely prevented. An essential problem inherent in all of these synchronisation methods is that correct, as well as incorrect, computations may be prevented.

Proponents of Time Warp would argue that Jefferson's temporal locality principle will limit the potential for instability without reducing optimism. Fujimoto (1990a) points out that this principle might not hold if incorrect work tended to spread with lower time stamp increment than correct computation. On the other hand practical systems in which this occurs have not been established.

There are performance overheads in optimistic mechanisms which are not shared by conservative approaches. All optimistic approaches must save state at intervals and restore a state after rollback. The degree to which this will affect performance depends on the average size of the LP state and the frequency of saving. The same factors also determine the memory requirements of Time Warp. In general, optimistic methods require greater memory than their conservative counterparts because of saving state. In Time Warp simulators if available memory is limited the mechanism by which memory usage is decreased, returned messages, may cause performance degradation. It should be noted that GVT calculation represents an additional overhead but the effect upon performance is limited by running a GVT calculation as a background processing task.

7.6.2 Empirical Evidence

Many parallel simulation implementations have been reported. Most papers have tended to measure the performance of conservative **or** optimistic synchronisation without making a direct performance comparison. Some recent studies of conservative synchronisation have looked at the simulation of queuing networks (Reed *et al.* 1988), performance under synthetic loads (Fujimoto 1989a), job-shops (Wing and Rayward-Smith 1991) and logic network simulation (Su and Seitz 1989). Optimistic synchronisation has been applied to such diverse areas as simulation of computer networks (Presley *et al.* 1989), biological systems (Ebling *et al.* 1989), combat situations (Wieland *et al.* 1989) and colliding pucks (Hontalas *et al.* 1989).

Studies comparing synchronisation methods have been carried out in the simulation of queuing network models (Fujimoto 1989b; Konas and Yew 1991). Both conservative null-message synchronisation and Time Warp have been implemented for traffic simulation (Roberts *et al.* 1990). Wieland and Jefferson (1989) compared time-driven synchronisation against Time Warp in simulation of a battle-field model.

Simulations of parallel systems have also been carried out to assess performance. Agre (1989) developed *Warpsim* which runs on a uniprocessor and models the Time Warp mechanism. The impact of algorithmic optimisation on Time Warp has been assessed using an independently developed model (Baezner *et al.* 1989). Clearly the accuracy of these models is heavily dependent upon the values used for parameters such as inter-processor communication delay.

Various studies provide strong evidence for the importance of lookahead in conservative methods. Reed, Malloney and McCredie (1988) simulated various queuing network models using deadlock avoidance and deadlock recovery algorithms. They deliberately did not attempt to exploit system specific information to generate lookahead. Although acyclic networks showed good speed-up, lack of lookahead in cyclic networks was disastrous. For deadlock avoidance a high proportion of null messages were generated, whilst deadlock resolution became a major overhead in the deadlock recovery algorithm. A central server model (see Figure 7.10) showed particularly bad performance with fractional speed-up, i.e. the parallel implementation took longer to execute than a sequential simulation. In a later study Fujimoto (1989a) proved that the use of lookahead could dramatically improve performance on these types of problem. Exploiting lookahead, the central server model was re-simulated and considerably higher speed-up generated.

A further queuing network study carried out by Fujimoto directly compared deadlock avoidance and recovery algorithms against a Time Warp implementation (Fujimoto 1989b). In general Time Warp seemed to outperform the conservative algorithms. The difference was very marked where lookahead was decreased by introducing prioritised customers and preemption of service. These changes caused a sharp drop in performance of the conservative algo-

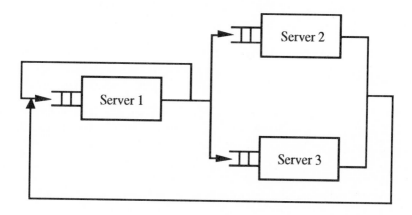

Figure 7.10: Central server model

rithms but the Time Warp implementation was only slightly affected and still produced a respectable speed-up.

Fujimoto's study gives strong support for using Time Warp in simulations where there is poor lookahead. However, the out-performance of the conservative algorithms (even with good lookahead) cannot be considered as indication that Time Warp is inherently superior. For instance, Roberts *et al.* (1990) report that a deadlock avoidance algorithm out-performed Time Warp on a traffic simulation.

In the simulation of queuing networks, Time Warp benefits from the generally small state vectors involved. This will minimise the cost of saving and restoring state. Fujimoto examined Time Warp performance with varying state sizes and found that state saving may be a major overhead. This may well provide justification for hardware support of Time Warp to limit the cost of state saving and restoration (Buzzell and Robb 1990). Fujimoto does not report wide-spread instability occurring in the simulation. In fact high efficiencies are generally achieved implying that correct computation is not being excessively delayed by erroneous work. Other experimental results do seem to show that incorrect work may significantly delay the progress of correct computation. Abrams (1989) reports instability in a Time Warp implementation of a queuing network simulation, a fractional speed-up of very much less than unity resulted. Less dramatically, in a Time Warp simulation of a battle-field with varying numbers of processors (Wieland *et al.* 1989) the increase in speed-up was inversely correlated to the increase in rollback count.

The Time Warp Operating System (Jefferson *et al.* 1987) has been successfully applied to the simulation of computer networks (Presley *et al.* 1989), biological systems (Ebling *et al.* 1989), and colliding pucks (Hontalas *et al.* 1989). In each case respectable speed-ups were reported, generally in excess of 50% of the processors used. There also seemed to be a low incidence of reverse messages which indicates that excessive memory usage was not a problem. Successful application of Time Warp to such diverse simulations seems to indicate the robustness of the algorithm. This has been further emphasised by Fujimoto who used synthetic workloads to measure performance of another Time Warp implementation (Fujimoto 1990b). Using 64 processors, speed-ups were as high as 54 under favourable conditions and 32 with adverse conditions.

In contrast conservative algorithms showed poor robustness in the queuing network studies carried out by Fujimoto (1989b) and by Konas and Yew (1991). Fujimoto found that the introduction of only one per cent prioritised jobs significantly degraded performance. Konas and Yew showed that different distributions of service times had a strong influence on performance. An exponential distribution (with a minimum service time of zero) produced poorer speed-up than the biased or deterministic cases.

For optimistic approaches which do not advance logical time as aggressively as Time Warp, two studies have provided conflicting evidence as to effectiveness. Reiher *et al.* (1989) measured the performance of two modified versions of Time Warp. The first modification penalised those processors with a high incidence of causality errors. The second used a time windowing approach which impeded the progress of the more time advanced processes. Reported results were not significantly better than that for unrestrained optimism under a wide range of conditions. In some cases performance was very much poorer with correct computation being impeded. The Moving Time Window algorithm (Sokol *et al.* 1988) is significantly different from Time Warp in some respects, but does use rollback and an optimistic time window technique. On a battle-field simulation study, Sokol and Stucky (1990) produced an impressive speed-up of 8 using 9 processors. However, critics of this approach would note that speed-up was measured relative to the same algorithm on a single processor and that anything other than the optimal window size produced much lower performance.

It is worth commenting that no performance study can give definitive answers as to which synchronisation approach is best. A central problem in assessing synchronisation performance is the number of other factors which affect the success of a parallel simulation. These include the degree of communication between processes, granularity of the computation, the degree of potential concurrency, and the assignment of processes to processors. There is also a whole range of hardware issues such as whether a shared memory or message passing machine is used. However, performance studies can indicate the relative merits of conservative or optimistic approaches and whether a particular approach is viable or not in specific application areas. In no application where there was high process connectivity and poor lookahead did a conservative method perform at near-linear speed-up. Optimistic methods seemed to do much better in these circumstances. On the other hand, a high ratio of processes to processors seemed to favour conservative methods over optimistic.

7.6.3 Analytical Evidence

Any analytical examination of synchronisation performance will be made difficult by the complexity of behaviour involved. This is especially true of optimistic synchronisation mechanisms in which cascaded rollback may occur. In answer to this problem most authors attempting analytical studies make simplifying assumptions, limiting to some extent the credibility of the derived results. However, useful insights are still produced by the analytical models generated.

In general we may divide analytical studies into those that examine a specific type of synchronisation and those that attempt to compare the performance of different types. A study of conservative synchronisation in parallel algorithms was carried out by Greenbaum (1989). She derived upper and lower bounds on the costs of synchronisation barriers and on synchronisation between neighbouring processors. The former was shown to be very costly on performance as the number of elements was increased.

Early analytical work concentrating on Time Warp analysed models with only two logical processes. Lavenburg and Muntz (1983) produced an approximate solution for system behaviour assuming low process interaction. This assumption was relaxed by Mitra and Mitrani (1984) who developed an exact solution. In both cases the system remains stable with a rollback cost proportional to the rollback distance. However, Fujimoto (1990a) constructs a two process example in which instability is produced if rolling back by time t is more expensive than forward progression of t. The extension of the work by Mitra and Mitrani to cover an arbitrary number of processes (and thus cascaded rollback) has not been reported. In other studies which deal with the n process case, rollback cost has been either irrelevant, assumed to be zero, or fixed and independent of rollback distance.

Madisetti *et al.* (1990) assume a constant rollback cost. They start with an analysis of a two processor case to determine the progress of the simulation. This is then extended to the general case. From this analysis it is argued that additional control messages may be necessary to correct the spread of erroneous computation. A state saving and rollback cost of zero is assumed by Gupta *et al.* (1991). They model the overall state of a Time Warp simulation as a one dimensional markov chain with state changes occurring as a result of GVT calculation, processing of an event or rollback. By computing transition rates between states the behaviour of the system as a whole is predicted. The result of this analysis was validated by experimental results showing quite close correspondence to those expected.

Authors comparing synchronisation methods generally assume a one-to-one mapping of logical processes to processors which favours optimistic over conservative synchronisation. Although a single *super-process* executing on a processor may actually model the action of several aggregated LPs (Fujimoto 1990a), the efficiency of conservative synchronisation will be boosted by having many processes on a processor.

Felderman and Kleinrock (1990) find an upper bound on the increase in performance of Time Warp compared to synchronous simulation. Since the most restrictive form of synchronisation is likely to be time-driven this analysis also provides bounds on the improvement of Time Warp over asynchronous conservative methods (CMB). Rollback cost may be ignored in the analysis as there will be no causality errors if best case execution of Time Warp is achieved. Given P processors, a state saving cost of zero and exponentially distributed times to simulate an event, an upper bound on improvement of log P is produced. This is consistent with experimental results (Wieland and Jefferson 1989).

Lipton and Mizell (1990) also look at upper and lower bounds on simulation progression. They compare CMB synchronisation with Time Warp, constructing an example in which Time Warp outperforms CMB synchronisation by an amount related to the number of processes. They also derive an upper bound on the possible improvement of CMB synchronisation over Time Warp which is constant. The conclusion is that Time Warp may arbitrarily out-perform CMB but the converse is not true. The possible lag in progress of Time Warp behind a conservative algorithm is bounded. This analysis relies on the assumption of a constant rollback cost.

Lin and Lazowska (1990) consider the conditions under which Time Warp will out-perform conservative synchronisation. They assume a zero state saving/rollback cost and that correct computation is never rolled back by incorrect. In these circumstances it is shown that Time Warp will always execute as fast as the critical path and possibly faster given lazy cancellation. In addition Time Warp will outperform conservative synchronisation in all feed-forward networks and a majority of networks that include feedback loops. Unfortunately these conclusions do not have profound implications as they merely confirm what might have been guessed intuitively. The given assumptions remove all those costs which are associated with Time Warp and not shared by a conservative algorithm.

Nicol (1990) derives an upper bound on the progression of a simulation with no lookahead. Then assuming constant rollback and state saving costs with good lookahead, an upper bound on Time Warp and a lower bound on a conservative synchronisation protocol (later developed as YAWNS, see Section 7.3.1) are derived. This is used in finding conditions under which Time Warp will out-perform the conservative protocol. Based upon his analytical model, Nicol suggests general principles by which parallel simulations should be developed. Exploitation of lookahead is emphasised as important even in optimistic synchronisation. Interestingly the same point was made in a simulation of Time Warp (Baezner *et al.* 1989).

We may make a number of comments on the studies carried out. Analysis of the two processor case (such as that carried out by Mitra and Mitrani) is interesting, but not of much practicable importance unless it can be extended. Although restrictive assumptions are still required, progress has been made on the multiple processor case. Gupta, Akyildez and Fujimoto successfully validated their analysis of Time Warp. The work carried out by Nicol predicted the importance of lookahead and connectivity. It does seem that analytical studies may yet contribute to the debate over the use of different synchronisation methods.

7.7   Concluding Remarks

Simulation seems an obvious candidate for parallel execution. This computationally intensive task is used widely and has a high degree of inherent parallelism. However, a number of factors seem to have prevented the widespread acceptance of parallel simulation techniques, notably a lack of multiprocessor systems and people experienced in their use. This has been coupled with doubts about the viability of the approach and simple ignorance of its existence.

Many of these barriers have now been removed. A much wider availability and use of parallel MIMD machines has spawned users at home with parallel computing. The techniques of parallel simulation have been developed and can no longer be dismissed as exotic or experimental. This has been emphasised by performance studies confirming parallel simulation as viable and offering significant improvements on sequential execution times. Ease of use may also be improved with the prospect of "off-the-shelf" parallel simulation packages where minimal alterations would be necessary for a particular application. In this area Time Warp has great potential.

The remaining obstacle to the widespread use of parallel simulation lies in ignorance and the inertia of the computing community. Although successfully applied in a diverse range of applications, parallel simulation has been used as a test of viability rather than to solve actual problems. A consequence is that results have been reported to the parallel simulation community rather than to those working in particular application fields. The final barrier of ignorance will only be eroded as parallel simulation is reported to be a practical solution for practical problems.

Acknowledgements

The invaluable advice of V. J. Rayward-Smith in preparing this chapter is gratefully acknowledged. My thanks are also due to S. A. Rush for his help with LateX.

References

Abrams M. (1989) The Object Library for Parallel Simulation (OLPS). *Proc. of the 1989 Winter Simulation Conference*, pp. 210–9.

Agre J. (1989) Simulation of Time Warp Distributed Simulations. *Proc. of the SCS Multiconference on Distributed Simulation*, **21**(2), pp. 85–90.

Axelrod T. (1985) Effects of Synchronisation Barriers on Multiprocessor Performance. *Parallel Computing*, **3**, pp. 129–40.

Ayani R. (1989) A Parallel Simulation Scheme Based on Distances Between Objects. *Proc. of the SCS Multiconference on Distributed Simulation*, **21**(2), pp. 113–8.

Ayani R. and Rajaei H. (1990) Parallel Simulation of a Generalised Cube Multistage Interconnection Network. *Proc. of the SCS Multiconference on Distributed Simulation*, **22**(1), pp. 60–3.

Baezner D., Cleary J., Lomow G. and Unger B. (1989) Algorithmic Optimisations of Simulations on Time Warp. *Proc. of the SCS Multiconference on Distributed Simulation*, **21**(2), pp. 73–8.

Ball D. and Hoyt S. (1990) The Adaptive Time-Warp Concurrency Control Algorithm. *Proc. of the SCS Multiconference on Distributed Simulation*, **22**(1), pp. 174–7.

Barnetson P. (1969) *Critical Path Analysis*. Butterworth and Co. Ltd.

Bellenot S. (1990) Global Virtual Time Algorithms. *Proc. of the SCS Multiconference on Distributed Simulation*, **22**(1), pp. 122–7.

Berry O. and Jefferson D. (1985) Critical Path Analysis of Distributed Simulation. *Proc. of the 1985 SCS Conference on Distributed Simulation*, **15**(2), pp. 57–60.

Brooks E. (1984) A Multitasking Kernel for the C and FORTRAN Programming Languages. *Technical Report UCID–20167*, Lawrence Livermore National Laboratory.

Bryant R. (1977) Simulation of Packet Communications Architecture Computer Systems. *Technical Report MIT–LCS–TR–188*, Massachusetts Institute of Technology.

Buzzell C. and Robb M. (1990) Modular VME Rollback Hardware for Time Warp. *Proc. of the SCS Multiconference on Distributed Simulation*, **22**(1), pp. 153–6.

Cai W. T. and Turner S. (1990) An Algorithm for Distributed Discrete-Event Simulation-The 'Carrier Null Message' Approach. *Proc. of the SCS Multiconference on Distributed Simulation*, **22**(1), pp. 3–8.

Chandy K. and Misra J. (1979) Distributed Simulation: A Case Study in Design and Verification of Distributed Programs. *IEEE Transactions on Software Engineering*, **SE–5**(5), pp. 440–52.

Chandy K. and Misra J. (1981) Asynchronous Distributed Simulation via a Sequence of Parallel Computations. *Communications of the ACM,* **24**(11), pp. 198–206.

Chandy K. and Sherman R. (1989a) The Conditional Event Approach to Distributed Simulation. *Proc. of the SCS Multiconference on Distributed Simulation,* **21**(2), pp. 93–9.

Chandy K. and Sherman R. (1989b) Space-Time and Simulation. *Proc. of the SCS Multiconference on Distributed Simulation,* **21**(2), pp. 53–7.

Cota B. and Sargent R. (1990) A Framework for Automatic Lookahead Computation in Conservative Distributed Simulations. *Proc. of the SCS Multiconference on Distributed Simulation,* **22**(1), pp. 56–9.

De Vries R. (1990) Reducing Null Messages in Misra's Distributed Discrete Event Simulation Method. *IEEE Transactions on Software Engineering,* **16**(1), pp. 82–91.

Dickens P. and Reynolds P. (1990) SRADS with Local Rollback. *Proc. of the SCS Multiconference on Distributed Simulation,* **22**(1), pp. 161–4.

Ebling M., DiLoreto M., Presley M., Wieland F., and Jefferson D. (1989) An Ant Foraging Model Implemented on the Time Warp Operating System. *Proc. of the SCS Multiconference on Distributed Simulation,* **21**(2), pp. 21–6.

Felderman R. and Kleinrock L. (1990) An Upper Bound on the Improvement of Asynchronous versus Synchronous Distributed Processing. *Proc. of the SCS Multiconference on Distributed Simulation,* **22**(1), pp. 131–6.

Fujimoto R. (1989a) Performance Measurements of Distributed Simulation Strategies. *Transactions Society for Computer Simulation,* **6**(2), pp. 89–132.

Fujimoto R. (1989b) Time Warp on a Shared Memory Multiprocessor. *Transactions Society for Computer Simulation,* **6**(3), pp. 211–139.

Fujimoto R. (1990a) Parallel Discrete Event Simulation. *Communications of the ACM,* **33**(10), pp. 31–53.

Fujimoto R. (1990b) Performance of Timewarp under Synthetic Workloads. *Proc. of the SCS Multiconference on Distributed Simulation,* **22**(1), pp. 23–8.

Greenbaum A. (1989) Synchronisation Costs on Multiprocessors. *Parallel Computing,* **10**, pp. 3–14.

Gupta A., Akyildiz, I. and Fujimoto R. (1991) Performance Analysis of Time Warp with Homogeneous Processors and Exponential Task Times. *Proc. of the 1991 ACM Sigmetrics Conference on Measurement and Modeling of Computer Systems*, pp. 101–10.

Hontalas P., Beckman B., DiLoreto M., Blume L., Reiher P., Sturdevant K., Van Warren L., Wedel J., Wieland F. and Jefferson D. (1989) Performance of the Colliding Pucks Simulation on the Time Warp Operating Systems. *Proc. of the SCS Multiconference on Distributed Simulation*, **21**(2), pp. 3–7.

Jefferson D. (1985) Virtual Time. *ACM Transactions on Programming Languages and Systems*, **7**(3), pp. 404–25.

Jefferson, D. and Reiher P. (1991) Supercritical Speedup. *Proc. of the 24th Annual Simulation Symposium*, pp. 159–68.

Jefferson D. and Sowizral H. (1985) Fast Concurrent Simulation Using the Timewarp Mechanism. *Proc. of the 1985 Conference on Distributed Simulation*, pp. 63–9.

Jefferson D., Beckman B., Wieland F., Blume L., DiLoreto M., Hontalas P., Laroche, P., Sturdevant K., Tupman J., Warren V., Wedel J., Younger H. and Bellenot S. (1987) Distributed Simulation and the Time Warp Operating System. *11th Symposium on Operating Systems Principles*, **21**(5), pp. 77–93.

Jones D. (1986) Concurrent Simulation: An Alternative to Distributed Simulation. *Proc. of the 1986 Winter Simulation Conference*, pp. 417–23.

Jones D., Chou C., Renk D., and Bruell S. (1989) Experience with Concurrent Simulation. *Proc. of the 1989 Winter Simulation Conference*, pp. 756–63.

Konas P. and Yew P.-C. (1991) Parallel Discrete Event Simulation on Shared-Memory Multiprocessors. *Proc. of the 24th Annual Simulation Symposium*, pp. 134–48.

Krishnamurthi M., Chandrasekaran U., and Sheppard S. (1985) Two Approaches to the Implementation of a Distributed Simulation System. *Proc. of the 1985 Winter Simulation Conference*, pp. 435–43.

Lavenburg S. and Muntz R. (1983) *Performance Analysis of a Rollback Method for Distributed Simulation*, pp. 117–32. Performance '84. Elsevier Science Publications, North Holland.

Lin Y.-B. and Lazowsky E. (1990) Optimality Considerations of 'Time Warp' Parallel Simulation. *Proc. of the SCS Multiconference on Distributed Simulation*, **22**(1), pp. 29–34.

Lin Y.-B., Lazowsky E., and Baer J. (1990) Conservative Parallel Simulation for Systems with No Lookahead Prediction. *Proc. of the SCS Multiconference on Distributed Simulation,* **22**(1), pp. 144–9.

Lipton R. and Mizell D. (1990) Time Warp vs. Chandy-Misra: A Worst Case Comparison. *Proc. of the SCS Multiconference on Distributed Simulation,* **22**(1), pp. 137–43.

Lo V. (1988) Heuristic Algorithms for Task Assignment in Distributed Systems. *IEEE Transactions on Computers,* **37**(11), pp. 1384–97.

Lubachevsky B. (1989a) Efficient Distributed Event Driven Simulations of Multiple-Loop Networks. *Communications of the ACM,* **32**(1), pp. 111–23.

Lubachevsky B. (1989b) Scalability of the Bounded Lag Distributed Discrete Event Simulation. *Proc. of the SCS Multiconference on Distributed Simulation,* **21**(2), pp. 100–7.

Lubachevsky B. (1990) Simulating Colliding Rigid Disks in Parallel Using Bounded Lag Without Time Warp. *Proc. of the SCS Multiconference on Distributed Simulation,* **22**(1), pp. 194–202.

Lubachevsky B., Shwartz A., and Weiss A. (1989) Rollback Sometimes Works . . . If Filtered. *Proc. of the 1989 Winter Simulation Conference,* pp. 630–9.

Madisetti V., Walrand J., and Messerschmitt D. (1988) WOLF: A Rollback Algorithm for Optimistic Distributed Simulation Systems. *Proc. of the 1988 Winter Simulation Conference,* pp. 296–305.

Madisetti V., Walrand J., and Messerschmitt D. (1990) Synchronisation in Message-Passing Computers—Models, Algorithms and Analysis. *Proc. of the SCS Multiconference on Distributed Simulation,* **22**(1), pp. 35–48.

Misra, J. (1983) Detecting Termination of Distributed Computations using Markers. *Proc. of the 2nd ACM Principles of Distributed Computing,* pp. 290–3.

Misra J. (1986) Distributed Discrete-Event Simulation. *ACM Computing Surveys,* **18**(1), pp. 39–65.

Mitchell D., Thompson J., Manson G., and Brookes G. (1990) *Inside the Transputer.* Blackwell Scientific Publications, Oxford.

Mitra D. and Mitrani I. (1984) *Analysis and Optimum Performance of two Message-Passing Parallel Processors Synchronised by Rollback,* pp. 35–50. Performance '84. Elsevier Science Publications, North Holland.

Mitrani I. (1982) *Simulation Techniques for Discrete Event Systems.* Cambridge University Press.

Nevison C. (1990) Parallel Simulation of Manufacturing Systems: Structural Factors. *Proc. of the SCS Multiconference on Distributed Simulation,* **22**(1), pp. 17–9.

Ni L., Xu C.-W., and Gendreau T. (1985) A Distributed Drafting Algorithm for Load Balancing. *IEEE Transactions on Software Engineering,* **SE-11**(10), pp. 1153–61.

Nicol D. (1988) Parallel Discrete-Event Simulation of FCFS Stochastic Queuing Networks. *ACM SIGPLAN Notices,* **23**(9), pp. 124–37.

Nicol D. (1990) Analysis of Synchronisation in Massively Parallel Discrete-Event Simulations. *ACM SIGPLAN Notices,* **25**(3), pp. 89–98.

Nicol D. and Saltz J. (1988) Dynamic Remapping of Parallel Computations with Varying Resource Demands. *IEEE Transactions on Computers,* **37**(9), pp. 1073–87.

Nicol D., Micheal C., and Inouge P. (1989) Efficient Aggregation of Multiple LPs in Distributed Memory Parallel Simulation. *Proc. of the 1989 Winter Simulation Conference,* pp. 680–5.

Peacock J., Wong J., and Manning E. (1979a) A Distributed Approach to Queuing Network Simulation. *Proc. of the 1979 Winter Simulation Conference,* pp. 399–405.

Peacock, J., Wong J., and Manning E. (1979b) Distributed Simulation Using a Network of Processors. *Computer Networks,* **3**, pp. 44–56.

Prakash A. and Subramanian, R. (1991) Filter: An Algorithm for Reducing Cascaded Rollbacks in Optimistic Distributed Simulations. *Proc. of the 24th Annual Simulation Symposium,* pp. 123–32.

Presley M., Ebling M., Wieland F., and Jefferson D. (1989) Benchmarking the Time Warp Operating System with a Computer Network Simulation. *Proc. of the SCS MultiConference on Distributed Simulation,* **21**(2), pp. 8–13.

Rajagopal R. and Comfort, J. (1989) Contrasting Distributed Simulation with Parallel Replication: A Case Study of a Queuing Simulation on a Network of Transputers. *Proc. of the 1989 Winter Simulation Conference,* pp. 746–54.

Reed D., Maloney A., and McCredie B. (1988) Parallel Discrete Event Simulation Using Shared Memory. *IEEE Transactions on Software Engineering,* **14**(4), pp. 541–53.

Reiher P. and Jefferson D. (1990) Virtual Time Based Dynamic Load Management in the Time Warp Operating System. *Proc. of the SCS Multi-Conference on Distributed Simulation,* **22**(1), pp. 103–11.

Reiher P., Bellenot S., and Jefferson D. (1991) Temporal Decomposition of Simulations under the Time Warp Operating System. *Parallel and Distributed Simulation (PADS).*

Reiher P., Wieland F., and Jefferson D. (1989) Limitation of Optimisation in the Time Warp Operating System. *Proc. of the 1989 Winter Simulation Conference,* pp. 765–70.

Reiher, P., Fujimoto R., Bellenot S., and Jefferson D. (1990) Cancellation Strategies in Optimistic Execution Systems. *Proc. of the SCS Multiconference on Distributed Simulation,* **22**(1), pp. 112–21.

Reynolds P. (1982) A Shared Resource Algorithm for Distributed Simulation. *Proc. of the 9th Annual Symposium on Computer Architecture,* **10**(3), pp. 259–66.

Reynolds P. (1988) A Spectrum of Options for Parallel Simulation. *Proc. of the 1988 Winter Simulation Conference,* pp. 325–32.

Roberts J., Baker S., Kirton M., Merrifeld B., Richardson S., and Youngman N. (1990) Concurrent Discrete Event Simulation on Supernode. *Text of paper presented at Esprit Workshop on Simulation and Prototyping.*

Shanker M. and Padman R. (1989) Adaptive Distribution of Model Components via Congestion Measures. *Proc. of the 1989 Winter Simulation Conference,* pp. 640–7.

Sokol L. and Stucky B. (1990) MTW: Experimental Results for a Constrained Optimistic Scheduling Paradigm. *Proc. of the SCS MultiConference on Distributed Simulation,* **22**(1), pp. 169–73.

Sokol L., Briscoe D. and Wieland A. (1988) MTW: A strategy for Scheduling Discrete Simulation Events for Concurrent Execution. *Proc. of the SCS Multiconference on Distributed Simulation,* **19**(3), pp. 34–42.

Steinman J. (1991) Interactive SPEEDES. *Proc. of the 24th Annual Simulation Symposium,* pp. 149–58.

Su W.-K. and Seitz C. (1989) Variants of the Chandy-Misra-Bryant Distributed Discrete-Event Simulation Algorithm. *Proc. of the SCS MultiConference on Distributed Simulation,* **21**(2), pp. 38–43.

Torpor R (1984) Termination Detection for Distributed Computations. *Information Processing Letters,* **18**, pp. 33–6.

Wagner D. and Lazowska E. (1989) Parallel Simulation of Queuing Networks: Limitations and Potentials. *Proc. of the International Conference on Measurement and Modelling of Computer Systems.*

Wagner D., Lazowska E. and Bershad B. (1989) Techniques for Efficient Shared-Memory Parallel Simulation. *Proc. of the SCS Multiconference on Distributed Simulation,* **21**(2), pp. 29–37.

Wieland F. and Jefferson D. (1989) Case Studies in Serial and Parallel Simulation. *Proc. of the 1989 International Conference on Parallel Processing,* pp. 255–8.

Wieland F., Hawley L., Feinberg A., DiLoreto M., Blume L., Reiher P., Beckman B., Hontalas P., Bellenot S. and Jefferson D. (1989) Distributed Combat Simulation and Time Warp: The Model and its Performance. *Proc. of the SCS Multiconference on Distributed Simulation,* **21**(2), pp. 14–20.

Wing A. and Rayward-Smith V. (1991) Distributed Discrete Event Simulation for the Printed Circuit Board Industry. *Applied Simulation and System Dynamics,* pp. 31–40.

CHAPTER 8

Genetic Algorithms in Optimisation and Adaptation

Philip Husbands
University of Sussex, UK

8.1 Introduction

This chapter provides a wide ranging and up to date survey of the field of genetic algorithms. Genetic algorithms are adaptive search techniques based on an abstract model of biological evolution. They can be used as an optimisation tool or as the basis of more general adaptive systems. The basic idea is to maintain a population of candidate solutions which evolve under a selective pressure that favours the better solutions. Parent solutions are combined in various ways to produce offspring solutions which then enter the population, are evaluated and may themselves produce offspring. The technique has attracted a great deal of interest because it has been shown to be highly robust and to perform well without recourse to special domain specific heuristics.

The chapter starts by providing some background in the form of a brief discussion of natural evolution. This makes the more abstract computational material easier to follow. The basic technique is described in some detail by referring to the early simple sequential algorithms. This is done as a matter of convenience and to give some historical perspective. Most of the underlying mechanisms are the same whether the implementation is sequential or parallel. It will become clear that the method is highly suitable for parallelisation and this topic is discussed in detail. A number of implementation techniques are presented and it is shown how genetic algorithms can extend the parallel paradigm and reach into uncharted areas of problem solving. A range of applications in function optimisation and combinatorial optimisation is described.

When genetic algorithms are used as an optimisation method, selection is controlled by some well defined and very specific fitness (objective) function. However, it is misleading to think of natural evolution in these terms: in general there is no well defined goal, there is no specific fitness function; the whole process is open-ended. It can be argued that if we are to understand natural intelligence, and perhaps build intelligent machines, we should consciously work against an evolutionary backdrop: intelligence was not designed, it evolved in response to environmental pressures. Models in which we attempt to understand more about the emergence of various sorts of organism behaviours will have to allow for this undirected open-endedness. Some form of genetic algorithm is likely to play an important role. Pioneering work in this direction is described towards the end of the chapter, along with earlier GA based research in machine learning. A radical argument is developed that suggests that GAs provide us with an opportunity to learn how to evolve complex systems rather than design them, our record at designing them being very poor.

8.2 Natural Evolution, Optimisation and Adaptation

Most of us are familiar with Darwin's notion of natural selection (Darwin 1859): the fitter an organism is the more likely it is to produce offspring and pass on those characteristics that made it fit. Darwin postulated that selection, acting on small variations within a species, provides the basic mechanism of natural evolution. When it first appeared, this theory sparked a fierce debate on its scientific plausibility. The physicist Kelvin showed that estimates of the age of the earth, based on the physics of the day, suggested that sufficient time had not yet elapsed to account for the higher life forms, assuming they had evolved as Darwin claimed. Fleeming Jenkins, the first professor of Engineering at Edinburgh University, argued that none of the theories put forward for the transmission of characteristics from parents to offspring explained how useful mutations could be maintained (pre-Mendelian theories assumed a kind of 'blending of characteristics' mechanism, which implies a reduction in variance by a factor of 1/2 each generation). These matters were not resolved until the advent of nuclear theory allowed far more accurate estimates of the earth's age and the development of modern genetic theory provided an understanding of the mechanism of inheritance and variation within a species.

Natural evolution is now understood in terms of those genes which promote greater fitness being maintained within a species. The genetic code (genotype) of an organism is expressed as the phenotype (developed or developing organism) which must be well adapted to its environment in order to survive. The fit, or well adapted, organism has a good chance of producing offspring and passing on its genes. Conversely, the poorly adapted organism is less likely to produce offspring and its genes disappear from the population. Mutations providing greater fitness will propagate relatively quickly throughout the population. Darwin's theory of natural selection plus the genetic theory of heredity is generally known as Neo-Darwinism (Maynard-Smith 1972).

It should be noted that Neo-Darwinism is not accepted as gospel in biological circles. The main debate is between the gradualists (Maynard-Smith 1987) and the punctuationists (Eldredge and Gould 1972; Gould 1989). Gradualism tends to be taken as a basic tenet of Neo-Darwinism. It states that adaptation is achieved, little by little, by selection fine tuning from a series of small variations. Large changes are unlikely to improve fitness as they will not provide a fine enough grading to closely fit the current environment; they are like random leaps into the void, being much less likely than a small change to result in a move up the evolutionary slope. On the other hand, the punctuationists believe macroevolution must be decoupled from microevolution within an alternative theoretical framework. The source of their belief: the fossil record. They claim that the fossil record reveals long periods of little or no change (i.e. stasis) punctuated by periods of rapid change. A number of authors have speculated that a better understanding of the issues (and, indeed, whether there is any contradiction) will come out of a complete theory of ecosystem evolution (Stenseth 1985). Such a theory would take into account inter-species dynamic: after all, the majority of most species' environments is made up of the effects of other species. These issues will be returned to later where it will be seen that they have a bearing on genetic algorithm problem solving models.

Real environments tend to be complex, dynamic and varied which helps to account for the diversity of life forms found across time and space. In particular, as the environment changes, the set of characteristics required for fitness will also vary. In the process of adaptation the goal posts are forever moving. However, if an organism's environment were to remain stable for long enough, the species should evolve to a highly adapted state. This may be to a state of equilibrium with other coevolved species in an ecosystem or, in the case of a relatively simple environment of which it is the only inhabitant, to a highly efficient way of life. In the latter unusual case we can make a strong connection between evolution and optimisation. The space of possible genotypes is being searched for the best solution. A solution is rated in terms of the performance of the phenotype. In the light of all this, it is tempting to imagine that extremely powerful techniques for optimisation, and for the design of adaptive systems, can be abstracted from the logic of natural evolution. Over the past forty or so years a number of researchers have tried to do just that. The most powerful and successful methods emerged in the late 1960s and early 1970s and are based on Holland's genetic algorithm (Holland 1975). Rechenberg and Schwefel's independently developed evolution strategies (Rechenberg 1965 and Hoffmeister and Back 1991) are useful methods for a more restricted class of problems; they will not be discussed further here.

The genetic algorithm will now be described in some detail.

8.3   Genetic Algorithms

8.3.1   Introduction

As already indicated, genetic algorithms (GAs) are adaptive search strategies based on a highly abstract model of biological evolution. They can be used as an optimisation tool or as the basis of more general adaptive systems. The fundamental idea is as follows. A population of structures, representing candidate solutions to the problem at hand, is produced. Each member of the population is evaluated according to some fitness function. Fitness is equated with goodness of solution. Members of the population are selectively interbred in pairs to produce new candidate solutions. The fitter a member of the population the more likely it is to produce offspring. Genetic operators are used to facilitate the breeding; that is, operators which result in offspring inheriting properties from both parents (sexual reproduction). The offspring are evaluated and placed in the population, quite possibly replacing weaker members of the last generation. The process repeats to form the next generation. This form of selective breeding quickly results in those properties which promote greater fitness being transmitted throughout the population: better and better solutions appear. Normally some form of random mutation is also used to allow further variation. A simple form of this algorithm is illustrated in Figure 8.1. This population based survival of the fittest scheme has been shown to act as a powerful problem solving method over a wide range of complex domains (Grefenstette 1985, 1987; Schaffer 1989; Belew and Booker 1991; Schwefel and Manner 1991; Davis 1990).

The analogies between GAs and natural evolution should be clear. The structures encode a solution to the problem; these are the genotypes. There will be some process for interpreting the structure as a solution: the phenotype. The interpretation is often implicitly embedded in the evaluation function and can be complex. When the encoding and the evaluation function are static, we are in the realms of optimisation. When they are not, the GA can be used to build adaptive systems; systems that are able to cope with a changing environment.

There are many different implementations of this idea, varying markedly in their specifics. Before giving the details of a commonly used model and then going on to explain why it works, it is worth stressing that the two most important elements to carry forward from this somewhat abstract description are selection and sexual reproduction.

8.3.2   Early Sequential Algorithms

The basic technique will now be described in some detail by referring to the early simple sequential algorithms. This is done as a matter of convenience and to give some historical perspective. Most of the underlying mechanisms are the same whether the implementation is sequential or parallel. It will become clear that the method is highly suitable for parallelisation and this topic is discussed in detail in Sections 8.5 and 8.6.

Holland's early work was concerned with developing a powerful abstract general formalism for adaptive systems (Holland 1962). This lead to his notion of a 'general reproductive plan' (Holland 1966) which, slightly modified, was christened a genetic algorithm by Bagley (1967). All subsequent GA research has taken as its starting point the deceptively simple algorithm developed by Holland and his group at the University of Michigan during this period. The following subsections describe the technique's essential elements.

The Population of Structures

The population of structures to undergo adaptation generally consists of strings (chromosomes) of a fixed length. Each element (gene) of the string represents some aspect of the solution and will have a set of possible values (alleles)

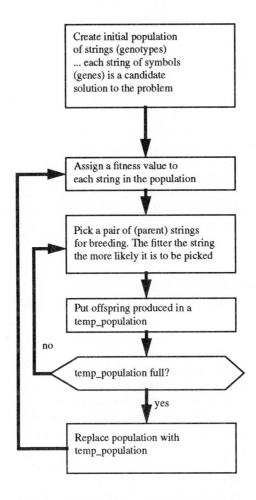

Figure 8.1: A simple genetic algorithm

mapped to various attributes. The fitness of such a string is measured by some objective function which costs the particular combination of attributes present. Hence the chromosomes may be, for instance, strings of real numbers, strings of integers, bit strings (string of 1s and 0s to be decoded into a set of parameter values), a permutation of some set of elements, a list of rules or some combination of these representations. Usually each gene has a fairly small set of alleles. In fact there are theoretical arguments (Holland 1975; Goldberg 1989a, 1989b) which suggest that very low cardinality alphabets should be used. This goes a long way to explain why bit strings, with the lowest possible cardinality of two, are so often employed. Indeed, a large part of the GA literature precludes discussion of any other representation scheme. However, the analyses on which the result stands are by no means complete. If a low cardinality alphabet string representation is natural to the problem in hand, fine. If not, there is much empirical evidence that high cardinality alphabet strings, especially when used with non-standard genetic operators, can be effective (Davis 1985, 1987; Davis and Coombs 1987 and Husbands and Mill 1991). A number of specific example applications are described later in this chapter. The encodings used are fully elucidated.

Population size is an important parameter to consider when applying a GA to a particular problem. If the population is too small premature convergence of the whole population onto a local optima is very likely to occur. Population sizes less than fifty tend to cause difficulties. In practice, more complex problems require larger populations. Figures of hundreds or thousands are not uncommon. Certain parallel implementations require very large populations, as will be seen later. For more detailed discussions of this topic see Goldberg (1989b).

The Genetic Operators

The set of genetic operators developed by Holland, and the one generally used, consists of three operators: crossover, inversion and mutation. Simple crossover involves choosing at random a crossover point (some position along the string) for two mating chromosomes, then two new strings are created by swapping over the sections lying after the crossover point. Inversion is simply a matter of reversing a randomly chosen section of a single string. Mutation changes the value of a gene to some other possible value. The genetic operators are applied at the breeding stage according to a routine like the following. When two strings are selected for breeding, first apply crossover (with some high probability) and randomly choose one of the two new strings thus formed. Next apply inversion (with a medium probability) to this string. Each gene on the resulting string undergoes mutation (with a very low probability) and the outcome is taken as the offspring. The basic operators and the breeding process are illustrated in Figure 8.2. Note the stochastic nature of this process. All operators are applied probabilistically and crossover and inversion points are chosen randomly.

The overall effect is to emphasise combinations of basic building blocks (groups of genes) which produce maximum fitness.

In some problem domains it may be beneficial to allow dynamic length strings. This can be achieved by randomly selecting different crossover points on each parent rather than forcing them to be the same. Other operators, such as translocation (moving a section of the string to a new location), may also be useful.

Action of the Operators

Crossover is the most important of the operators and serves two complimen-

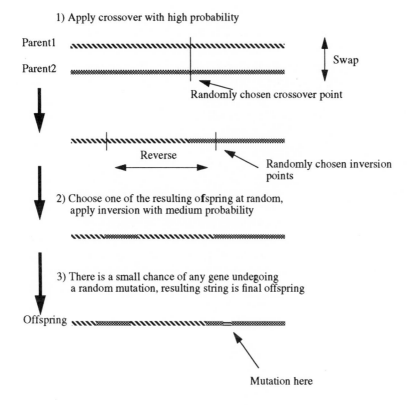

1) Apply crossover with high probability

Parent1

Parent2

Swap

Randomly chosen crossover point

Reverse

Randomly chosen inversion
points

2) Choose one of the resulting ofspring at random,
apply inversion with medium probability

3) There is a small chance of any gene undegoing
a random mutation, resulting string is final offspring

Offspring

Mutation here

Figure 8.2: Application of the genetic operators

tary search functions. It provides new points for further testing within the
hyperplanes, defined by gene combinations, already present in the population.
It also introduces instances of new gene combinations by exchanging material between strings. The power of this operator will be analysed in the next
subsection.

Alternative crossover operators such as two point crossover and uniform
crossover (each gene is taken from the two parents with respective probabilities
p and $1 - p$ (Ackley 1987b; Syswerda 1991)) are often used.

Inversion cannot always be easily used as it may produce illegal structures
when applied in combination with simple crossover. This is particularly likely if
the genetic coding is order sensitive. The purpose of this operator is to increase
the linkage of gene combinations exhibiting good performance. It does this
by reducing their length and hence making their disruption by crossover less
likely. It can also bring previously widely separated alleles into close proximity.
Holland has shown that its action is, like that of crossover, intrinsically parallel

(see Section 8.3.2).

Mutation is very much a background operator, used to guarantee that the probability of searching a particular subspace of the problem space is never zero. It provides the crossover operator with the full range of alleles which should prevent the algorithm from becoming trapped on local optima. The probability of any gene undergoing mutation must be low enough to ensure that this process does not become overactive and start disrupting the advances made by crossover. Typically all genes on the string have equal but independent probabilities of mutation.

Selection Mechanisms

Selection, whereby more credence is given to fitter population members, provides the dynamo that powers the algorithm. However, this survival of the fittest scheme is more subtle than it at first seems. In this context survival of the fittest does not equate to the fittest will survive, but the fittest are *more likely* to survive. In more detail, the fittest are more likely to pass on some of their genes to later generations. This probabilistic element—which is found in other parts of the method, for instance the genetic operator mechanisms—helps to account for the technique's power and robustness.

Four selection mechanisms are dealt with here:

- Breeding Pool

- Roulette Wheel

- Ranking

- Local rules

Each of these usually requires an explicit measure of fitness. In optimisation applications, the fitness function is directly related to the objective function of the problem. Fitness must be a non-negative merit value. If the problem is to maximise some objective function $g(x)$, fitness may well just be taken as the value of $g(x)$. More commonly optimisation problems involve the minimisation of the objective function. In this case the fitness function, $f(x)$, is very often taken as:

$$f(x) = \begin{cases} Cost_{max} - g(x) & \text{if } g(x) < Cost_{max}, \\ 0 & \text{otherwise.} \end{cases} \tag{8.1}$$

$Cost_{max}$ is usually either the highest value of g observed so far, the highest value of g in the current population, or the highest value observed over the last k generations. Sometimes it relates to more complex population statistics.

Breeding pool selection works as follows. The fitness of each member of the current population is calculated. The relative fitness of each member is then calculated as follows:

$$rel f_i = \frac{f_i}{\sum_{i=0}^{i=N} f_i} \tag{8.2}$$

where f_i is the fitness of the ith member of the population and N is the population size. The expected number of offspring for each individual is then calculated using equation 8.3, $round$ rounds to the nearest integer:

$$num\_offspring_i = round(rel f_i \times N). \tag{8.3}$$

This number typically varies between zero and four or five. Each member of the population then has the appropriate $num\_offspring_i$ copies of itself put into a temporary population, or breeding pool. Pairs for mating are then chosen at random from this pool. The offspring produced are used to replace the current population to form the next generation. Clearly the greater the number of copies of an individual in the pool, the more chance it has of contributing towards the next generation. Note some individuals, those with $num\_offspring_i = 0$, are excluded from breeding.

On each cycle the whole population might be replaced or some predefined proportion of it. In the latter case, it is usual to use a stochastic process which biases replacement towards the weaker members of the current population. An intermediate scheme, the so called *elitist strategy* is often practised. In this scheme the whole population is replaced each generation, except the fittest individual which is carried through unscathed.

Roulette wheel selection is slightly different. Again, each member of the population has their relative fitness, $rel f_i$, calculated. This value can be thought of as assigning the correct sized slice of pie, or an appropriately sized sector of a roulette wheel. Individuals are chosen for mating by generating a random number between 0 and 1 and then moving through the population an individual at a time until the cumulative relative fitness is greater than the random value. This is directly analogous to spinning the roulette wheel; the individual associated with the sector that comes to a halt opposite the pointer is selected. The bigger the relative fitness the more likely the individual is to be selected for breeding. Note that with this scheme no member of the population is excluded from breeding; they all have some chance of contributing to the next generation.

Ranking schemes for selection are particularly straightforward. Using this strategy the population is ranked, or ordered, according to the fitness values of its members. Selection is then performed by following a pre-determined probability distribution function, such as the ones shown in Figure 8.3. This may be a simple linear function that constrains the first ranked (fittest) individual to be twice as likely to be selected as the median ranked individual. This scheme tightly controls the selective pressure and allows strong differentiation of the population, even at later stages when their fitness values are very close. It has been argued that this can help to prevent premature convergence. Indeed, Whitley has developed a sequential variant of the genetic algorithm that

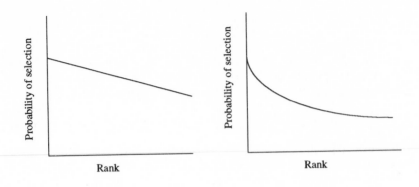

Figure 8.3: Rank-based selection probability distribution functions

incorporates ranked-based selection in a steady state algorithm (Whitley 1989). That is, offspring are introduced into the population one at a time, rather than whole population at a time, which allows a more gradual, and supposedly more robust, search. Whitley's algorithm can be summarised as follows:

1. Randomly generate population

2. Rank the population by fitness

3. Pick pair of parents by using a pre-determined ranked-based distribution

4. Breed to produce offspring

5. Insert in correct position in population (no repeats), push off bottom ranked member of population

6. Unless stopping criteria met, go to 3.

These selection schemes are intended for use with sequential GAs. An alternative form of selection, that only really makes sense in the context of parallel GAs, is governed by local rules of interaction. Briefly, the idea is that a population is somehow split up into many subpopulations, either explicitly or implicitly, and selection occurs *locally*. That is, with reference only to the subpopulation, not to the global population. Local schemes may be based on the methods described earlier or may be simpler. For instance, the population may be effectively spread out over a 2D grid and selection occurs among groups of very close neighbours. In this scheme, any individual has some small number of potential mates. One is chosen according to a probability distribution based on their fitnesses. The advantages of these sorts of schemes, and of parallel GAs in general, are discussed in Section 8.5, which deals with parallel GAs in detail.

Before closing this section there should be some discussion of fitness scaling. Relative fitness based selection, whether used in sequential or parallel GAs, introduces two problems, particularly when small populations are used. First, if a few good individuals appear early on in a run, relative fitness selection rules can easily allow an almost complete take over by these individuals, leading to premature convergence. Second, later in a run, although there may still be significant diversity in the population, the average fitness may be close to the best fitness. In these circumstances, relative fitness selection is not sensitive enough to push the search along. In other words, selective pressure has disappeared. Rank-based selection is one solution to these problem, fitness scaling is another. Linear scaling is often used; the scaled fitness, f', is related to the raw fitness, f, as follows:

$$f' = af + b \qquad (8.4)$$

where the coefficients a and b are chosen such that the average scaled fitness equals the average raw fitness. A more detailed discussion of these matters can be found in Goldberg (1989a).

Schemata Analysis, Implicit Parallelism

So far there have only been intuitive hand waving arguments to explain why GAs work: selection identifies useful building blocks and the genetic operators slot them together in more and more powerful combinations. This section provides a brief overview of Holland's original formal analytic work which goes a long way to explaining the power of GAs and gives further insight into the mechanics of the technique.

If a search strategy is to test for and incorporate structural properties associated with better performance, there must be some method of identifying these useful properties. This is the function of schemata: providing a means of comparing the properties of population strings. A schema describes a subset of strings with similarities at certain positions. Suppose we have a population of strings of length 6 where for each string position, i, there are three possible values $\{a_i, b_i, c_i\}$. If we add the symbol $*$ to mean don't care—i.e. it is irrelevant which value is taken at that point—we can represent schemata. The schema $a_1 a_2 * *b_5 c_6$ matches 9 strings, those with the fixed values shown in the 1st, 2nd, 5th and 6th positions, but any of the three legal values in both the 3rd and 4th positions. In this example, with an alphabet of cardinality 3 for each string position, the number of possible schemata is $4^6 = 4096$. For a population of N strings of length l, there will be an upper bound of $N2^l$ different schemata present. 2^l is the number of schemata contained in any string—each position may take on its actual value or the $*$ symbol. It can be seen that contained within these schemata there is a very large amount of information about the similarities between strings in a given population.

During the reproduction phase the fittest strings provide the most copies for the breeding pool, so it follows that schemata associated with high fitness are sampled more often than those that are not. To understand the effect of crossover we need another simple definition. The distance between the first and last specified (fixed value) string positions in a schema is known as the schema's defining-length. For instance, the schema $b_1 * c_3 a_4 *$ has defining-length $4 - 1 = 3$ and $a_1 a_2 * **$ has defining-length $2 - 1 = 1$. Clearly, schema with short defining-lengths are less likely to be disrupted by crossover than those with long defining-lengths. It can be seen intuitively, then, that the combined effect of selection and crossover is to propagate highly fit short defining-length schemata throughout the population. These schemata are generally known as 'building blocks'. It is these building blocks which combine to produce fitter and fitter strings.

To put things on a firmer basis, Holland (1975) has shown that a GA gives exponentially increasing sampling to the observed best schemata. This is an extremely important result and goes a long way to explaining the power of the method. The result Holland gave, for a GA using crossover and mutation (a commonly used algorithm), was:

$$P(S, t+1) \geq \left[1 - P_c \frac{\delta(S)}{l-1}(1 - P(S,t))\right] (1 - P_m)^{O(S)} \frac{f(S)}{\overline{f}} P(S,T),$$

(8.5)

where, $P(S, t)$ is the proportion of schema S in the population at generation t, P_c and P_m are the probabilities of crossover and mutation, l is the population string length, $\delta(S)$ is the defining length of S, $f(S)$ is the fitness of S, $\overline{f}$ is the average schema fitness, $O(S)$ is the order of S - number of defined (fixed value) positions in S.

Assuming P_m is very small (see previous section), and ignoring negligible cross products, this can be rewritten as:

$$P(S, t+1) \geq P(S,t)\frac{f(S)}{\overline{f}} \left[1 - \frac{P_c \delta(S)}{l-1} - O(S)P_m\right]. \qquad (8.6)$$

Although Holland's derivation of Equation 8.5 is not particularly difficult to follow, it is too long to reproduce here. However, by examining Equation 8.6, we can see how the result implies exponentially increasing sampling for fit schemata. If we assume S is a constant amount fitter than average, i.e. $f(S) = C\overline{f}, C > 1$, then Equation 8.6 can be rewritten as Equation 8.7, where K is a constant greater than 1.

$$P(S, t+1) \geq P(S,t)K \qquad (8.7)$$

Tracing this recursive relation back to generation 0, we readily obtain Equation 8.8:

$$P(S, t) \geq P(S,0)K^t \qquad (8.8)$$

that is, fitter than average schemata receive exponentially increasing sampling as the GA runs from generation to generation. Similarly, schema less fit than average $(f(S)/\overline{f} < 1)$ will receive exponentially decreasing sampling. It can be seen from Equations 8.2 and 8.3 that K will be larger for low defining-length $(\delta(S))$ schemata. This supports the building blocks explanation of the efficacy of GAs.

In the same work Holland also showed that in a population of N strings, although a GA processes N strings per generation, it processes in the order of N^3 schemata. He refers to this equally important result as the *implicit parallelism* of genetic algorithms.

These two fundamental results follow from the very straightforward mechanics of a GA; they do not require any special accounting procedures or complicated transformations, just a population of strings and the application of the simple operators described.

Other Analyses

The formal analysis of genetic algorithms is an active area of research. Most of the research is aimed at understanding under what circumstances a GA will perform well and when it will perform badly. There are a number of interrelated questions on how to encode the problem, how to set the algorithm control parameters, such as population size and genetic operator probabilities, and how to control selection. A discussion of this work is outside the scope of this chapter. However, it is worth noting that most analyses take Holland's schema work as their starting point. The non-linear dynamics of a typical GA system makes analysis very hard. This, coupled with the diversity of application details, means that empirical results tend to hold more weight. Some references on GA analysis worth pursuing are Goldberg (1989c, 1989d), Goldberg and Bridges (1990), Liepins and Vose (1990), Radcliffe (1991) and Rawlins (1990).

Advantages of GAs

Because Holland and his students developed genetic algorithms to serve as adaptive problem solving strategies able to operate over a large range of environments, they have qualities that make them suitable for many large combinatorial problems and string representable search tasks. By a combination of selection and reproduction via genetic operators, they are able to find very fit structures by searching only a tiny proportion of the whole problem space. As long as the string representations and the cost function are accurate, GAs can conduct a successful search without recourse to any special domain specific heuristics. The subtlety of their action prevents them from getting stuck on local optima and ensures that they simultaneously search widely separated parts of the problem space. This is largely due to the random elements in the action of the genetic operators. No assumptions need to be made about the search space, often in contrast to the situation with branch and bound and various heuristic search techniques. These qualities make genetic algorithms an extremely robust problem solving method. It is this robustness that makes them an attractive and useful search method. GAs work by combining candidate solutions together to produce new candidate solutions. They identify building blocks which promote greater fitness and propagate these through the population producing better and better solution strings. Of course this implies that the genotypic encoding describes a search space that is amenable to this process: that contains fairly independent building blocks that can be combined in useful ways. An example of a domain in which this is difficult to achieve is the generation of computer programs. A naive encoding might represent the series of instructions as genes on a simple string chromosome. However, a single random mutation in a working program is likely to render the whole thing useless. This means that the genetic operators would be highly unlikely to be successful. Added to this is the problem of evaluation. How is it decided that a program 'almost' works? For an interesting approach to overcoming some of these problems and showing how to evolve specialist programs see Koza (1990).

In general, GAs do not provide a magic solution. A great deal of ingenuity is often needed to derive a suitable encoding and provide it with an appropriate set of genetic operators. However, when this is achieved, especially in fields where existing techniques are weak, GAs can be a remarkably powerful tool.

8.4   A Selection of Applications

By now there are hundreds of successful applications of GAs. A few will be mentioned here, but consult the following for many more examples: Grefenstette (1985, 1987), Schaffer (1989), Belew and Booker (1991) and Davis (1987, 1990).

GAs have been widely used in the optimisation of multimodal continuous functions (Goldberg 1989a; Richardson *et al.* 1989; Muhlenbein *et al.* 1991). They have advantages over traditional gradient descent techniques in terms of robustness and avoidance of local optima. Typically, a complex function of several variables, $f(x_1, x_2, .., x_n)$, is to be minimised. The genotype used is a simple bit string, 1001001010...001, where each gene can take the value 0 or 1. The string is decoded in chunks, where each chunk is taken as the binary value of one of the parameters $x_1..x_n$. Fitness is measured in terms of the value of f, as described in Section 8.3.2. Much of the early empirical evidence for the usefulness of GAs comes from De Jong's 1975 study of their performance as function optimisers (De Jong 1975). Function optimisation applications include pipeline and structure optimisation (Goldberg 1989a), machine design (Karr 1990), and systems identification (Johnson and Husbands 1991).

GAs have also been applied to many combinatorial optimisation problems, e.g. Husbands and Mill (1991), Whitley (1990) and Jones and Beltramo (1991). Some would say this is the area where they are most appropriate. Genotypes for these sorts of problems tend to be lists of integer codes or permutations of a set of elements.

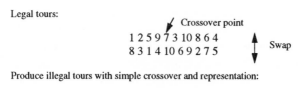

Figure 8.4: Illegal TSP offspring

An interesting combinatorial optimisation exemplar is the travelling sales-man problem (TSP). The task is to find the shortest route through a set of cities, visiting each once only and returning to the starting point. An obvious genotype is a permutation of a list of integers representing the cities. Using this representation, simple crossover would produce illegal tours most of the time, as shown in Figure 8.4. Early efforts (Goldberg and Lingle 1985; Grefen-stette 1985) overcame this problem by correcting the offspring tours so that the duplicate cities were replaced by the omitted cities or otherwise eliminated. Reasonable results were found like this. Improvements were later made by in-corporating heuristics (Liepins *et al.* 1987; Suh and van Gucht 1987). Whitley *et al.* (1991) produced better results by developing a representation and recom-bination operator that manipulated edges (links between cities) rather than the cities themselves (Whitley and Starkweather 1990). Their edge recombination operator uses an 'edge map' to construct an offspring that inherits as much information as possible from the parent structures. This map stores all the connections from the two parents that lead into and out of a city. An offspring is started by choosing at random one of the two initial cities from its parents. A tour is then built up by adding a city at a time while favouring those cities with the fewest unused edges (to avoid a city becoming isolated). Candidate 'next' cities will be taken from those connected to the 'current' city in either of the parents (this is the information the edge map holds). This is a good example of using a problem-appropriate representation and recombination operator within the logical framework of the GA. Gorges-Schleuter has also produced very good results for large TSP problems by incorporating heuristics into a parallel GA (Gorges-Schleuter 1989); this work is discussed later in Section 8.5.

GAs have been applied to such practical combinatorial optimisation prob-lem as production plan optimisation (Husbands *et al.* 1990; Vancza and Markus 1991), scheduling (Davis 1985; Whitley and Starkweather 1990; Clevland and Smith 1989; Husbands and Mill 1991; Nakano and Yamada 1991), bin packing (Smith 1985), compaction of symbolic layout (Fourman 1985), circuit parti-tioning (Hulin 1991), and timetabling (Colorni *et al* 1991; Ling 1991). Two further examples of GA solutions to combinatorial optimisation problems are described in detail in Section 8.6.

An application, falling between function optimisation and combinatorial optimisation, that has stirred up a lot of interest is the optimisation of component designs. Powell *et al.* (1989) have developed a domain independent design optimisation tool that integrates expert systems and genetic algorithms. Bramlett and Cusic (1989) have applied a related technique to the parametric design of aircraft.

A number of researchers have applied GAs to control problems by representing the control strategy as a set of rules. The genotype is then a string representing the whole set of rules. A number of different coding schemes have been used (Nordvik and Renders 1991; Grefenstette 1990).

To close this section it is worth mentioning two application principles that have been found useful:

1. If there are domain search algorithms available consider hybridizing them with a GA.

2. If there are useful domain heuristics incorporate these into a GA by developing appropriate encodings and recombination operators.

8.5   Parallel Genetic Algorithms

It should be clear from the preceding sections that the genetic algorithm is highly parallel in form. From the very early days of its development its potential for parallelisation, with all the attendant benefits of efficiency, have been noted (Holland 1962). Mainly because of the lack of availability of hardware, it is only recently that significant work has been done in this direction. The more successful parallel implementations are more than just simple translations of the sequential algorithms into a multi-processor environment. After all, the early algorithms were sequential because of hardware limitations; they were inevitably impoverished in the way they handled population-distributed processing and information.

In the early 1980s Grefenstette studied four proposed parallel implementations (Grefenstette 1981). These were:

• Synchronous master-slave

• Semi-synchronous master-slave

• Distributed, asynchronous concurrent

• Network

In the first of these implementations, the master processor controls selection and mating while the slave processors do the evaluations of population members. This makes some sense given that evaluation is very often the bottleneck in the sequential algorithm. However, it is rather wasteful if there is much difference in evaluation times and it relies on the health of the master processor. It is suitable for implementation on SIMD machines and could potentially allow the use of very large populations. The second model overcomes the first of the drawbacks encountered with the synchronous model by relaxing the strict synchronous requirement. The third model is fully asynchronous and concurrent; each processor performs selection, mating and evaluations independently. The processors access a common shared memory storing population information. There is a requirement that population members cannot be simultaneously processed by different processors. This implementation is highly robust, if not as subtle as some of the later versions described below. The final, network, model involves a number of independent simple GAs each with its own memory. Every so often the best individuals discovered by each population are broadcast to the other populations. This requires far less communication than the other schemes and is suitable for MIMD or LAN architectures. It was some time before results from real implementations of these models appeared (Pettey *et al.* 1987).

Cohoon *et al.* (1987) presented a model inspired by the theory of punctuated equilibria (Eldredge and Gould 1972). Punctuated equilibria was briefly described earlier in Section 8.2. Cohoon's algorithm is similar to Grefenstette's fourth model but with some refinements. Their model involves N subpopulations of size n, each residing on an individual processor. Each processor executes a sequential genetic algorithm for some fixed number of generations (an epoch). At the end of an epoch the population will be approaching equilibrium, but will still have some diversity. Next, each subpopulation copies a random subset of itself to neighbouring subpopulations. Each processor now has a surplus of solutions and so makes a probabilistic selection of n of them to serve as its initial population for the next epoch. The cycle continues. This process mimics the stasis-catastrophe-rapid evolution-stasis cycle of Eldredge and Gould's theory. Cohoon applied this algorithm to the Optimal Linear Arrangement problem and reported:

> In our experiments, the result was more than just a hardware acceleration, rather better solutions were found with less total work. (Cohoon *et al.* 1987)

Clearly the division of the population into subpopulations which evolve largely independently of each other, is well suited to MIMD machines or networks of individual processors. Robertson (1987) did some early work on large GA models using the fine-grained massively parallel SIMD Connection Machine. In this work, members of a very large homogeneous population were evaluated in parallel. However, mechanisms such as selection were treated in a sequential way on a front-end computer (essentially Grefenstette's first model). Since evaluation is the most time consuming part of a GA, this scheme produced huge speed ups. Connection Machines are now fairly standard platforms for the larger US GA research groups. Many of them now employ more subtle extended varieties of parallelism, some of which are discussed later in this section.

Sannier and Goodman (1987) developed a very interesting model (AS-GARD) which simulated the effects of a changing spatial distribution of the population. Simple simulated organisms roamed a 2D toroidal world. The organisms' genomes coded for their behaviour. Enough food had to be eaten to stay alive and a certain minimum energy level had to be achieved before mating was possible. Mate selection and offspring placement was biased by distance from the organism involved. Conditions in the world varied in time and space; the emergence of different subpopulations, taking advantage of different properties of the world, was observed. In the same paper the authors briefly report work to extend this model to develop co-adapted process schedulers capable of cooperating to improve performance in a LAN. One of the reasons the ASGARD model is important is that it was the first to explore geographically-based selection schemes; many researchers now see these as important in situations where evaluation criteria are implicit or ill defined.

Jog and Gucht (1987) and Tanese (1987) give further empirical evidence for the thesis that a well designed parallel GA will give more than just speed up.

By 1989 more sophisticated parallel GAs had started to appear. Muhlenbein and Gorges-Schleuter developed their parallel asynchronous ASPARAGOS algorithm (Muhlenbein *et al.* 1988; Muhlenbein 1989; Gorges-Schleuter 1989) which employs a number of interesting extensions derived from population genetics theory. Individuals live on a 2D grid with local selection operating. An important addition is that when each individual comes into existence it does local hill climbing to improve its fitness. Experiments were performed on producing offspring using polysexual voting recombination. The basic algorithm is represented below; it is applied asynchronously and in parallel to all individuals:

1. Randomly generate N individuals

2. Each individual does local hill climbing to increase fitness

3. Each individual chooses partner(s) for mating in its neighbourhood, according to their rank in the neighbourhood.

4. Create new offspring with genetic operators

5. Replace individual

6. If not finished, go to 2.

In their neighbourhood model there is a notion of subpopulations 'isolated by distance', but neighbourhoods do overlap. The population is viewed as a continuous structure, with local interactions only, making it quite different from the isolated subpopulation models. The justification for this model comes from the work of the population geneticist Wright. Specifically it is based on his claim (Wright 1982) that the most favourable population structure is one

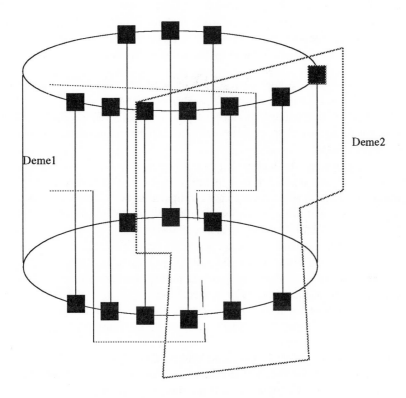

Figure 8.5: ASPARAGOS population structure

subdividing the species into local populations (demes) sufficiently
small and isolated that accidents of sampling may overwhelm many
weak selection pressures. There must, however, be enough diffusion that a deme that happens to acquire a favourable interaction
system may transform its neighbours to the point of autonomous
establishment of the same peak, and thus ultimately transform the
whole species or at least that portion of it in which the new system
actually is favourable.

Implementations were on a MIMD machine with configurable inter-
processor communication. The population structure is represented in Figure
8.5. A sparse graph (ring, ladder and torus) representation is used in which
overlapping neighbourhoods (demes) can be seen. In this example individuals
are given a 'mobility radius' of 2, yielding a connectivity of 7, and hence a
deme of size 8. The nodes represent processors, one for each individual.

The polysexual voting recombination operators, found to work very well
with certain problems, perform as follows. In p-sexual voting recombination p

parents, from the same neighbourhood, are used to produce a single offspring. For each gene on the genotype, if a particular allele occurs more times than some threshold in the group of parents, then it is placed on the offspring genotype. The remaining genes are filled in by mutations. This idea was first proposed by Ackley (1987a).

ASPARAGOS was applied to the quadratic assignment problem and found a new optimum for the largest published problem (Muhlenbein 1989). The algorithm was also found to perform very well on large TSP problems and to be highly robust with respect to settings of the GA control parameters (Gorges-Schleuter 1989).

More recently, Muhlenbein and his group have concentrated on testing the limits of their algorithm by applying it to a range of complex problems (Muhlenbein et al 1991). In each case the algorithm was applied to the largest published problem instance. Reported solutions are comparable with or better than those found by other methods. Experiments also revealed that the algorithm consistently outperformed other less parallel GAs.

Manderick and Spiessens (1989, 1991) have reported a number of experiments with a fine grained parallel genetic algorithm implemented on massively parallel machines such as the Connection Machine and the DAP. Like Muhlenbein and Gorges-Schleuter, they use a population distributed over a planar grid with selection restricted to small neighbourhoods on the grid. Their neighbourhoods are larger than Muhlenbein's and their selection mechanisms are different (simple relative fitness, rather than ranking). Their algorithm is quite different to Robertson's early attempt at a fine grained model (Robertson 1987) mentioned near the beginning of this section. Robertson's algorithm used a front-end computer to do selection and mating, thereby introducing a communications bottleneck. Local selection is a far more efficient solution and, according to empirical evidence, leads to more powerful robust algorithms. It also brings the GA closer to the natural model which inspired the early sequential algorithms. It should be remembered that the early models were sequential by necessity, and employed various tricks to model an essentially distributed parallel process as a sequential one.

More recent work has refined some of these models and explored the application of parallel GAs to highly complex problems requiring very large populations. Some of this work will be discussed in detail in the next section. Here we will mention two of the more interesting general studies. Collins and Jefferson (1991) have explored a number of selection mechanisms in massively parallel GAs. Their motivation is neatly summarised in the following quote:

> Genetic algorithms that use panmictic selection and mating (where any individual can potentially mate with any other) typically converge on a single peak of multimodal functions, even when several solutions of equal quality exist (Deb and Goldberg 1989). Genetic convergence is a serious problem when the adaptive landscape is constantly changing as it does in both natural and artificial ecosys-

tems. Crowding, sharing, and restrictive mating are modifications to panmictic selection schemes that have been proposed to deal with the problem of convergence, and thus allow the population to simultaneously contain individuals on more than one peak on the adaptive landscape (De Jong 1975; Goldberg and Richardson 1987; Deb and Goldberg 1989). These modifications are motivated by the natural phenomena of niches, species, and assortative mating, but they make use of global knowledge of the population, phenotypic distance measures, and global selection and mating, and thus are not well suited for parallel implementation. Rather than attempting to *directly implement these natural phenomena [as the aforementioned works did], we exploit the fact that they are emergent properties of local mating* (my italics).

The results of their empirical study were quite definite and in keeping with suspicions welling up in the GA community. Although they were careful to point out that further studies should be carried out, they concluded that local selection and mating was far more appropriate for Artificial Life studies (see Section 8.8) than panmictic schemes, and that local schemes also appear to be superior in function optimisation applications. More specifically, local selection and mating

1. finds optimal solutions faster (in terms of solutions evaluated);

2. typically finds multiple optimal solutions in the same run;

3. is much more robust than global schemes.

On top of these, local selection and mating schemes produce more efficient parallel implementations by cutting down on interprocessor communications and by largely eliminating sequential bottle necks, such as relative fitness calculations. They are most naturally implemented on fine grained SIMD machines.

Similar conclusions have been drawn by Davidor (1991).

8.6 Extending the Parallel Paradigm

8.6.1 Introduction

The previous section described recent advances in the parallel implementation of simple genetic algorithms. It has now been established that parallelisation has more to offer than just the speed up provided by simultaneous evaluation of population members. Local selection and mating within a geographically distributed population has been shown to give a more powerful robust search. Such algorithms are not as prone to premature convergence as sequential GAs or the early naive parallel GAs. They can also handle very large populations much more efficiently. Another advantage is the ease with which these models can be extended in a number of interesting, natural and potentially useful directions. For instance, evaluation criteria could be varied over the 2D grid, and individuals could have varying degrees of mobility. All of these advances seem to push GAs closer to the biological mechanisms from which they were derived. However, all the work described so far has been concerned with a single species finding solutions to a single well defined problem. The remainder of this section deals with two very recent models which have shown how to extend the problem solving capabilities of GAs by introducing coevolution. In so doing they have shown how to further exploit potential for parallelisation and have highlighted the fact that natural evolution still has plenty to offer as a source of inspiration.

8.6.2   Parasites and Sorting Networks

Danny Hillis, who lead the team that developed the Connection Machine (Hillis 1985), was the first to significantly extend the parallel GA paradigm by showing how to develop a more powerful optimisation system by making use of coevolution (Hillis 1990). In his experiments Hillis takes full advantage of massive parallelism, routinely using populations of a million individuals evolving over tens of thousands of generations.

Hillis uses a diploid genetic algorithm. Individuals are represented as pairs of number strings, analogous to the chromosome pairs of diploid organisms. The solutions, or phenotypes, coded for by the string pairs, are constructed in the usual way by interpreting fixed regions of the strings as coding for particular parameter values, or phenotypic traits. Discrepancies between the two strings of a pair are resolved according to some rule of dominance. He has also found that locally controlled selection is more robust than the simple global variety. Specifically, in his experiments individuals evolve on a 2D toroidal grid. The x and y displacements of an individual from potential mates are taken as a binomial approximation of a Gaussian distribution. After a pair mate, the two offspring they produce replace them, in the same spatial locations, so the genetic material remains spatially local.

An interesting complex optimisation problem that he has tackled using GAs, is the problem of finding minimal sorting networks for a given number of elements. A sorting network (Knuth 1973) represents a sorting algorithm in which comparisons and exchanges take place in some predetermined order. Finding good networks is of significant practical interest, bearing on the development of optimal sorting algorithms, switching circuits and network routing algorithms. A sorting network is represented in Figure 8.6. The horizontal lines correspond to the elements to be sorted. The unsorted input is on the left and the sorted output is on the right. In between, comparison-exchanges of elements are indicated by arrows pointing from one element to another. A comparison-exchange of the ith and jth elements is indicated by an arrow from the ith to the jth line. Elements are exchanged if the element at the head of the arrow is strictly less than the element at the tail. The network shown in the

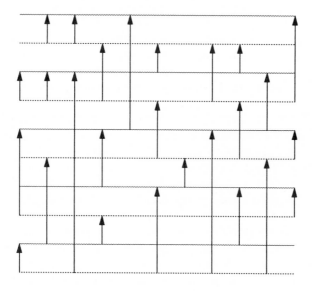

Figure 8.6: A sorting network

figure is random.

The genotype of each individual consists of a pair of bit string chromosomes. Each chromosome can be thought of as sixty eight-bit genes. Each gene consists of two four-bit numbers representing elements to be compared and possibly exchanged. The phenotype (sorting network) is generated by traversing the chromosomes in a fixed order. If a pair of chromosomes have the same gene (comparison-exchange elements) at a particular site, then only that comparison-exchange pair is generated in the sorting network. If the genes are different, then both pairs are generated. In this way the network can contain between sixty and one hundred and twenty comparison-exchanges.

The phenotype is scored according to how well it sorts. The measure used is the percentage of correct sorts performed on a sample of test cases. In the first set of experiments this sample was randomly generated. The best sorting network found contained sixty five exchanges. This compares favourably with solutions found by other methods. In fact, for a number of years the best know solution used sixty five exchanges. After seven years hard work this was reduced to sixty.

While this first set of experiments produced good results, closer analysis revealed two sources of inefficiency. The first was a tendency to get stuck in (rather good) local optima. The second was an inefficiency in the evaluation process. After the first few generations most of the test cases were sorted successfully by most networks, so they provided little information about differential fitness. Several methods for overcoming these problems were investigated. The most interesting and the most successful was the coevolution of test cases. This is analogous to the biological evolution of a host parasite. In this extended model there are two independent gene pools, each evolving according to local selection and mating. One population, the hosts, represents sorting networks, the other, the parasites, represents test cases. Interaction of the populations is via their fitness functions. The sorting networks are scored according to the test cases provided by the parasites in their immediate vicinity. The parasites are scored according to the number of tests the network fails on.

Coevolution provided two benefits. First, it helped to avoid the curse of local optima. If a large but imperfect subpopulation evolves, it becomes a target towards which the parasitic test cases will evolve, hence reducing the fitness of the networks and moving the subpopulation off the local optima. Second, the testing process became much more efficient. After a short while, only those test cases that show up weaknesses are widely represented in the population. Hence it is only necessary to apply a few test cases per individual per generation.

The runs with coevolving parasites were found to produce consistently better results faster than those without. The best result found was a network requiring sixty one exchanges, only one more than the best known. To quote Hillis (1990):

> It is ironic, but perhaps not surprising, that our attempts to improve simulated evolution as an optimisation procedure continue to take us closer to real biological systems.

8.6.3   Symbiosis and Emergent Scheduling

A New Approach to Combinatorial Optimisation

The underlying structure of many combinatorial optimisation problems of practical interest is highly parallel. However, traditional approaches to these problems tend to use mathematical characterisations that obscure this. By contrast, the use of biologically inspired models casts fresh light on a problem and may lead to a more general characterisation which clearly indicates how to exploit parallelism and gain better solutions. This section describes a model based on simulated coevolution that has been applied to a highly generalised version of the manufacturing scheduling problem, a problem previously regarded as too complex to tackle. Whereas Hillis's model is analogous to a host-parasite ecology, this model is closer to a symbiotic ecology, that is, a number of separate species interacting in ways that are to their mutual advantage. Further details can be found in Husbands and Mill (1991), Husbands *et al.* (1991) and Husbands (1992).

Although this section is largely focused on one particular optimisation problem, it should be noted that the model presented can be generalised. This work is concerned with using parallel GA search in a form of distributed problem solving. As we have already seen, most previous parallel GA work has been concerned with speed up and devising parallel implementations which provide a more robust search. In contrast, the work reported here is concerned with using parallel GAs to simultaneously solve interacting subproblems. From this emerges the solution to some wider more complex problem. The idea is to re-cast a highly complex application in terms of the cooperative and simultaneous solution of a number of simpler interacting subproblems. Using this variation of divide and conquer, the inherent parallelism is brought out and thoroughly exploited. The model involves a number of separate populations each evolving under the influence of a GA. The genotype for each population is different and represents a solution to one of the subproblems. Because the fitness of any individual in any population takes into account the interactions with members of other populations, the separate species coevolve in a shared world. In this model, possible conflicts between species (for instance, disputes over shared resources) are decided by a further coevolving species, the Arbitrators. The Arbitrators evolve under a pressure to make decisions that benefit the whole ecosystem. Without explicitly encoding the overall problem, the Arbitrators are used to try and adhere to its global constraints.

Referring to the application of the model, there is already a very large body of work on solving planning and scheduling problems, mainly emanating from the fields of Artificial Intelligence and Operations Research. However, traditional AI approaches have had limited success in real-world applications; indeed their shortcomings have been thoroughly explored and documented (Chapman 1985). The general resource planning, or scheduling, problem is well known to be NP-Complete (Garey and Johnson 1979). Consequently OR techniques have been developed to give exact solutions to restricted versions of the problem, but in general there is a reliance on heuristic-based methods. Because of the complexity and size of the search spaces involved, a number of simplifying assumptions have always been used in practical applications. These assumptions are now implicit in what have become the standard problem formulations. The author holds the view that in many instances this has led to the most general underlying optimisation problem being ignored or, more often, not even acknowledged as existing at all. It is claimed that the parallel GA model presented here, derived from a completely fresh way of looking at the application, runs counter to this tendency and opens up new areas of investigation.

Domain of Application: Manufacturing Planning and Scheduling

Consider a manufacturing environment in which n jobs or items are to be processed by m machines. Each job will have a set of constraints on the order in which machines can be used and a given processing time on each machine. The jobs may well be of different lengths and involve different subsets of the m machines. The job-shop scheduling problem is to find the sequence of jobs on each machine in order to minimise a given objective function. The latter will be a function of such things as total elapsed time, weighted mean completion time and weighted mean lateness under the given due dates for each job (Christophedes 1979). In the standard model process planning directly precedes the scheduling. A process plan is a detailed set of instructions on how to manufacture each part (process each job). This is when decisions are made about the appropriate machines for each operation and any constraints on the order in which operations can be performed (Chang and Wysk 1985). Very often completed process plans are presented as the raw data for the scheduler. Scheduling is essentially seen as the task of finding an optimal way of interleaving a number of fixed, or maybe slightly flexible, plans which are to be executed concurrently and which must share resources. However, in many manufacturing environments there are a vast number of legal plans for each component. These vary in the orderings between operations, the machines used, the tools used on any given machine and the orientation of the work-piece on any given machine. They will also vary enormously in their costs. Instead of just generating a reasonable plan to send off to the scheduler, it is desirable to generate a near optimal one. Clearly this cannot be done in isolation from the scheduling: a number of separately optimal plans for different components might well interact to cause serious bottle-necks. Because of the complexity of the overall optimisation problem, that is simultaneously optimising the individual plans and the schedule, and for the reasons outlined in the introduction, up until now very little work has been done on it. However, recasting the problem to fit an 'ecosystem' model of coevolving organisms has provided a solution. Success is partly due to the power of the central optimisation technique (genetic algorithms) and partly because the recasting has allowed many of the complex interactions inherent in the problem to be represented in a simple and natural way.

The Model

The idea behind the coevolving species model is shown in Figure 8.7. The genotype of each specie represents a feasible process plan for a particular component to be manufactured in the machine shop. Separate populations evolve under the pressure of selection to find near-optimal process plans for each of the components. However, their fitness functions take into account the use of shared resources in their common world (a model of the machine shop). This means that without the need for an explicit scheduling stage, a low cost schedule will emerge at the same time as the plans are being optimised.

The data provided by a *plan space generator*, whose operation is outlined later, is used to randomly generate populations of structures representing possible plans, one population for each component to be manufactured. An important part of this model is the population of Arbitrators, again initially randomly generated. The Arbitrators' job is to resolve conflicts between members of the other populations; their fitness depends on how well they achieve this. It is important to note that the environment of each population includes the influence of all the other populations.

In order to understand the interpretation of the genomes described later, a

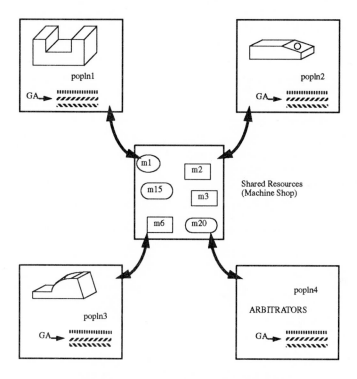

Figure 8.7: The coevolving species model

few words need to be said about the plan space generation. This is done by a knowledge-based system, which breaks down the manufacture of a component into a number of nearly independent operations. The entire space of possible plans can then be generated by finding all the possible ways to carry out each operation, along with ordering constraints. The execution of an operation is defined in terms of a (machine/process/ tool/setup) tuple. The first three fields indicate how to use the machine and the fourth refers to the orientation of the work-piece (partially completed component). The output from this process is a large number of interconnected networks like the one in Figure 8.8. A manufacturing process for the sub-goal described by the fragment of network shown is a route from the starting conditions node to the goal conditions node. Implicit in the representation are functional dependencies and ordering constraints between sub-operations. Further details can be found in Husbands *et al.* (1990).

The genotype of a process plan organism can be represented as follows:

$$op_1 m_1 s_1 op_2 m_2 s_2 G op_3 m_3 s_3 op_4 m_4 s_4 op_5 m_5 s_5 G \ldots \ldots$$

where op_i refers to the ith operation in a plan, m_i to the machine to use

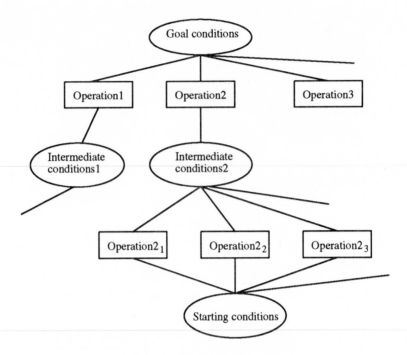

Figure 8.8: Part of a planning network

for that operation and s_i to the setup. Operations with interdependencies are grouped together, each group being terminated by a special symbol (G in above example). As long as the group terminators are the only legal crossover points, the crossover operation will always produce legal plans. If crossover were to occur within a group, data for dependent operations would be split up and illegal plans would probably occur on recombination. The mutation operator is also fairly involved because the gene values are context sensitive due to the dependencies. This encoding encapsulates the network structures of the data produced by the plan space generator. Each op_i, m_i and s_i have associated with them finite sets of possible integer coded values. Because these sets are all quite different, bit string representations would be awkward and unnatural, hence so called real valued codes are used. The genotype is transformed into another form for interpretation by the fitness function, to be described later. This is to take into account the ordering aspect of the problem. There is a network of partial ordering constraints associated with each genotype (specie); the operations must be ordered in accordance with these.

The Arbitrators' genotype is a bit string which encodes a table indicating which population should have precedence at any particular stage (defined later) of the execution of a plan, should a conflict over a shared resource occur. There is one bit for each possible population pairing at each possible stage. Hence the Arbitrator genome is a bit string of length $SN(N-1)/2$, where S = maximum number of stages in a possible plan and N = number of process plan organism populations. Each bit is uniquely identified with a particular population pairing and is interpreted according to the following function:

$$f(n_1, n_2, k) = g\left[\frac{kN(N-1)}{2} + n_1(N-1) - \frac{n_1(n_1+1)}{2} + n_2 - 1\right]$$

$$(8.9)$$

where n_1 and n_2 are unique labels for particular populations, $n_1 < n_2$, k refers to the stage of the plan and $g[i]$ refers to the value of the ith gene on the Arbitrator genome. If $f(n_1, n_2, k) = 1$ then n_1 dominates, else n_2 dominates. By using pair wise filtering the Arbitrator can be used to resolve conflicts between any number of different species.

The cost functions for plan organisms involve two stages, for arbitrators just one. The first stage involves population specific criteria and the second stage takes into account interactions between populations. The first stage cost function for the process plan organisms, COST1 shown below, is applied to the genotype shown above after it has first been translated into a linearised format that can be interpreted sequentially.

$$COST1(plan) = \sum_{i=1}^{i=N} (M(m_i, i) + S(s_i, i, m_i)) \qquad (8.10)$$

where s_i = setup used while processing ith operation, m_i = machine used for processing ith operation, $S(s_i, i, m_i)$ = setup cost for ith operation, $M(m_i, i)$ = machining cost for ith operation, N = number of operations to be processed and $M(m_i, i)$ has been previously calculated and is looked up in a table. Note that a setup cost is incurred every time a component is moved to a new machine or its orientation on the same machine changes. This function performs a basic simulation of the execution of the plan. Its input data is an ordered set of (machine, setup) pairs, one for each operation. The operations must be ordered in such a way that none of the constraints laid down by the planner are violated. Ordered sets of operations to be processed using a particular machine/setup combination are built up on a 2D grid. $S(s_i, i, m_i)$ governs the way in which the sets are built up on the grid. The operations in any set can be performed in isolation from those in any other set. Such a set will be referred to as a *stage* of a job in the rest of this chapter. These sets themselves are ordered and the outcome is a process plan like the one shown below, where the integers in the sets refer to particular operations.

```
1) machine:  6  setup:  5   [0,3,5,7]
2) machine:  2  setup: 21   [1,8,12,19]
3) machine: 11  setup:  4   [2,4,6,9,13,15]
...etc
```

In fact COST1 provides a mapping from the process plan genotype to its phenotype: one of the plans illustrated above. Note that the setup cost is often considerably more (orders of magnitude) than the basic machining costs. The essential working of COST1 is to sequentially process the transformed genome in order to group operations together in clusters which can then be scheduled as single units (stages). At the same time the final executable ordering of the operations is found, as well as the basic machining costs. The mechanics of the function are fairly complex and will not be dealt with here, but see Husbands and Mill (1991) and Husbands (1992). The function involves a complicated interpretation of the genotype, more complex than at first sight seems necessary. However, it should be noted that a carefully chosen genetic encoding plus a complex, but *computationally cheap*, interpretation and costing function, allow very simple and cheap genetic operators to be employed which are capable of searching the combined ordering and machine selection space. The alternative would have been to use a simple costing function in conjunction with *computationally expensive* reordering genetic operators (Goldberg and Lingle 1985).

The second phase of the cost function involves simulating the simultaneous execution of plans derived from stage one. Additional costs are incurred for waiting and going over due dates. What happens when two plans want the same resource at the same time? Fixed precedences would be far too inflexible and random choices would be of no help. As already indicated, the most general and powerful solution developed was to introduce a new species, the Arbitrators, whose genetic code holds a table indicating which population had precedence at any stage. The Arbitrators are costed according to the amount of waiting and the total elapsed time for a given simulation. The smaller these two values, the fitter the Arbitrator. Hence the Arbitrators, initially randomly generated, are allowed to coevolve with the plan organisms. Each individual's fitness is calculated according to its total cost. This means that selection pressure takes account of both optimisation problems: interactions during phase two that increase an individual's cost will reduce its chances of reproduction, just as will a poor result from phase one of the costing. In general, a population of co-evolving Arbitrators could be used to resolve conflicts due to a number of different types of operational constraint.

The first implementation of this model, on a MIMD machine, had the various populations on separate processors and involved a complicated ranking mechanism to allow coevolution to produce useful results (Husbands and Mill 1991); global selection was employed. The second, more satisfactory, implementation spreads each population over a 2D toroidal grid. Selection is local, very similar to the schemes used by Hillis; interactions are also local. The second phase of the costing involves individuals from each population at the same location on the grid. This provides a highly parallel model which consistently provided better results faster than the first, less parallel, implementation.

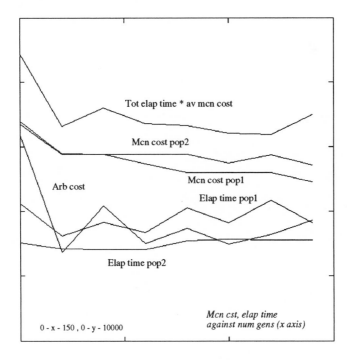

Figure 8.9: Results of coevolution model

Results

Results of typical runs are shown in Figures 8.9 to 8.12. The graphs show
how the machining costs (COST1) of the best individual in each population
reduce with time, and also how the Arbitrator costs reduce. It also shows
how the total elapsed time reduces. The gantt charts show how the emergent
schedule evolves. The vastly reduced number of stages after a few tens of
generations reflects the fact that machining costs can be decreased by putting
more operations into a single stage. Clearly both optimisation problems have
been tackled simultaneously. Note that there is some tension between the
various objectives; one cost may momentarily rise while others drop, but the
overall trend is down. A model of a real job-shop is used and the components
planned for are of medium to high complexity needing 25–60 operations to
manufacture. Each job has a number of internal partial ordering constraints but
is by no means strongly constrained. Typically each operation has 8 candidate
machines and each of these machines has 6 possible setups. To simplify
matters, tool changes and machine transfer costs have not been modelled in
great detail. However, it is a simple matter to include them and future versions
of the model will be complete in that respect. Experiments with up to 10

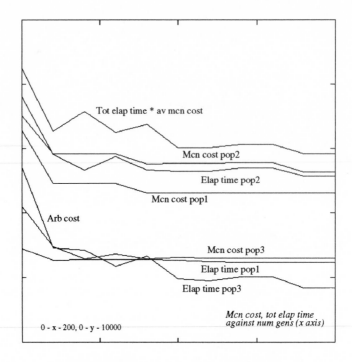

Figure 8.10: Results of coevolution model

jobs have been conducted. Very promising results have been obtained for this extremely complex optimisation problem, never before attempted. The search spaces involved are unimaginably huge, greater than 10^{60}, but this model has exploited parallelism sufficiently to produce good results.

Davis has done some work on using GAs to solve job-shop scheduling problems (Davis 1985), but his solution was for the simplified problem that does not take into account the proper relationship between planning and scheduling. Each genome represented an entire schedule. That approach cannot exploit the inherent parallelism of the problem in the same way that the work described here has. Hilliard *et al.* (1987) have used a classifier system to discover scheduling heuristics. That work may possibly tie in with ongoing research on enabling the Arbitrators to learn how to resolve a number of different type of conflicts. There is no reason why the Arbitrators should not become fully blown classifier systems. Because this system runs on a powerful parallel machine (500 transputers) very good solutions are found within a few minutes; because of this not much effort has yet been put into making the system react to sudden changes in the manufacturing environment. However, this is an area for future research. One possible scenario that is envisaged is that the system will run in

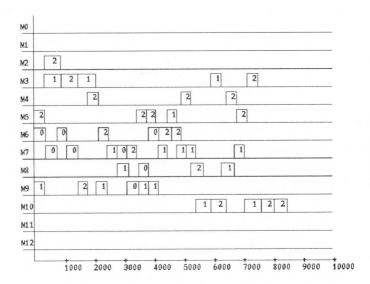

Figure 8.11: Gantt chart, generation 0, 3 jobs

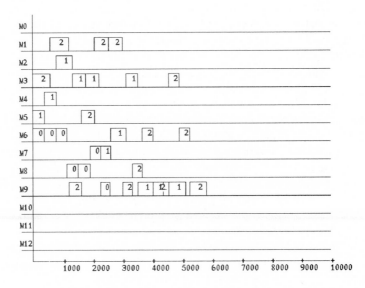

Figure 8.12: Gantt chart, generation 180, 3 jobs

the background and be continuously updated with feedback from the job-shop; in other words the simulated environment will dynamically mirror the actual manufacturing environment. Various local selection and interaction schemes are to be investigated on a new Connection Machine implementation of the model.

8.7 GAs and Machine Learning: Classifier Systems

8.7.1 Overview

In this section classifier systems, GA-based machine learning systems, are introduced. Their highly parallel nature soon becomes obvious and will not be stressed further.

Holland's development of the genetic algorithm was part of a wider effort to design robust adaptive systems capable of autonomous learning and decision making. Although he set the theoretical foundations in the early 1960s (Holland 1962), it was not until the late 1970s that the first genetics based learning system was implemented (Holland and Reitman 1978).

The object of a learning system, natural or artificial, can be described as the expansion of its knowledge in the face of uncertainty. One important way that such a system can improve its performance is by generalising on past experience. However, if a system is faced with more or less perpetual novelty, experience can only guide future actions if there are useful regularities in the environment. Here 'useful' refers to something that is relevant to the system's goals. For an animal these will be highly survival oriented. For a machine learning system, then, the problem is to construct relevant categories (representing the regularities) and discover what actions are appropriate to each category. Very likely this process will include an element of ongoing refinement. This is a difficult problem, especially if we wish to construct robust systems untethered by strong a priori assumptions about category representation. On top of this, in most realistic situations, information is very noisy and there are continual demands to act, rather than to sit and process ad infinitum. An equally difficult problem is that of giving sufficient credit to intermediate stage setting actions. In many environments only the final goal achieving action is directly rewarded. However, very often a long chain of actions is necessary. So a second major problem for machine learning systems is to be able to correctly value these stage setting actions, which generally only have implicit worth. This problem is only made worse when the goals cannot be precisely defined; e.g. an organism's search for food in a complex environment.

Holland's Classifier System (Holland 1986; Goldberg 1989a; Booker *et al.* 1989) is a machine learning system that goes some way towards tackling these fundamental problems. It consists of three main parts:

- A rule and message system

- An apportionment of credit system

- A genetic algorithm

The rule and message system is a special kind of production system (Davis and King 1976). A production system uses production rules of the form: if [condition(s)] then [action(s)]. When the condition is satisfied, as long as there are no conflicts, the action is taken (the rule fires). Many early symbolic AI systems and the later expert systems used production rules with particularly simple forward or backward chaining pattern matching algorithms (Waterman and Hayes-Roth 1978). Production rule based systems were long ago shown to be computationally complete (Post 1943). The classifier system uses much lower level rules (sub-symbolic) than expert systems and its algorithms are quite different in order to cope with the demands of learning.

To quote Holland (1986):

> The research that has culminated in the design of expert systems is a solid achievement for AI: given appropriately restricted domains, expert systems display the reasoned consideration that one expects of an expert. The source of this success, the domain-specific character of the systems, is also a source of limitations. The systems are brittle in the sense that they respond appropriately only in narrow domains and require substantial human intervention to compensate for even slight shifts in domain. This problem of brittleness and ways to temper it ...[are a primary concern of classifier system].

And then later in the same chapter he sketches a classifier system thus:

> Message passing, rule-based production systems in which many rules are active simultaneously offer attractive possibilities for the exploitation of general-purpose machine learning algorithms. In such systems each rule can be looked upon as a tentative hypothesis about some aspect of the task environment, competing against other plausible hypotheses being entertained at the same time. In this context there are two major tasks for machine learning algorithms: 1) apportionment of credit and 2) rule discovery.

A classifier system uses a population of syntactically very simple rules (classifiers) which are activated in parallel; this permits the coordination of multiple simultaneous actions. The key piece of information for the system to learn is the value of any given rule in a particular circumstance. To allow this, classifier systems have the rules coexisting in what Holland has described as "an information based service economy". A competition is held between classifiers where the right to respond to messages, from the environment or other classifiers, goes to the highest bidder. The bids are redistributed to previously active (message sending) classifiers. In this way a chain of actions can be rewarded. The competition ensures that good rules survive and bad ones die out.

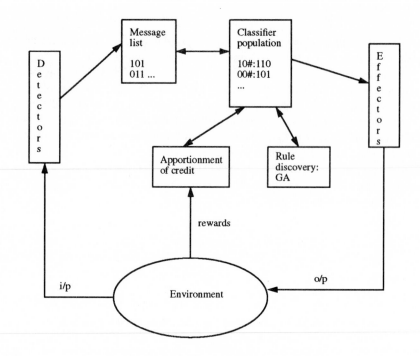

Figure 8.13: Classifier system architecture

A genetic algorithm is used for rule discovery. To use a population of rules that are fixed for all time is very restrictive. Instead the system can adapt, can improve itself, by replacing its least useful rules by newly discovered ones. The simple string nature of classifiers allows the use of the standard GA operators for this purpose. Selection is performed based on the rules' strengths (measure of usefulness). Only a fraction of the population is replaced after breeding to allow gradual and useful adaptation. The properties of the GA mean that the discovery procedure is biased by the system's experience: rule building blocks found useful in the past are recombined to form new rules.

8.7.2   A Closer Look

Figure 8.13 shows a typical classifier system architecture. Information flows from the environment through the detectors and is decoded to a finite length message. This is posted to a finite length message list. Messages on the list may activate classifiers. When activated, a classifier posts a message to the list. This may cause further classifiers to be activated or an action to be taken through the effectors.

A message is a finite length string over some finite alphabet, usually binary. Using BNF formalism and assuming length k,

$$[\text{message}]::=\{0,1\}^k$$

A classifier is a production rule:

$$[\text{classifier}]::=[\text{condition}]:[\text{message}]$$

where,

$$[\text{condition}]::=\{0,1,\#\}^k.$$

Note that # is a wildcard character that can match either 1 or 0. A condition is matched by a message if at every position in the condition there is a match with the corresponding character in the message. Clearly, the implementor will specify how the rules are decoded so that they are appropriate to the domain of application. Some examples of classifier system applications are given in the next section. Once a classifier condition is matched it becomes a candidate for activation; an 'auction' determines whether or not it is. The apportionment of credit mechanism is bound up with this 'auction'. There are a number of algorithms used but most are variations on *The Bucket Brigade Algorithm* (Holland 1986).

In outline, the workings of this algorithm are as follows. Each classifier has a strength which initially will be some pre-defined value. Each matched classifier makes a 'bid' proportional to its strength. The highest bidders are activated and their bids are divided among those classifiers responsible for sending the messages that matched their conditions. This division of payoff ensures the emergence of coadapted sets of rules to cover a range of situations. Classifiers are usually taxed by some small percentage at each cycle to prevent unproductive rules surviving.

So, for any classifier we can write:

$$S(t+1) = S(t) - C_{bid}S(t) - C_{tax}S(t) + P(t) \qquad (8.11)$$

where, $S(t)$ is the strength at time t C_{bid} is the bid coefficient, $C_{bid}S(t)$ is the bid, C_{tax} is the tax coefficient, $C_{tax}S(t)$ is the tax, and $P(t)$ is the payment from other classifiers or the environment. The bid coefficient is typically 20–30%, the tax coefficient much lower at about 5%.

A genetic algorithm is used to produce new, possibly better, rules. Using a straightforward GA like the one described earlier, new rules are created by the usual reproduction, crossover and mutation processes. Selection is made according to rule strengths. Because it takes some time to evaluate a classifier properly, it is no longer desirable to use a non-overlapping population model (i.e. one in which the whole population is replaced at each generation). Instead only a proportion (maybe 10–20%) of the population is replaced. Another way in which the classifiers are exposed to a sufficiently wide range of situations for accurate assessment of their value, is to call up the GA with period longer than one rule and message matching cycle. The GA period may be fixed or may be treated stochastically. The GA may be additionally called upon if the overall performance of the system has fallen below some threshold.

In summary, a typical classifier system algorithm is:

1. randomly generate a population of classifiers each with initial strength, $S(0)$.

2. Decode environmental message, if there is one, and post it to message list.

3. Form the set of matched classifiers.

4. Hold an auction to determine which classifiers are activated (message list is finite and cannot be overfilled). Each classifier bids according to its strength; the bids are distributed to previously active classifiers who cause current ones to activate by posting their messages; tax each classifier.

5. Activate winning classifiers by posting their messages.

6. If action is to be taken (through effectors) do so, if this results in payoff from environment distribute this to the classifiers who caused action.

7. If a genetic event is to take place, produce new offspring, giving them some initial strength; replace weaker classifiers.

8. Increment time, goto 2.

8.7.3   The Pitt Alternative

An alternative classifier system approach, know as the Pitt approach because the work was done in Pittsburgh, was developed by Smith (1980, 1983). In this approach a population of structures, each of which represents an entire rule set, is maintained. Rule sets are crossed and mutated under a selective pressure to form new rule sets for evaluation. Smith's system uses a standard production rule inference engine which employs a working memory to store the current state, and matching production rules are fired in parallel. Using this approach, it is possible to almost completely sidestep the apportionment of credit problem and avoid schemes like the bucket brigade algorithm. Another interesting feature of this system is the use of variable length rule sets and non-standard crossover to handle this. Smith used poker as his study domain, and his system was able to consistently beat Waterman's adaptive poker player system (Waterman 1970).

8.7.4   Example Classifier System Applications

Holland and Reitman's original implementation (Holland and Reitman 1978) learned a series of maze running tasks. The study demonstrated that the system was capable of transferring experience on simpler problems to accelerate learning on harder ones. Booker (1982) simulated the adaptive behaviour of a simple animal by giving it a classifier system brain. Goldberg (1983) developed classifier systems for control applications, specifically the pole balancing problem and the control of a gas pipeline compressor system. In both cases a critic presented payoff on each cycle. Wilson used a classifier system to focus and centre a moveable video camera on an object placed in its field of vision (Wilson 1983). Later he developed an interesting stripped-down system that controlled an artificial creature (animat) in a simulated environment (Wilson 1985). Classifier systems have been used for learning models of consumer choice (Greene and Smith 1987) and for discovering scheduling heuristics (Hilliard *et al.* 1987). Bonelli and Parodi (1991) have used the technique in a medical diagnostic system.

8.8 Genetic Algorithms and Artificial Life

8.8.1 Simulating Adaptive Behaviour

Recently there has been a surge of interest in the possibilities provided by the study of animats—simulated animals or autonomous robots (Wilson 1987)—in helping to make fundamental breakthroughs in our understanding of adaptive behaviour (Meyer and Wilson 1991). The idea is that the study of such systems, whose innards the experimenter has direct control over, is a powerful and flexible approach to uncovering the underlying mechanisms that allow animals to adapt and survive in uncertain environments. A closely related strand of research aims to devise mechanisms, not necessarily biologically inspired, to control autonomous robots capable of sophisticated behaviour in unstructured and unfamiliar surroundings. It seems highly likely that the two areas have much to offer each other.

This kind of research has found itself under the umbrella of the Artificial Life movement, whose aims are best summed up by its founder, Chris Langton:

> Artificial Life is the study of man-made systems that exhibit behaviours characteristic of natural living systems. It complements the traditional biological sciences concerned with the *analysis* of living organisms by attempting to *synthesize* life-like behaviours within computers and other artificial media. By extending the empirical foundations upon which biology is based *beyond* the carbon-chain life that has evolved on Earth, Artificial Life can contribute to theoretical biology by locating *life-as-we-know-it* within the larger picture of *life-as-it-could-be* [from Langton 1989].

This vision may seem extreme to many. However, if simulations go on proving useful to our understanding of the behaviour of natural systems, it seems likely that simulated evolution, in the form of a genetic algorithm, will play an increasingly important role. The sophisticated adaptive behaviour we observe in the higher animals evolved. Indeed, all behaviour, intelligent or otherwise, should be viewed against an evolutionary backdrop, against a backdrop of adapt or die. We are what we are, in the main, because that way we have survived.

An increase in our understanding of such issues in natural systems can then feed into our development of artificial ones. The following sections explore one aspect of this, namely the use of genetic algorithms to evolve, rather than design, significant parts of systems. They will mainly be concerned with systems that are likely to play an important role in future developments in artificial intelligence.

8.8.2   GAs and Neural Nets

Over the last decade there has been growing interest and activity in the field variously know as parallel distributed processing, connectionism and artificial neural networks. This paradigm borrows from what we know of real brains and performs computations using a network of simple interconnected processors (Rumelhart and McClelland 1986). Various mechanisms by which networks can learn, or adapt, have been devised, numerous applications now exist, the formal foundations have been strengthened, biologically plausible models are entering the field; all in all the adaptive powers of these networks point to their increasing importance in a number of areas, particularly artificial intelligence. However, one serious problem remains. Neural networks are difficult to design.

It is almost impossible to predict the behaviour of many large multi-layer networks. The kinds of parallel processing and distributed representations characteristic of neural nets are very hard for us to picture. As a result, trying to hand-design useful new network architectures is a daunting task indeed.

However, if the design problem is to find the architecture which produces the best results for a particular problem according to a particular criteria, genetic algorithms can be of use. A neural network's structure can be reasonably easily encoded on a genotype, although there are still many research issues involved in developing the most appropriate, concise and flexible encodings. A number of investigations into this use of GAs have been successful. They have typically being concerned with small networks and use representation schemes ranging from the rather inflexible string encoding of the network's connection matrix (Miller *et al.* 1989) to weak specifications requiring a developmental mapping from genotype to phenotype (Harp *et al.* 1989). The first type of scheme is good for optimising the structure of small networks while the second is a more promising approach for representing large complex nets and for optimising other properties of the network as well as its topology.

Neural nets operate by adjusting connection weights between their units. Learning in a network can be viewed as finding the optimal distribution of weights. Whitley has used genetic algorithms as the learning mechanism in a fixed topology (Whitley and Hanson 1989; Whitley *et al.* 1991). He reports results that compare favourably with other techniques, indeed his algorithms often outperform the others.

Ackley and Littman (1990) have done some interesting work on combining evolutionary and lifetime learning. Their *evolutionary reinforcement learning* makes use of a genetic algorithm to allow effective learning based solely upon natural selection. Their simulations involve animats, governed by a neural network, involved in open ended survival situations. Lifetime learning is via a special reinforcement algorithm. Both the network and the evaluation function used by the reinforcement technique are coded on the genotype. They evolve under selective pressure. This technique has been shown to give better performance than either just evolution or just lifetime learning. This result is closely related to recent discussion of the Baldwin effect—whereby lifetime learning speeds up evolution (Hinton and Nolan 1987; Maynard-Smith 1987; Schull 1991). Littman and Ackley (1991) have also shown that their technique produces better adaptation in dynamic environments.

8.8.3 Evolution Versus Design

Neural networks are not alone; in general we find it very hard to design any complex system well, particularly those involving interactions between many parts. We have evolved in such a way that it is difficult for us to consciously hold many things in our heads at the same time. This effectively makes it impossible for us to foresee many of the interactions inherent in complex non-linear systems. Trial and error and the use of accumulated experience is one way forward. But some would claim that this process becomes more and more infeasible as more complex tasks are attempted (Husbands and Harvey 1992) and that artificial evolution is a more powerful and general technique, applicable to a range of tasks, not just the design of neural networks.

Evolving the design of physical objects having many control points has already been mentioned in Section 8.4. The problem of evolving robot control systems will now be briefly discussed. A useful and truly autonomous robot must be able to perform in uncertain environments—this requires a sophisticated coordination of perception and action. In short, its control programs must, to a large degree, conquer many of the problems of artificial intelligence. Traditional approaches to the development of autonomous robot control systems have made only modest progress, with fragile and computationally very expensive methods. A large part of the blame for this can be laid at the feet of an implicit assumption of functional decomposition—the assumption that perception, planning and action can be analysed independently of each other. This failure has led to recent work at MIT which bases robot control architectures instead around behavioural decomposition (Brooks 1991). However, it is accepted that the design of robust mobile robot control systems is highly complex, whichever decomposition is used, because of the extreme difficulty of foreseeing all possible interactions with the environment, and because of the interactions between separate parts of the robot itself (Brooks 1991; Moravec 1983). The design by hand of such a cognitive architecture inherently becomes more complex much faster than the number of layers or modules within the architecture—the complexity can scale with the number of possible interactions between modules.

If, however, some objective fitness function can be derived for any given architecture, and it is possible to specify a set of (probably extensible) base units, perhaps neural networks, for the architecture, there is the possibility of automatic evolution of the architecture without explicit design. Natural evolution is the existence proof for the viability of this approach, given appropriate resources. So, it is suggested that GAs, suitably extended in their application, are a means of evading the problems mentioned above. There is no longer any need to make strong assumptions about the means to achieve a particular behaviour, as long as this behaviour is directly or indirectly included in the evaluation process.

A methodology for developing cognitive architectures along these lines is described in Husbands and Harvey (1992). In particular it is argued that a gradual building up of competence is necessary. This requires a dynamic developmental genotype as discussed in Harvey (1991).

8.8.4   The Future of AI?

The discussion of the previous section can be taken further. A possible conclusion is that we will never be able to design an artificial intelligence but we might be able to artificially evolve one. Of course this begs the question of what intelligence actually is. The higher levels of human intelligence have emerged very recently on an evolutionary time scale. By higher levels I mean those capable of proving theorems in logic, or playing chess, or solving puzzles, or interpreting sophisticated English sentences—those with which the field of artificial intelligence has largely been obsessed for the past three decades. This recent emergence suggests that the essence of intelligence, the largest part common to all higher life forms, is the ability to react appropriately to the environment, to be able to coordinate perception and action to good advantage. Different levels of intelligence are bound to different sorts of adaptive behaviour, different survival tools.

All this suggests that trying to understand in detail the most recent and advanced evolutionary developments from the top down is an impossibly naive and ambitious task. A much better approach might be to try and understand the evolutionary progression by which they arrived. Attempting to simulate evolution, but simulating with hindsight, seems a promising way forward. Lack of space forbids further discussion of this topic, but I predict that it will become of central importance in future AI work. Related views, although not explicitly involving simulated evolution, have been expressed by others (Brooks 1991; Cliff 1991).

8.9   Conclusions

This chapter has provided a wide survey of the field of genetic algorithms. The parallel nature of the technique has been stressed and a number of parallel implementations have been discussed in depth. It has been concluded that parallel implementations using local selection and breeding with some genotype mobility are superior to others. They provide greater robustness and reliability.

The use of GAs as function optimisers has been dealt with. It has been shown how GA-based models can cast fresh light on a problem and may lead to a more general characterisation which clearly indicates how to exploit parallelism. Lastly, after a discussion of the uses of GAs in various areas of AI, it has been suggested that they will have a fundamental role to play as we learn to evolve, rather than design, complex systems.

References

Ackley D. (1987a) An empirical study of bit vector function optimisation. In Davis L. (1987).

Ackley D. (1987b) *A connectionist machine for genetic hillclimbing*. Kluwer Academic Publishers.

Ackley D. and Littman M. (1990) Learning from natural selection in an artificial environment. In Langton *et al.* (Eds.), *Proc. 2nd Artificial Life Conf.*. Addison-Wesley.

Back T., Hoffmeister F. and Schwefel H. (1991) A Survey of Evolution Strategies. In *Proc. 4th Int. Conf. on Genetic Algorithms* (Ed. by R. Belew and L. Booker), pp. 2–9. Morgan Kaufmann.

Bagley J. (1967) The behaviour of adaptive systems that employ genetic and correlation algorithms. PhD Thesis, University of Michigan.

Belew R. and Booker L. (Eds) (1991) *Proc. 4th Int. Conf. on GAs*. Morgan Kaufmann.

Bonelli P. and Parodi A. (1991) An efficient classifier system and its experimental comparison with two representative learning methods on three medical domains. In Belew R. and Booker L. (1991), pp. 288–295.

Booker L. (1982) Intelligent behaviour as an adaptation to the task environment. PhD. thesis, University of Michigan.

Booker L., Goldberg D. and Holland, J. (1989) Classifier systems and genetic algorithms. *Artificial Intelligence* **40**, 1–3, pp. 235–282.

Bramlett M. and Cusic R. (1989) A Comparative Evaluation of Search Methods Applied to Parametric Design of Aircraft. In Schaffer D. (1989), pp. 213–218.

Brooks R. (1991) Intelligence without representation. *Artificial Intelligence* **47**, pp. 139–159.

Chang T. and Wysk R. (1985) *An Introduction to Automated Process Planning Systems.* Prentice-Hall.

Chapman D. (1985) Planning for conjunctive goals. *Technical Report* AI–TR–802, MIT AI Lab.

Christophedes N. (1979) *Combinatorial Optimisation.* Wiley.

Clevland G. and Smith S. (1989) Using Genetic Algorithms to Schedule Flow Shop Releases. In Schaffer D. (1989), pp. 160–169.

Cliff D. (1991) Computational Neuroethology: A provisional manifesto. In Meyer J. and Wilson, S. (1991), pp. 29–39.

Cohoon J., Hegde S., Martin W. and Richards D. (1987) Punctuated equilibria: A parallel genetic algorithm. In Grefenstette (1987), pp. 148–154.

Collins R. and Jefferson D. (1991) Selection in massively parallel genetic algorithms. In Belew R. and Booker L. (1991), pp. 249–256.

Colorni A., Dorigo M. and Maniezzo V. (1991) Genetic Algorithms and Highly Constrained Problems: The Time-Table Case. In Schwefel H. and Manner R. (Eds.) (1991), pp. 55–59.

Darwin C. (1859) *On the Origin of Species.* John Murray, London.

Davidor Y. (1991) A naturally occurring niche and species phenomenon: the model and first results. In Belew R. and Booker L. (1991), pp. 257–263.

Davis L. (1985) Job Shop Scheduling with Genetic Algorithms. In Grefenstette, J. (1985), pp. 136–140.

Davis L. (1987) *Genetic Algorithms and simulated annealing.* Pitman, Research notes in AI.

Davis L. and Coombs S. (1987) Genetic Algorithms and Communication Link Speed Design: Theoretical Considerations. In Grefenstette J. (1987), pp. 252–256.

Davis L. (1990) *The Handbook of Genetic Algorithms.* Van Nostrand Reinhold.

Davis R. and King J. (1976) An overview of production systems. In *Machine Representation of Knowledge*, (Ed. by Elcock E. and Michie D.) Wiley.

Deb K. and Goldberg D. (1989) An investigation of niche and species formation in genetic function optimization. In Schaffer D. (1989), pp. 42–50.

De Jong K. (1975) An analysis of the behaviour of a class of genetic adaptive systems. PhD thesis, Dept. Computer and Communication Sciences, University of Michigan.

Eldredge N. and Gould S. J. (1972) Punctuated equilibria: an alternative to phyletic gradualism. In *Models in Paleobiology*, (Ed. by T. Schopf) W. H. Freeman, pp. 82–115.

Fourman M. (1985) Compaction of Symbolic Layout Using Genetic Algorithms. In Grefenstette J. (1985), pp. 141–153.

Garey M. R. and Johnson D. S. (1979) *Computers and intractability*. Freeman, San Francisco.

Gould S. J. (1989) *Wonderful Life*. Hutchinson, London.

Goldberg D. (1983) Computer aided gas pipeline operation using genetic algorithms and rule learning. PhD. thesis, University of Michigan.

Goldberg D. (1989a) *Genetic Algorithms*. Addison-Wesley.

Goldberg D. (1989b) Sizing populations for serial and parallel genetic algorithms. In Schaffer D. (1989), pp. 70–79.

Goldberg D. (1989c) Genetic algorithms and Walsh functions: part I, a gentle introduction. *Complex Systems*, **3**, pp. 129–152.

Goldberg D. (1989d) Genetic algorithms and Walsh functions: part II, deception and its analysis. *Complex Systems*, **3**, pp. 153–171.

Goldberg D. and Bridges C. (1990) An analysis of a reordering operator on a GA-hard problem. *Biological Cybernetics*, **62(5)**, pp. 397–405.

Goldberg D. and Lingle R. (1985) Alleles, Loci and The Travelling Salesman Problem. In J. Grefenstette (Ed), *Proc. Int. Conf. on GAs and their Applications*, Lawrence Erlbaum.

Goldberg D. and Richardson J. (1987) Genetic algorithms with sharing for multimodal function optimisation. In Grefenstette J. (1987), pp. 41–49.

Gorges-Schleuter M. (1989) ASPARAGOS An Asynchronous Parallel Genetic Optimisation Strategy. In Schaffer D. (1989), pp. 422–427.

Greene D. and Smith S. (1987) A genetic system for learning models of consumer choice. In Grefenstette, J. (1987), pp. 217–223.

Grefenstette J. (1981) Parallel adaptive algorithms for function optimisation. Tech. Rep. CS–81–19, Vanderbilt University Compter Science Dept.

Grefenstette J. (Ed) (1985) *Proc. of an Int. Conf. on GAs and their applications*. Lawrence Erlbaum.

Grefenstette J. (Ed) (1987). *Proc. 2nd Int. Conf. on GAs*. Lawrence Erlbaum.

Grefenstette J. (1990) Strategy acquisition with genetic algorithms. In Davis L. (1990).

Harp S., Samad T. and Guha A. (1989) Towards the genetic synthesis of neural networks. In Schaffer D. (1989), pp. 360–369.

Harvey I. (1991) Species adaptation genetic algorithms: a basis for a continuing SAGA. In *Proc. 1st European Conf. on Artificial Life*. MIT Press.

Hilliard M., Liepins G., Palmer M., Morrow M. and Richardson J. (1987) A Classifier based system for discovering scheduling heuristics. In J. Grefenstette (Ed), *Proc. 2nd Int. Conf. on GAs*. Lawrence Erlbaum.

Hillis W. D. (1985) *The Connection Machine*. MIT Press.

Hillis W. D. (1990) Co-Evolving parasites improve simulated evolution as an optimization procedure. *Physica D*, **42**, pp. 228–234.

Hinton G. and Nolan S. (1987) How learning can guide evolution. *Complex Systems* **1**, pp. 495–502.

Hoffmeister F. and Back T. (1991) Genetic Algorithms and Evolution Strategies—Similarities and Differences. In *Parallel Problem Solving from Nature*, (Ed. by Schwefel, H.-P. and Manner R.), Springer-Verlag, LNCS Series, pp. 455–470.

Holland J. (1962) Outline for a logical theory of adaptive systems. *Journal of ACM*, **3**.

Holland J. (1966) Universal Spaces: A basis for studies of adaptation. In *Automata Theory*, (Ed. by Caianiello E.), Academic Press, pp. 218–231.

Holland J. (1975) *Adaptation in Natural and Artificial Systems*. University of Michigan Press.

Holland J. (1986) Escaping Brittleness. In *Machine Learning* (Ed. by Michalski *et al.*). Morgan Kaufmann.

Holland J. and Reitman J. (1978) Cognitive systems based on adaptive algorithms. In *Pattern directed inference systems* (Ed. by Waterman and Hayes-Roth). Academic Press.

Hulin M. (1991) Circuit Partitioning with Genetic Algorithms Using Code Schemes to Preserve the Structure of a Circuit. In Schwefel H. and Manner R (Eds.) (1991), pp. 75–79.

Husbands P. (1992) An ecosystem model for integrated production planning. *Int. Journal of Computer Integrated Manufacturing*.

Husbands P. and Harvey I. (1992) Evolution versus Design: Controlling Autonomous Robots. Submitted to *Proc. AIS 92*, IEEE Publications.

Husbands P. and Mill F. (1991) Simulated co-evolution as the mechanism for emergent planning and scheduling. In Belew R. and Booker L. (1991), pp. 264–270.

Husbands P., Mill F. and Warrington S. (1990) Generating Optimal Process Plans from First Principles. In *Expert Systems for Management and Engineering* (Ed. by Balagurasamy E. and Howe J.). Ellis Horwood, pp. 130–152.

Husbands P., Mill F. and Warrington S. (1991) Genetic algorithms, production plan optimisation and scheduling. In Schwefel H. and Manner R. (1991), pp. 80–84.

Jog P. and Gucht D. (1987) Parallelisation of Probabilistic Sequential Search Algorithms. In Grefenstette (1987), pp. 170–176.

Johnson T. and Husbands P. (1991) Systems Identification using genetic algorithms. In Schwefel H. and Manner, R. (1991), pp. 85–89.

Jones D. and Beltramo M. (1991) Solving partitioning problems with genetic algorithms. In Belew and Booker (1991), pp. 442–449.

Karr C. (1990) Air injected hydrocyclone optimisation via genetic algorithms. In Davis L. (1990).

Knuth D. (1973) *Sorting and searching, Vol. 3, The art of computer programming*. Addison-Wesley.

Koza J. (1990) Genetic programming: A paradigm for genetically breeding populations of computer programs to solve problems. Tech. Report STAN–CS–90–1314, Dept. Compt. Sci., Stanford University.

Langton C. (1989) Artificial Life. In *Artificial Life* (Ed. by Langton C.), Addison-Wesley, pp. 1–48.

Liepins G., Hilliard M., Palmer M. and Morrow M. (1987) Greedy genetics. In Grefenstette J. (1987), pp. 90–99.

Liepins G. and Vose M. (1990) Representational issues in genetic optimization. *Journal of Experimental and Theoretical Artificial Intelligence*, **2**, pp. 101–115.

Ling S. (1991) Constructing a college timetable by integrating two approaches: logic programming and genetic algorithms. MSc. Thesis, School of Cognitive and Computing Sciences, University of Sussex.

Littman M. and Ackley D. (1991) Adaptation in constant utility non stationary environments. In Belew R. and Booker L. (1991), pp. 136–142.

Manderick B. and Spiessens P. (1989) Fine-grained parallel genetic algorithms. In Schaffer (1989), pp. 428–433.

Maynard-Smith J. (1972) *On Evolution*. Edinburgh University Press.

Maynard-Smith J. (1987) When learning guides evolution. *Nature*, **329**, pp. 761–762.

Meyer J. and Wilson S. (Eds.) (1991) *From animals to animats*. MIT Press.

Miller G., Todd P. and Hegde S. (1989) Designing neural networks using genetic algorithms. In Schaffer D. (1989), pp. 379–384.

Moravec H. (1983) The Stanford cart and CMU rover. *Proc. IEEE*, **71**, pp. 872–884.

Muhlenbein H. (1989) Parallel Genetic Algorithms, Population Genetics and Combinatorial Optimisation. In Schaffer D. (1989), pp. 416–421.

Muhlenbein H., Gorges-Schleuter M. and Kramer O. (1988) Evolution Algorithms in Combinatorial Optimisation. *Parallel Computing*, **7(1)**, pp. 65–88.

Muhlenbein H., Schomisch M. and Born J. (1991) The parallel genetic algorithm as function optimizer. *Parallel Computing*, **17**.

Nakano R. and Yamada T. (1991) Conventional Genetic Algorithms for Job Shop Problems. In Belew R. and Booker L. (1991), pp. 474–479.

Nordvik J. and Renders J. (1991) Genetic algorithms and their potential for use in process control: a case study. In Belew R. and Booker L. (1991), pp. 480–486.

Pettey C., Leuze M. and Grefenstette J. (1987) A Parallel Genetic Algorithm. In Grefenstette (1987), pp. 155–161.

Post E. L. (1943) Formal reductions of the general combinatorial decision problem. *American Journal of Mathematics*, **vol. 65**.

Powell D., Tong S. and Skolnick M. (1989) EnGENEous: Domain Independent, Machine Learning for Design Optimisation. In Schaffer D. (1989), pp. 151–159.

Radcliffe N. (1991) Equivalence class analysis of genetic algorithms. *Complex Systems*, **5(2)**, pp. 183–205.

Rawlins G. (Ed.) (1990) *Foundations of genetic algorithms*. Morgan Kaufmann.

Rechenberg I. (1965) Cybernetic solution path of an experimental problem. Royal Aircraft Establishment Translation No. 1122, Ministry of Aviation, Farnborough.

Richardson J., Palmer M., Liepins G. and Hilliard M. (1989) Some guidelines for genetic algorithms with penalty functions. In Schaffer D. (1989), pp. 191–197.

Robertson G. (1987) Parallel implementation of genetic algorithms in classifier systems. In Grefenstette (1987), pp. 140–147.

Rumelhart D. and McClelland J. (1986) *Parallel Distributed Processing*. MIT Press.

Sannier A. and Goodman E. (1987) Genetic Learning Procedures in Distributed Environments. In Grefenstette (1987), pp. 162–169.

Schaffer D. (Ed) (1989) *Proc. 3rd Int. Conf. on GAs*. Morgan Kaufmann.

Schull J. (1991) The view from the adaptive landscape. In Schwefel H. and Manner R. (1991), pp. 415–428.

Schwefel H. and Manner R. (Eds) (1991) Parallel Problem Solving from Nature. Springer Verlag, *Lecture Notes in Computer Science*, **496**.

Smith D. (1985) Bin packing with adaptive search. In Grefenstette, J. (1985), pp. 202–206.

Smith S. (1980) A learning system based on genetic adaptive algorithms. PhD. thesis, University of Pittsburgh.

Smith S. (1983) Flexible learning of problem solving heuristics through adaptive search. *Proc. 8th Int. Joint Conf. on AI*, pp. 422–425.

Stenseth N. C. (1985) Darwinian evolution in ecosystems: the Red Queen view. In *Evolution: essays in honour of John Maynard-Smith*. Cambridge University Press, pp. 55–72.

Suh J. and Van Gucht D. (1987) Incorporating heuristic information into genetic search. In Grefenstette, J. (1987), pp. 100–108.

Syswerda G. (1991) Uniform crossover in genetic algorithms. In Belew R. and Booker L. (1991), pp. 2–9.

Tanese R. (1987) Parallel Genetic Algorithm for a Hypercube. In Grefenstette (1987), pp. 177–183.

Vancza J. and Markus A. (1991) Genetic Algorithms in Process Planning. *Proc. first CIRP Workshop of Intelligent Manufacturing Systems Seminar on Learning in IMS*, Budapest, pp. 329–348.

Waterman D. (1970) Generalization learning techniques for automating the learning of heuristics. *Artificial Intelligence*, **1**, pp. 121–170.

Waterman D. and Hayes-Roth F. (Eds) (1978) *Pattern-Directed Inference Systems*. Academic Press.

Whitley D. (1989) The GENITOR algorithm and selection pressure: why rank-based allocation of reproductive trials is best. In Schaffer D. (1989), pp. 116–121.

Whitley D. and Hanson T. (1989) Optimizing neural networks using faster, more accurate genetic search. In Schaffer D. (1989), pp. 391–397.

Whitley D. and Starkweather T. (1990) GENITOR II: A distributed genetic algorithm. *Journal of Experimental and Theoretical Artificial Intelligence*, **2**, pp. 189–214.

Whitley D., Dominic S. and Das R. (1991) Genetic reinforcement learning with multilayer neural networks. In Belew R. and Booker L. (1991), pp. 562–570.

Wilson S. (1983) On the retinal-cortical mapping. *Int. J. Man-Mach. Stud.*, **18**, pp. 361–389.

Wilson S. (1985) Knowledge growth in an artificial animal. In Grefenstette J. (1985), pp. 16–23.

Wilson S. (1987) Classifier systems and the animat problem. *Machine Learning*, **2**, pp. 199–228.

Wright S. (1982) Character changes, speciation, and the higher taxa. *Evolution*, **36(3)**, pp. 427–443.

CHAPTER 9

Randomised Algorithms for Packet Routing on the Mesh

Sanguthevar Rajasekaran

University of Pennsylvania, USA

9.1 Introduction

9.1.1 Randomised Algorithms

Classical approaches to introducing randomness in algorithms typically assume a distribution on possible inputs and compute the expected performance of various (deterministic) algorithms. Quicksort is a good example. If one assumes that each input permutation is equally likely to occur, Quicksort runs in an expected $O(n \log n)$ time to sort n numbers. The credibility of such an approach critically depends on the assumption made on the inputs. There may be applications where the input distribution is quite different from the one used in the probabilistic analysis.

As an attractive alternative, Rabin and Solovay and Strassen proposed introducing randomness in the algorithm itself. A randomised algorithm is one where certain decisions are made based on the outcomes of coin flips. No matter what the input is, a large fraction of all possible outcomes for the coin flips will ensure 'good performance' of the algorithm. Thus the two approaches differ in the probability space used for analysis. In the former one considers the space of all possible inputs and in the latter one employs the space of all possible outcomes for coin flips.

Since the introduction of randomised algorithms in 1976, a wide variety of computational problems have been solved (both sequentially and in parallel) using this technique. In this chapter we study randomised algorithms for packet routing.

9.1.2 Packet Routing

Fixed connection machines are some of the most practical models of parallel computing, as inferred from the parallel computers available today. A fixed connection machine is usually represented as a directed graph whose nodes correspond to processing elements, and whose edges correspond to communication links. The speed of a parallel computer is determined by 1) the computing power of component processors, and 2) the speed of inter-processor communication. Nowadays the computing power of individual processing elements can be made arbitrarily high owing to the decline in hardware costs. Thus the speed of any parallel machine crucially depends on how fast the inter-processor communication is.

A single step of inter-processor communication in a fixed connection network can be thought of as the following task (also called *packet routing*): each node in the network has a packet of information that has to be sent to some other node. The task is to send all the packets to their correct destinations as quickly as possible such that at the most one packet passes through any wire at any time.

A special case of the routing problem is called the *partial permutation routing*. In partial permutation routing, each node is the origin of at the most one packet and each node is the destination of no more than one packet. A packet routing algorithm is judged by 1) its *run time*, i.e., the time taken by the last packet to reach its destination, and 2) its *queue length*, which is defined as the maximum number of packets any node will have to store during routing. Contentions for edges can be resolved using a *priority scheme*. Furthest destination first, furthest origin first, etc. are examples of priority schemes. We assume that a packet not only contains the message (from one processor to another) but also the origin and destination information of this packet. An algorithm for packet routing is specified by 1) the path to be taken by each packet, and 2) a priority scheme.

9.1.3   Different Models of Packet Routing

How large a packet is (when compared with the channel width of the communication links) will determine whether a single packet can be sent along a wire in one unit of time. If a packet is very large it may have to be split into pieces and sent piece by piece. On this criterion many models of routing can be derived. A packet can be assumed to be either atomic (this model is known as the *store and forward model*), or much larger than the channel width of communication links (thus necessitating splitting).

In the latter, if each packet is broken up into k pieces (also called *flits*), where k depends on the width of the channel, the routing problem can be studied under two different approaches. We can consider the k flits to be k distinct packets, which are routed independently. This is known as the *multipacket routing approach*. Each flit will contain information about its origin and destination.

Alternatively, one can consider the k flits to form a *snake*. All flits follow the first one, known as the head, to the destination. A snake may never be broken, i.e., at any given time consecutive flits of a snake are at the same or adjacent processors. Only the head has to contain the origin and destination addresses. This model is called the *cut through routing with partial cuts* or simply the *cut through routing*.

Efficient algorithms for store and forward, multipacket, and cut through routing are presented in this chapter.

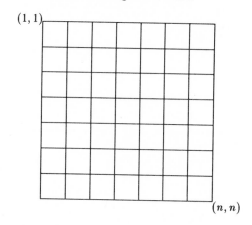

Figure 9.1: An $n \times n$ mesh connected computer

9.1.4 Mesh Connected Computers

The fixed connection machine assumed in this chapter is the *Mesh Connected Computer*. The basic topology of a two dimensional mesh is an $n \times n$ square grid with one processor per grid point (see Figure 9.1). Except for processors at the boundary, every other processor is connected to its neighbours to the *left*, *right*, *above*, and *below* through bidirectional links. Variations in this topology are possible depending on whether one or more of the following connections are allowed : 1) vertical wrap arounds, 2) horizontal wrap arounds, and 3) connections to diagonal neighbours. In this chapter we only consider the mesh with the basic topology. The instruction stream assumed is MIMD. This in particular means that each node can send and receive a packet (or a flit) from all its (four or less) neighbours in one unit of time.

Mesh connected computers (MCCs) have drawn the attention of computer scientists in recent times because of their many special properties. Some of the special features of MCCs are: 1) they have a simple interconnection pattern, 2) many problems have data which map naturally onto them, and 3) they are linear-scalable.

9.1.5 Some Definitions

We say a randomised algorithm uses $\widetilde{O}(g(n))$ amount of any resource (like time, space etc.) if there exists a constant c such that the amount of resource used is no more than $cg(n)$ with probability $\geq 1 - n^{-\alpha}$ on any input of size n.

By *high probability* we mean a probability of $\geq 1 - n^{-\alpha}$ for any $\alpha \geq 1$ (n being the input size of the problem at hand).

Let $B(n, p)$ denote a binomial random variable with parameters n and p.

9.1.6 Chernoff Bounds

One of the most frequently used facts in analyzing randomised algorithms is *Chernoff bounds*. These bounds provide close approximations to the probabilities in the tail ends of a binomial distribution. Let X stand for the number of heads in n independent flips of a coin, the probability of a head in a single flip being p. X is also known to have a binomial distribution $B(n, p)$. The following three facts (known as Chernoff bounds) are now folklore:

$$\text{Prob.}[X \geq m] \leq \left(\frac{np}{m}\right)^m e^{m-np},$$

$$\text{Prob.}[X \geq (1 + \epsilon)np] \leq exp(-\epsilon^2 np/2), \text{ and}$$

$$\text{Prob.}[X \leq (1 - \epsilon)np] \leq exp(-\epsilon^2 np/3),$$

for any $0 < \epsilon < 1$, and $m > np$.

9.2 The Queue Line Lemma

In the process of packet routing in a network, the time taken by any packet to reach its destination is dictated by two factors: 1) the *distance* between the packet's origin and destination, and 2) the number of steps (also called the *delay*) the packet waits in queues. The *Queue Line Lemma* enables one to compute an upper bound on the delay of any packet.

Consider the set of paths $\mathcal{P}$ taken by the packets. Two packets are said to *overlap* if they share at least one edge in their paths. The set of paths is said to be *nonrepeating* if for any two paths in $\mathcal{P}$, the following statement holds: if these two paths meet, share some successive edges, and diverge, then they will never meet again.

Lemma 9.2.1 *The amount of delay any packet q suffers waiting in queues is no more than the number of distinct packets that overlap with q, provided the set of paths taken by packets is nonrepeating.*

Proof. Let π be an arbitrary packet. If π is delayed by each of the packets that overlap with π no more than once, the lemma is proven. Else, if a packet (call it q) overlapping with π delays π twice (say), then it means that q has been delayed by another packet which also overlaps with π and which will never get to delay π. $\square$

9.3 Routing on a Linear Array

In this section we study different routing problems on a linear array. These results will help us analyze routing algorithms on the mesh. As will be shown, routing on a mesh can be broken into a constant number of phases, where in each phase routing is performed either along the rows or along the columns.

Problem 1. Each node of a linear $n-$array has $k \geq 1$ packets initially and each node is the destination of exactly k packets. Send all the packets to their destinations sending at the most one packet along any edge in a single step.

Lemma 9.3.1 *If we use the furthest origin first priority scheme, problem1 can be solved in time* $\leq \frac{k(n+1)}{2} + \frac{n}{2}$.

Proof. Let the nodes of the array be numbered $1, 2, \ldots, n$ from left to right. Each packet takes the shortest path between its origin and destination. In order to compute an upper bound on the delay that any packet suffers, it suffices to compute the number of distinct packets that can potentially delay this packet (see the queue line lemma 9.2.1). Also notice that since each node can communicate with all its neighbours in one unit of time, the flow of packets in one direction (say left to right) does not delay the flow in the other direction (right to left). And hence, for the worst case analysis we can assume without loss of generality that all the packets are traversing in the same direction (say from left to right).

Consider a packet originating from node i. Without Loss Of Generality assume that its destination is to the right. This packet can be delayed by at the most ki (the number of packets at and to the left of i) steps. Also, the delay is at the most $(n - i + 1)k - 1$ since only these many packets have destinations to the right of i. Therefore, the maximum delay for the packet at i is $\leq \min[ki, (n - i + 1)k]$ (This is an upper bound on the number of packets that overlap with the packet at i). Thus, the maximum delay any packet suffers is $\leq \max_i[\min(ki, (n - i + 1)k)] = \frac{k(n+1)}{2}$ (for $i = \frac{n+1}{2}$). Therefore, all the packets can be routed in time $\frac{k(n+1)}{2} + \frac{n}{2}$. $\square$

Problem 2. On an n-array, node i has k_i ($1 \leq k_i \leq n$ and $\sum_{i=1}^{n} k_i = n$) packets initially (for $i = 1, 2, \ldots, n$). Each node is the destination for exactly one packet. Route the packets.

Lemma 9.3.2 *If furthest destination first priority scheme is used, the time needed for a packet starting at node i to reach its destination is no more than the distance between i and the boundary in the direction the packet is moving. That is if the packet is moving from left to right then this time is no more than $(n - i)$ and if the packet is moving from right to left the time is $\leq (i - 1)$.*

Proof. Consider a packet q at node i and destined for j. Assume (without loss of generality) it is moving from left to right. q can only be delayed by the packets with destinations $> j$ and which are to the left of their destinations. Let $k_1, k_2, \ldots, k_n$ be the number of such packets (at the beginning) at nodes $1, 2, \ldots, n$ respectively. (Notice that $\sum_{l=1}^{n} k_l \leq n - j$).

Let m be such that $k_{m-1} > 1$ and $k_{m'} \leq 1$ for $m \leq m' \leq n$. Call the sequence $k_m, k_{m+1}, \ldots, k_n$ *the free sequence*. Realise that a packet in the free sequence will not be delayed by any other packet in the future. Moreover, at every time step at least one new packet joins the free sequence. Thus, after

$n - j$ steps, all the packets that can possibly delay q would have joined the free sequence. q needs only an additional $j - i$ steps, at the most, to reach its destination. The case the packet moves from right to left is similar.

One could also use the queue line lemma to prove the same result. □

Problem 3. In a linear array with n nodes more than one packet can originate from any node and more than one packet can be destined for any node. In addition, the number of packets originating from the nodes $1, 2, \ldots, j$ is no more than $j + f(n)$ (for any j and some function f). Route the packets.

Lemma 9.3.3 *Under the furthest origin first priority scheme, Problem 3 can be solved within $n + f(n)$ steps.*

Proof. Let q be a packet originating from node i and destined for node j (to the right of i). q can potentially be delayed by at the most $i + f(n)$ packets (since only these many packets can originate from the nodes $1, 2, \ldots, i$ and hence have a higher priority than q). q only needs an additional $j - i$ steps to reach its destination. Thus the total time needed for q is $\leq j + f(n)$. The maximum of this time over all the packets is $n + f(n)$. □

9.4 Store and Forward Routing on the Mesh

In this section an optimal algorithm is presented for packet routing on an $n \times n$ mesh. The algorithm has a run time of $2n + \widetilde{O}(\log n)$ and a queue length of $\widetilde{O}(1)$. First, a $3n + o(n)$ step algorithm with a queue size of $\widetilde{O}(\log n)$ is presented. This algorithm is modified in stages to obtain the stated bounds.

9.4.1 A $3n + o(n)$ Step Algorithm

Consider an $n \times n$ mesh where there is a packet at each node initially. Processors are named $(i, j), i = 1, 2, \ldots, n, \ j = 1, 2, \ldots, n$ with $(1, 1)$ at the left top corner. Let q be any packet and let (i, j) and (r, s) be its origin and destination respectively. What follows is a three phase algorithm with a run time of $3n + o(n)$ with high probability.

In phase I send q along the column j to a random node (k, j) in column j (each node in column j being equally likely). In phase II send q to (k, s) along the row k. And finally, in phase III send q to (r, s) along the column s. In phase II employ the furthest origin first priority scheme, and in phase III employ the furthest destination first priority scheme(with ties broken arbitrarily). Notice that each phase of routing corresponds to routing on a linear array.

Since there is one packet at each node at the beginning, phase I can be completed in $\leq n$ steps without any packet suffering any delay whatsoever by a continuous flow of packets along the columns. In phase II, the number of packets that can start from an arbitrary node (k, l) in row k $(1 \leq k \leq n)$ is $B(n, 1/n)$, since in phase I each of the n packets in column l would have

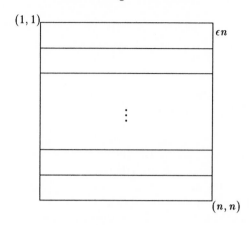

Figure 9.2: Partition of columns for the $(2 + \epsilon)n$ algorithm

chosen (k, l) with probability $1/n$. Therefore, the total number of packets starting their phase II from any one of the nodes $(k, 1), (k, 2), \ldots, (k, j)$ is $B(nj, 1/n)$. Using Chernoff bounds (equation 1), this number is no more than $j + n^\delta$ (for any constant $\delta > 1/2$) with high probability. Now, an application of lemma 9.3.3 shows that phase II can be completed in $n + n^\delta$ steps with high probability. Phase III is nothing but Problem 2 of Section 9.3 and hence can be performed within n steps. Put together, the algorithm has a run time of $3n + n^\delta = 3n + o(n)$.

9.4.2   A $(2 + \epsilon)n$ Routing Algorithm

The run time of the above algorithm can be improved to $(2 + \epsilon)n, \epsilon \geq 1/\log n$ if in phase I, each packet is sent to a random position (along its column) within a distance of ϵn from its origin.

Assume without loss of generality, ϵn divides n. Divide each column of the mesh into slices of length ϵn as shown in Figure 9.2.

Modify phase I as follows. Every processor (i, j) chooses a random position in column j within the slice it is in and sends its packet to that position along column j. Phases II and III remain the same.

Lemma 9.4.1 *The above modified algorithm runs in time $(2 + \epsilon)n + c\alpha n^\delta$, for any $\delta > 1/2$ and some fixed constant c, with probability $\geq 1 - n^{-\alpha}$.*

Proof. Phase I can now be implemented in time ϵn without any additional delay since there is one packet at each node initially.

Consider a packet that starts phase II at (k, j). Without Loss Of Generality, assume it is moving to the right. The number of packets starting this phase in $(k, 1), (k, 2), \ldots, (k, j)$ is $B(j\epsilon n, \frac{1}{\epsilon n})$. The mean of this variable is j. Using Chernoff bounds (equation 1), this number is no more than $j + n^\delta$ (for any constant $\delta > 1/2$) with high probability. Therefore lemma 9.3.3 applies here to imply that phase II can be completed in $n + n^\delta$ steps with high probability.

In phase III there are n packets starting from any column and each node is the destination of exactly one packet. Thus in accordance with lemma 9.3.2, phase III terminates within n steps. $\square$

One could choose as small an ϵ as desired in order to decrease the run time. However, the smaller the value of ϵ, the larger will be the queue size. Note that if in phase I, packets randomise over slices of length ϵn, then for the worst case input, queue size at the end of phase II will be $\Omega(1/\epsilon)$. (An example is when all the packets that have to be routed to a particular column appear in the same row in the input). The queue size of this algorithm can be shown to be $O(1/\epsilon + \log n)$.(The proof is similar to the one given for the next algorithm and hence is omitted here.) Thus ϵ has to be greater than $1/\log n$ if only $O(\log n)$ queue size is allowed.

9.4.3   An Optimal Routing Algorithm

In the $(2 + \epsilon)n$ algorithm above, say we overlap the three phases. That is, a packet that finishes phase I before ϵn steps can start phase II without waiting for the other packets to finish phase I and so on. We expect some packets to finish faster now.

The only possible conflict between phases is between phases I and III. A packet that is doing its phase I and a packet that is doing its phase II might contend for the same edge. Always, under such cases, we give precedence to the packet that is doing its phase I. If a packet q doing its phase III contends for an edge with a packet that is doing its phase I, it means that q has completed its phases I and II within ϵn steps. Since after ϵn steps from the start of the algorithm every packet will be doing either its phase II or phase III, q will reach its destination within $(1 + \epsilon)n$ steps. Packets like q are not interesting to our analysis. Thus we will not mention them hereafter in any analysis. A packet doing its phase II cannot be delayed by packets doing their phase I or phase III and vice versa.

In summary, no packet suffers additional delays due to overlap of phases. It is also easy to see that the maximum queue length at any node does not increase due to overlap. Can some of the packets finish faster now? We answer this question next.

Consider a packet q initially at node (i, j) with (r, s) as its destination. Assume without loss of generality, (r, s) is below (i, j) and to the right (the other cases can be argued in exactly the same lines). If in phase I q chooses a position (k, j) (in the slice (i, j) is in), then q will finish phase A in $|i - k|$ steps. Also q will take at the most $(s - 1) + n^\delta$ steps to finish phase II (with high probability) and $(n - k)$ steps to finish phase III (see Figure 9.3 and the proof of lemma 9.4.1).

Therefore q takes $|i - k| + (s - 1) + (n - k)$ (plus lower order terms) total steps to complete all the three phases. Call this sum the *normal path length* of

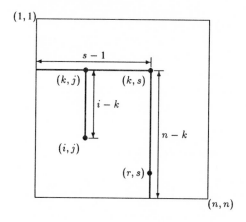

Figure 9.3: Normal path length of a packet: $(i - k) + (s - 1) + (n - k)$

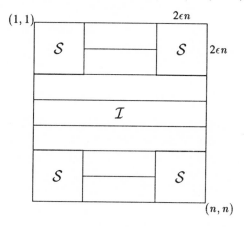

Figure 9.4: Superior and inferior packets

q. We require the normal path length of every packet to be $\leq 2n$. This can be ensured if every packet moves in the direction of its destination in phase I, because then $|i - k| + (n - k)$ will always be $\leq n$.

The normal path length of a packet is $\leq 2n$ also when the following condition holds: 'the origin of the packet is at least $2\epsilon n$ distance away from its nearest row or column boundary'. In the example of Figure 9.3 if $i \geq 2\epsilon n$, even if the packet is sent ϵn away from its destination (in phase I), $|i - k| + (n - k)$ will still be $\leq n$. Packets that satisfy the above condition are the ones that have their origins in the I region of Figure 9.4.

Call these packets the *inferior packets*. Packets that do not satisfy this condition are those that have their origins in the S region of Figure 9.4. Call these packets the *superior packets*. Superior packets have the potential of having a normal path length of $> 2n$.

Our routing algorithm will route the inferior packets using the $(2 + \epsilon)n$ algorithm given above. Superior packets are given special treatment. We will route the superior packets such that those packets with a **potential** normal path length of $\geq 2n$ move in the direction of their destinations in phase I. Overlap of phases is assumed for both the types of packets. Details of the algorithm follow.

Algorithm for inferior packets

> Inferior packets use the algorithm of Section 9.4.2. Every column is divided into slices of length ϵn. In phase I, a packet at (i, j) with a destination (r, s) is sent to a random position (k, j) in its slice along the column j. In phase II, the packet is sent to (k, s) along row k and finally in phase III, the packet is sent to (r, s) along the column s.

Algorithm for superior packets

> Each node (i, j) in the upper two S squares chooses a random k in $\{2\epsilon n + 1, 2\epsilon n + 2, \ldots, 2\epsilon n + (1/4)n\}$ and sends its packet, q, to (k, j) along column j. If (r, s) is the destination of q, q is then sent to (k, s) along row k and finally along column s to (r, s).

> Each node (i, j) in the lower two S squares chooses a random k in $\{n - 2\epsilon n - 1, n - 2\epsilon n - 2, \ldots, n - 2\epsilon n - (1/4)n\}$ and sends its packet to (k, j) along column j. The packet is then sent to (k, s) along row k and to (r, s), the packet's destination, along column s.

Queue Disciplines

- In phases I and III no distinction is made between superior and inferior packets.

- If a packet doing its phase I and a packet doing its phase III contend for the same edge, the packet doing its phase I takes precedence.

- Furthest destination first priority scheme is used for phase III.

- In phase II, superior packets take precedence over inferior packets. Among superior (inferior) packets furthest origin first priority scheme is used.

Lemma 9.4.2 *The above algorithm runs in time* $2n + \tilde{O}(\log n)$. *Further, the maximum queue length at any node at any time is* $\tilde{O}(\log n)$.

Proof
Superior packets

A superior packet does not suffer any delay in phase I.

A superior packet q starting its phase II at (k, j) can be delayed by at the most the number of superior packets that will ever start their phase II from a node to the left of (k, j) in row k (assuming without loss of generality, q is moving from left to right). The number of such packets is upper bounded by $B(j2\epsilon n, (4/n))$. The expected value of this binomial is $8\epsilon j$. If $j \geq \log n$, the Chernoff bounds affirm that this delay will be $\leq j$ with probability $\geq 1 - n^{-\alpha}$ for any $\alpha \geq 1$ if we choose a proper constant $\epsilon < 1/8$. The delay will be $\widetilde{O}(\log n)$ for $j < \log n$. Therefore, superior packets will complete phase II in $n + \widetilde{O}(\log n)$ steps.

A superior packet that moves in the direction of its destination in phase I spends at the most n steps in total in phases I and III (as was shown before). A packet that was moving in a direction opposite to its destination in phase I will complete phases I and III in a total of $2((1/4)n + 2\epsilon n)$ steps. Since ϵ is chosen to be less than $1/8$, this total is $\leq n$.

And hence all the superior packets will complete all the three phases in $2n + \widetilde{O}(\log n)$ steps.

Inferior packets

Let q be a packet at (i, j) initially, that moves to (k, j) at the end of phase I, to (k, s) at the end of phase II, and to (r, s) at the end of phase III.

q spends at the most ϵn steps in phase I.

In the following case analysis ignore the presence of superior packets.

case1: $k \leq 2\epsilon n$ or $k > n - 2\epsilon n$.

Delay for q in phase II will be $\leq j - 2\epsilon n + n^\delta$ with high probability. This is because q is delayed by at the most the packets that will ever do their phase II starting from a node to the left of (k, j) in row k (assuming q moves from left to right), and in row k packets in columns $1, 2, \ldots, 2\epsilon n$ have been vacated. Using an analysis similar to the one in Section 9.4.2, q completes phase II in $\leq n - 2\epsilon n + n^\delta$ steps.

q takes $\leq n$ steps to finish phase III. Thus, q takes $\leq \epsilon n + n - 2\epsilon n + n^\delta + n = 2n - \epsilon n + n^\delta$ steps to complete all the three phases.

case2: $n - 2\epsilon n \geq k > 2\epsilon n$.

q takes $\leq n + n^\delta$ steps to complete phase II with high probability.

In phase III q spends $\leq n - 2\epsilon n$ time since $n - 2\epsilon n$ is the maximum distance from row k to any row boundary. Therefore q completes all the three phases in $\epsilon n + n + n^\delta + n - 2\epsilon n = 2n - \epsilon n + n^\delta$ steps.

The only unaccounted delay so far is the delay of inferior packets by the superior packets in phase II. The number of superior packets that will ever do their phase II in a given row k is $B(8\epsilon^2 n^2, 4/n)$. The expected value of this number is $32\epsilon^2 n$. The Chernoff bounds imply this number is $\leq g\alpha\epsilon^2 n$ with probability $\geq 1 - n^{-\alpha}$ for some fixed constant $g(> 32)$. The number of superior packets that will delay an inferior packet doing its phase II in row k is therefore at the most $g\alpha\epsilon^2 n$. Hence, the inferior packet will complete all the three phases in time $\leq 2n - \epsilon n + n^\delta + g\alpha\epsilon^2 n \leq 2n$, for small enough ϵ.

Thus all the inferior packets will complete all the three phases in $\leq 2n$ steps with high probability.

The queue length of the above algorithm is $\widetilde{O}(\log n)$ as proven next. During any phase of routing, note that the queue length at any node is no more than the maximum of 1) the number of packets at the beginning of the phase in this node, and 2) the number of packets in this node at the end of the phase. Consider any node (i, j) in the mesh. During phase I, only one packet starts from any node and $\widetilde{O}(\log n)$ packets will end up in any node. During phase II, $\widetilde{O}(\log n)$ packets start from any node and $\widetilde{O}(\log n)$ packets will end up in any node. And in phase III, $\widetilde{O}(\log n)$ packets start from any node and only one packet will end up in any node. Therefore, the queue length of the whole algorithm is $\widetilde{O}(\log n)$. $\square$

Thus we have the following:

Theorem 9.4.1 *Any permutation routing of n^2 elements on an $n \times n$ MIMD MCC can be completed in $2n + \widetilde{O}(\log n)$ steps, the queue length being $\widetilde{O}(\log n)$.*

9.4.4 Reducing the Queue Length to $\widetilde{O}(1)$

We shall now modify the previous algorithm in order to reduce the size of the queues to constant. In the $2n + O(\log n)$ algorithm of the above section we chose ϵ to be a constant. One can see that the expected queue size at any node at the end of phase I and II is a constant.

The idea is instead of considering individual processors, to "divide" each column into consecutive blocks of $\log n$ nodes each. Then the expected number of packets per block is $\Theta(\log n)$, and using Chernoff bounds we get that with high probability we have $\Theta(\log n)$ packets per block, or, $\Theta(1)$ packets per processor.

We use therefore the same algorithm as in Section 9.4.2. However, we do not store packets at their target row and column at the end of phases I and II respectively, rather we redistribute packets within each block so as to obtain a constant number of packets per node. Redistributing packets introduces at most $O(\log n)$ delay. The results are summarised in the following theorem.

Theorem 9.4.2 *There is a randomised algorithm for routing on an $n \times n$ mesh with the property that any partial permutation can be realised within $2n + K\alpha \log n$ steps with probability at least $1 - n^{-\alpha}$ for some constant K and any $\alpha > 1$. Further, the queue length is $\tilde{O}(1)$.*

9.5 Cut Through Routing

9.5.1 Cut Through Routing on a Linear Array

In any network, if the width of the communication links is less than the size of a packet, it is more practical to break up each packet into k flits, where k depends on the width of the link. In cut through routing the k flits corresponding to each packet form a snake. A snake can never be broken, i.e., all the flits of a snake should be in adjacent nodes at any time.

Formally, a snake of k flits, s, at any time t, is defined by a $(k + 1)$-tuple, $(s_1, s_2, \cdots, s_i, t)$, where $s_1, s_2, \cdots, s_i, i \le k$, are consecutive processors that contain at least one flit of the snake. The *length* of a snake s at time t, $length(s, t)$, is defined to be the number of processors over which the snake is distributed. A snake is in *full-extension* if $length(s, t) = k$.

As in the case of store and forward routing, here also routing on a mesh can be broken into a constant number of phases of routing along columns and rows. Thus we first consider some cut through routing problems on a linear array.

Lemma 9.5.1 *Permutation routing can be performed on a linear array of n processors in $\frac{n(k+1)}{2}$ steps under the cut through model. Here k is the number of flits in each packet.*

Proof. The algorithm used by the processors is quite simple. At each time unit, every processor transmits the flit in the head of its queue. At the same time it receives a flit from each neighbour and appends these flits to appropriate queues.

The proof of run time will be given on a more restricted model of routing called the 'restricted cut through model'. An upper bound obtained under this model will clearly be an upper bound on the (more powerful) cut through model as well. In the restricted cut through model, once a packet gains full extension, it will not be compressed again. Transmission of a packet starts at time instances that are multiples of k.

An immediate consequence of the restricted cut through model and the algorithm described above is that at least one packet gains full extension every k steps. In the case of any cut through routing algorithm, the natural priority scheme to be assumed is the first in first out (FIFO) scheme. But to simplify the presentation, we assume throughout the furthest destination first priority scheme or the furthest origin first priority scheme. Similar analysis could be used to prove the same bounds for the FIFO scheme as well. Consider a packet q that originates in node i ($1 \le i \le n$) and whose destination is j. W.l.o.g. assume that j is to the right of i and all the packets are traversing from left to right. Also assume the furthest destination first priority scheme.

Packet q can possibly be delayed by at the most $(n - j)$ other packets (with higher priority). Also realise that q can only be delayed by $(j - 1)$ other packets. Therefore, the number of **distinct** packets that can delay q is $\min\{(n - j), (j - 1)\}$. If q were not delayed by any other packet, it needs only $k + (j - i)$ steps to reach its destination. Therefore, applying the queue line lemma (lemma 9.2.1), the time needed for q to reach its destination is no more than $\min\{(n - j)k, (j - 1)k\} + k + (j - i)$. The maximum of this quantity over all possible is and js is $\frac{n(k+1)}{2}$. $\square$

Using similar arguments we can also prove the following lemma:

Lemma 9.5.2 *If there are m packets on a linear array such that each processor has possibly more than one packet to start with and each processor is the destination of exactly one packet, routing can be completed in $(k - 1)m + n$ steps.*

9.5.2   Algorithm for $n \times n$ Mesh

The algorithm that we shall present for cut through model on the mesh is for permutation routing and it resembles the algorithm presented in Section 9.4.2. The run time of this algorithm is $kn + o(kn)$ and the queue length is $O(k)$ with high probability. The queue length is asymptotically optimal and the run time is within a factor of 2 of the lower bound presented in Section 9.6. We will first describe a $2kn + O(k \log n)$ algorithm which will be modified to run in $kn + O(k \log n)$ steps, where k is the number of flits in a packet.

The grid is divided into two regions **S** and **I**, and packets with origins in these two regions are called *superior* packets and *inferior* packets respectively (see Figure 9.4).

Inferior packets are routed using the following algorithm. The rows are divided up into $1/\epsilon$ strips of ϵn rows each. The algorithm has three phases. A packet at processor (x, y), destined for processor (r, s), is first routed along the column y to (w, y), a processor chosen at random in the same column and strip as (x, y). The packet is then sent to (w, s) along row w, after which it is routed to its destination along column s.

The superior packets are routed using a slightly different algorithm. A processor (x, y) in the upper two **S** squares, with destination (r, s) sends its packet to (w, y) along column y, where w is chosen at random from $\{2\epsilon n + 1, 2\epsilon n + 2, \cdots, 2\epsilon n + (1/4)n\}$. The packet is then sent to (w, s) along row w, and then, to (r, s) along column s. ϵ is chosen to be less than $1/8$, so that superior packets in the upper half remain in the upper half after the randomisation phase. The algorithm for the packets in the lower two S squares is symmetric.

During the first and third phases, no distinction is made between superior and inferior packets with respect to the queuing disciplines. If they contend for the same edge, a packet performing in its first phase takes precedence over one in its third phase. In phase II, superior packets take precedence over inferior packets and among the inferior (superior) packets the furthest origin first priority scheme is used. In phase III, packets that have further to go have higher priority.

9.5.3 Routing Time Analysis

Superior Packets

A superior packet q starting its phase II at (u, v) can be delayed by at the most $2\epsilon nv$ packets, each with probability $4/n$. Therefore, the number of packets that can potentially delay q is $m = B(2\epsilon nv, \frac{4}{n})$. Using Chernoff bounds (equation 1), we can show that m is $v + O(\log n)$, with high probability. q needs to traverse a distance of $\leq n - v$ in phase II. Thus, superior packets will complete phase II in at the most $kn + O(k \log n)$ steps, with high probability. Superior packets will complete phases I and III in at the most kn steps. Therefore, all superior packets will complete all three phases in $2kn + O(k \log n)$ steps with high probability.

Inferior Packets

For an inferior packet that starts phase II at row w, we have two possible cases.

1. $w \leq 2\epsilon n$ or $w \geq n - 2\epsilon n$

   Suppose an inferior packet q starts phase II at row w, $w \leq 2\epsilon n$, and w.l.o.g., it moves from left to right. Suppose, the packet starts the phase at column $2\epsilon n + t$. Then with high probability, the delay q suffers will be at the most $kt + kn^\delta$, for some $\delta < 1$. (This follows from the fact that we use the furthest origin first priority scheme in this phase, and that the number of inferior packets that can delay any inferior packet in phase II is $B(t\epsilon n, \frac{1}{\epsilon n})$, and an application of the Chernoff bounds equation 2.) Since q has to travel a distance of no more than $n - t - 2\epsilon n$, it will complete the phase II in $\leq kn - 2k\epsilon n + kn^\delta$ steps with high probability. Since the packet spends $\leq k\epsilon n$ steps in phase I and at the most kn steps in phase III (from lemma 9.5.2), it takes no more than $2kn - k\epsilon n + kn^\delta$ steps to complete all phases. This is $\leq 2kn$ for appropriate ϵ.

2. $2\epsilon n < w < n - 2\epsilon n$

   For such a packet q, phase I completes in $k\epsilon n$ steps, and phase III in

$kn - 2k\epsilon n$ steps. Suppose a packet starts phase II in column t. The number of inferior packets that delay q is $t + n^\delta$ (for some $\delta < 1$), with high probability. The number of superior packets that will delay the packet during phase II is $B(8\epsilon^2 n^2, 4/n)$ whose expectation is $32\epsilon^2 n$. Using Chernoff bounds equation 2, we can show that the number of packets delaying our packet is no more than $\alpha\epsilon^2 n$, with high probability, for some $\alpha > 32$. Thus, the total routing time of the packet, in this case, is $k\epsilon n + kn - 2k\epsilon n + kn + kn^\delta + k\alpha\epsilon^2 n \leq 2kn$, for small enough ϵ.

Therefore, all inferior packets will complete all three phases in at the most $2kn$ steps with high probability.

9.5.4   Modification to the Algorithm

We can reduce the number of steps taken by the algorithm by making the following modifications. Initially, each inferior processor flips a coin and colours its packet black or white depending on the result. The mesh is partitioned into both vertical and horizontal slices of ϵn columns and rows respectively.

In phase I, all the white packets choose a random node in the same column and horizontal slice as their origin and go there along the column of origin. Also in phase I, the black packets choose a random node in the same row and vertical slice as their origin and go there along the row of origin. During phase II, all white packets are routed along rows till they reach their column destination, while black packets are routed along columns till they reach their row destination. In phase III, white packets are routed along columns to their destinations, while black packets are routed along rows. There is no change in the algorithm for superior packets.

It is likely that white and black inferior packets contend for the same edge. For instance, a white inferior packet in phase I may compete for an edge with a black inferior packet performing its phase II. Whenever there is such a conflict between black and white packets, preference is given to packets in lower phases. In the above example, the white packet will be given priority.

Theorem 9.5.1 *Using a randomised colouring scheme, routing can be performed in $kn + O(k \log n)$ steps on an $n \times n$ mesh with queue size of $O(k)$ flits, with high probability.*

Proof. As a result of the colouring, the number of inferior packets that will perform their phase II along any row (column) and the number of inferior packets that will perform their phase III along any column (row) is no more than $n/2 + n^{3/4}$, with high probability. This is due to the fact that the above number is a binomial, $B(n, 1/2)$. Using analysis similar to that shown in the previous section, we find that, if we use a randomised colouring scheme, routing can be completed in $kn + O(k \log n)$ steps with high probability.

The conflict between white and black packets does not affect the run time for the following reason. Consider the example of a white packet in phase I conflicting with a black packet in phase II. If such a conflict occurs, it means that the black packet has completed its phase I well within ϵkn steps, and even if it waits for all the (black and white) packets to complete their phase I, it will start its phase II at the latest by step ϵkn, thus unaffecting the analysis. Conflicts of other kinds can also be argued similarly. $\square$

Queue Length Analysis

The queue size of the above algorithm in any phase is no more than the maximum queue size at the beginning and end of the phase. In phase I, only one packet starts from any node and the number of packets that will end up in any node is upper bounded by $B(\epsilon n, \frac{1}{\epsilon n})$. Using Chernoff bounds (equation 2), this number is $\widetilde{O}(\log n)$. In a similar way we also see that the queue sizes of phase II and phase III are $\widetilde{O}(\log n)$ packets (i.e., $\widetilde{O}(k \log n)$ flits).

Using ideas similar to ones given in Section 9.4.4, we can reduce the queue size to $O(k)$ flits (i.e., a constant number of packets). The crucial fact used is that the queue size of any $\log n$ successive processors in the array is still $O(k \log n)$, with high probability. Each column (as well as row) is partitioned into slices of $\log n$ successive nodes. Packets that have to be stored in each such slice are distributed among the nodes in the slice. That is, if a node in a slice has to store more than ck flits (for some constant $c > 1$), it will send the additional packets to its neighbour in the slice, and the neighbour will do the same thing. With high probability, this redistribution will be local to each slice.

9.6 A Lower Bound for Run Time

The run time of any cut through routing algorithm on an $n \times n$ mesh will have to be $\Omega(kn/2)$ where k is the number of flits in each packet. This can be proved by constructing a worst case permutation to be routed.

Consider a permutation where all the packets starting from the left half of the mesh have destinations in the right half, and all the packets originating from the right half have destinations in the left half. $kn^2/2$ flits from the left have to cross column $n/2$ to reach their destinations. There are only n nodes in column $n/2$ and each node can transmit only one flit in one unit of time from the left half to the right half. Therefore, even if all the nodes in column $n/2$ are busy all the time, they need at least $kn/2$ steps before they transmit all the $kn^2/2$ flits from the left to the right.

The above lower bound also holds for $k - k$ routing that we consider next.

9.7 Multipacket Routing

9.7.1 Preliminaries

In Multipacket routing, each packet is broken up into k flits, and these flits behave as though they are independent entities. In particular, each flit carries along with it, its source and destination addresses. The problem of $k - k$ routing is the problem of routing where $\leq k$ packets originate from any node and $\leq k$ packets are destined for any node. It need not be the case that if one of the k packets originating from a node (say i) is destined for a node (say j), then the other $k - 1$ packets originating from i will also be destined for j. In this section we present a $kn + O(k \log n)$ time, $O(k)$ queue length randomised algorithm for the general $k - k$ routing problem.

Lemma 9.7.1 $k - k$ *routing can be completed on an n-node linear array in* $\frac{n(k+1)}{2}$ *steps under the multipacket model.*

Proof. Very similar to that of lemma 9.5.1 □.

Lemma 9.7.2 *If there are m packets on an n-node linear array with zero or more packets originating from any node and zero or more packets destined for any node, routing can be performed within* $m + n - 1$ *steps.*

Proof. An immediate consequence of the queue line lemma. □
    The proof of the following lemma is similar to those of lemmas 3.3 and 7.1:

Lemma 9.7.3 *Let there be* xn *packets in a linear array. If the number of packets with an address* $> j - 1$ *is* $\leq (n - j + 1)x + g(n)$, *and the number of packets with an address* $< j$ *is* $\leq (j - 1)x + g(n)$, *then routing can be completed within* $xn + g(n)$ *steps using the furthest destination first priority scheme.*

9.7.2 Algorithm

The algorithm for Multipacket routing is similar to the one presented in the last section. There are k packets initially at each node. Each processor flips a 2-sided coin k times and colours its packets black or white depending on the outcomes. The packets are then routed in exactly the same way as they are for the cut through model, except that now there are no snakes but only independent packets. Also, now the superior packets are also coloured black or white depending on the outcomes of coin flips. White and black superior packets execute symmetric but opposite algorithms (i.e., in phase I, a white packet chooses a random node in the column of its origin and goes there, whereas a black packet chooses a random node in the row of its origin and goes there, and so on.) Conflicts between white and black packets are resolved by assigning higher priority to packets in lower phases.

Theorem 9.7.1 *Using this algorithm,* $k - k$ *routing will take* $kn + O(k \log n)$ *steps, on an* $n \times n$ *mesh with queue size of* $O(k)$, *with high probability.*

Proof. Lemma 9.7.3 is crucial to the proof. The analysis is similar to the one in Section 9.5. A superior packet completes its phases I and III in no more than $kn/2$ steps. In phase II, a superior packet that starts from column v can only be delayed by $8\epsilon kv + O(k \log n)$ other packets with high probability. It needs to traverse a distance of at the most $n - v$. Put together, the time needed for a superior packet in phase II is no more than $8\epsilon kv + O(k \log n) + n - v$, which is $\leq \frac{kn}{2} + O(k \log n)$, for a proper ϵ. Thus a superior packet will complete all the three phases in $\leq kn + O(k \log n)$ steps with high probability.

An inferior packet spends at the most $k\epsilon n$ steps in phase I. In phase II, if an inferior white packet q starts from row w and if $w \leq 2\epsilon n$ or $w \geq n - 2\epsilon n$, it will complete phase II in at the most $\frac{kn}{2} - 2k\epsilon n + kn^\delta$ steps with high probability. Phase III can be completed in $\frac{kn}{2} + kn^\delta$ steps with high probability. This is because only $\frac{kn}{2} + kn^\delta$ packets will be performing their phase III along any column with high probability and an application of lemma 9.7.3. Thus q needs no more than $kn - k\epsilon n + O(kn^\delta)$ steps for all the three phases. This number of steps is $\leq kn$ for an appropriate ϵ.

The case of a white inferior packet starting from a row w such that $2\epsilon n < w < n - 2\epsilon n$ is similar. The same analysis applies to the black inferior packets as well. The time bound remains the same even after accounting for conflicts between white and black packets. The reason is, if a packet in a lower phase conflicts with a packet in a higher phase, priority is given to the packet in the lower phase. If such a conflict occurs at all, it implies that the packet in higher phase has completed its lower phases well ahead of time and hence even if it is delayed by the lower priority packet, it will reach its destination within the stated time. (See also Section 9.5.4). The queue length analysis is also similar to the one in Section 9.5.4. □

Let the size of each packet to be routed be s bits and let the transmission time per bit be t_r. Since each packet carries with it its source and destination addresses the size of the packet is $s + 2 \log n$ bits. Consider two scenarios: 1) Use the store and forward algorithm to perform routing treating each packet as atomic. The run time in this case will be $[2n + O(\log n)](s + 2 \log n)t_r$; 2) Split each packet into k independent flits and use the multipacket routing algorithm. In this case the time bound is $[kn + O(k \log n)](\frac{s}{k} + 2 \log n)t_r$. Assuming s is much larger than $2 \log n$, the latter approach is nearly twice as fast as the former.

9.8   A Class of Efficient Routing Networks

The reason why our store and forward routing algorithm runs optimally is the following. All the packets in the I region of Figure 9.4 are less than $2n$ distance away from their destinations, and most of the nodes of the mesh are in the I region. This suggests that if we somehow distribute the nodes of the S regions into the I region we will get an improved run time. This can be done by changing the topology of the mesh as shown in Figure 9.5. Mesh1 of Figure 9.5 is exactly like an $n \times n$ mesh. Each node here also is connected to its four neighbours to the *left*, to the *right, above*, and *below* (except for the nodes in the boundary). The only difference is Mesh1 has a boundary shown in Figure 9.5

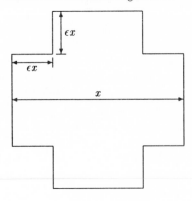

Figure 9.5: Mesh1 has a diameter of $\sqrt{3}n$

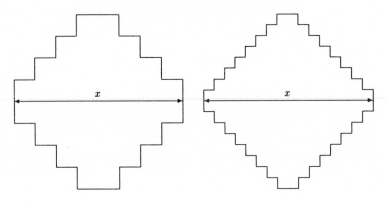

Figure 9.6: Mesh2 and Mesh3 have diameters $\sqrt{5/2}n$, and $1.5n$ respectively

rather than a square. If x is the maximum span along the vertical (or horizontal) direction of Mesh1, then the area of Mesh1 is $x^2 - 4\epsilon^2 x^2$. We require to place n^2 nodes in this mesh. Therefore, we get $x = \frac{n}{\sqrt{1-4\epsilon^2}}$. Diameter d of this mesh is $(2 - 2\epsilon)x$. Minimum of d (over all ϵ's) occurs for $\epsilon = 1/4$ and the minimum value is $\sqrt{3}n$.

The above idea can be extended further by choosing the boundaries shown in Figure 9.6 for a mesh. For these meshes also minimum diameter occurs when $\epsilon = 1/4$. These have diameters $\sqrt{5/2}n$ and $1.5n$ respectively. These meshes are all approximations to a rhombus inclined at 45° to the axes, which is a circle in the manhattan metric.

The limiting case in this class of networks is a rhombus with a diameter of $\sqrt{2}n$. This rhombus has $\sqrt{2}n$ rows. If the rows are numbered $-\frac{\sqrt{2}n}{2}, -(\frac{\sqrt{2}n}{2} - 1), \ldots, -1, 0, 1, \ldots, \frac{\sqrt{2}n}{2}$, then row 0 has $\sqrt{2}n$ nodes. Row $\pm i$ has $(\sqrt{2}n - 2i)$

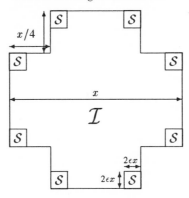

Figure 9.7: Routing on Mesh1

nodes, for $1 \le i \le \frac{\sqrt{2}n}{2}$.

Our store and forward routing algorithm runs in time $d + \tilde{O}(\log n)$ in a circular mesh of diameter d, the queue length being the same. Also, the cut through and multipacket routing algorithms run in $\frac{dk}{2} + \tilde{O}(k \log n)$ steps with a queue length of $\tilde{O}(k)$. The only modification needed to be made in the algorithms is to define the inferior and superior packets of the circular mesh accordingly. The following theorem concerns only store and forward routing. We invite the reader to prove analogous results for cut through and multipacket routings.

Lemma 9.8.1 *Store and forward routing algorithm of Section 9.4 runs in time $d + \tilde{O}(\log n)$ on a circular mesh of diameter d.*

Proof. We will prove this lemma for Mesh1. The same technique can be used for other circular meshes also. The region of inferior packets for Mesh1 will look like Mesh2. Let the size of the sliced out square be $2\epsilon x \times 2\epsilon x$, where x is the maximum vertical (or horizontal) span. The diameter of this Mesh1 is $1.5x$ (see Figure 9.7). Like the algorithm in Section 9.4.3, each column is divided into slices of length ϵx. There are three phases in the algorithm.

Each inferior packet chooses a random node in the slice it is in and traverses to that node in the first phase. In the second phase it goes to the correct column, and in the third phase it goes to the correct row.

Each superior packet in the top left square chooses a random row in the range $2\epsilon x + 1, 2\epsilon x + 2, \ldots, 2\epsilon x + (1/8)x$ in the first phase and reaches that row. Phases two and three are similar to inferior packets. Superior packets originating from other squares use a symmetric algorithm.

The analysis of the run time and the queue size for the above algorithm runs in exactly the same lines as the one given in Sections 9.4.3 and 9.4.4.

Any superior packet originating in the top left square will complete its phase II in $0.5x + O(\log n)$ steps with high probability. If this superior packet moved in the direction of its destination in phase I, then it needs at the most a total of x steps for phases I and III together. If it moved in a direction opposite to its destination in phase I, it will complete phases I and III in no more than $2((1/8)x + 2\epsilon x)$ steps. If ϵ is chosen to be less than $1/8$, this total is strictly less than x. Thus any superior packet originating in the top left square will complete all the phases in $\leq 1.5x + O(\log n)$ steps with high probability. The cases of other superior packets are similar.

Consider any inferior packet that starts its phase two from row k, where $2\epsilon x < k \leq (1/4)x$. This packet takes $\leq \epsilon x$ steps for phase I, and $\leq 0.5x + n^\delta$ steps (for any $\delta > 1/2$) for phase II with high probability. The expected delay for this packet in phase II due to superior packets is $64\epsilon^2 x$. Thus this delay is no more than $\alpha\epsilon^2 x$ (for any $\alpha > 64$) with high probability. Since this packet is at the most $x - 2\epsilon x$ distance away from any boundary in phase III, it will complete phase III in $x - 2\epsilon x$ steps or less.

Put together, the total time needed for this inferior packet is $\leq \epsilon x + 0.5x + n^\delta + \alpha\epsilon^2 x + x - 2\epsilon x$ with high probability. This total will be $\leq 1.5x$ for a small enough ϵ.

The cases of other inferior packets are similar. $\square$

Notes and References

Chernoff bounds given in Section 9.1.6 were derived by Chernoff (1952), and Angluin and Valiant (1979).

The study of randomised algorithms for packet routing on the mesh was initiated by Valiant and Brebner (Valiant 1982; Valiant and Brebner 1981). The queue line lemma and the $3n + o(n)$ algorithm in Section 9.4.1 are due to them. Store and forward algorithms (in Sections 9.4.2-9.4.4) were discovered by Krizanc, Rajasekaran, and Tsantilas (Rajasekaran and Tsantilas 1987; Krizanc *et al.* 1988). Kunde (1988) showed that one could avoid randomisation in the algorithm of Section 9.4.2, by sorting submeshes of size $\epsilon n \times \epsilon n$ (with respect to destination columns of packets in column major order). The resultant deterministic algorithm has a similar behaviour as that of the randomised algorithm, i.e., it has a run time of $(2 + 3\epsilon)n$ and a queue length of $\frac{2}{\epsilon}$.

Using the technique of classifying packets according to their origins (as superior and inferior) (Krizanc *et al.* 1988), and the sort and route paradigm of Kunde (1988), Leighton, Makedon, and Tollis (1989) were able to obtain a $2n - 2$ step deterministic algorithm for store and forward packet routing with a constant queue length. The queue length of this algorithm was over 600. Rajasekaran and Overholt (1990) reduced the queue length to 58 retaining the time bound. It still remains an open problem if one could perform permutation routing on the mesh within $2n - 2$ steps with a queue length of ≤ 5. Leighton (1990) analyzes the expected behaviour of certain greedy algorithms for packet routing.

Kunde and Tensi (1989) applied the sort and route paradigm of Kunde to obtain an efficient algorithm for multipacket routing. Their algorithm for multipacket routing has a run time of $\frac{5}{4}kn + O(kn/q)$ and a queue size of $O(kq)$ (for any $1 \leq q \leq n$). They also proved the lower bound in Section 9.6. The algorithm described in Section 9.5 is due to Rajasekaran and Raghavachari (1991). In a recent work, Kunde (1991) has provided a deterministic algorithm with a run time and queue length of $kn + o(kn)$ and $O(k)$ respectively.

Makedon and Simvonis (1990) proved that cut through routing on the mesh can be performed within $\frac{3}{2}kn + O(kn/q)$ steps, the queue length being $O(kq)$, for any $1 \leq q \leq n$. They also presented a randomised algorithm with a time bound of $kn + O(kn/q)$, the queue size being the same. Cut through routing algorithm given in Section 9.7 is due to Rajasekaran and Raghavachari (1991).

The problem of (full) permutation routing can be reduced to sorting. (We just sort the packets according to their destination addresses). A lot of work has been done in the area of sorting on the mesh, starting from the article of Thompson and Kung (1977). Thompson and Kung, and Kumar and Hirschberg (1983) obtained $O(n)$ algorithms for sorting. The algorithms of Thompson and Kung (1977), Schnorr and Shamir (1986), and Ma, Sen and Scherson (1986) need a queue length of only 1 and have a time bound of $3n + o(n)$ on an $n \times n$ mesh. Recently Kaklamanis, Krizanc, Narayanan, and Tsantilas (1991) showed that sorting can be performed using a randomised algorithm in $2.5n + o(n)$ steps, with a constant queue length of > 1. Later Kunde (1991) matched this time bound with a deterministic algorithm and a queue length of 2.

Various algorithms for off-line routing can be found in Nassimi and Sahni (1980), Raghavendra and Kumar (1986), Annexstein and Baumslag (1990), and Krizanc (1991). A survey of randomised parallel algorithms can be found in Rajasekaran (1988).

References

Angluin D. and Valiant L. G. (1979) Fast Probabilistic Algorithms for Hamiltonian Paths and Matchings. *Journal of Computer and Systems Science,* **18**, pp. 155–193.

Annexstein F. and Baumslag M. (1990) A Unified Approach to Off-Line Permutation Routing on Parallel Networks. *Proc. Symposium on Parallel Algorithms and Architectures,* pp. 398–406.

Chernoff H. (1952) A Measure of Asymptotic Efficiency for Tests of a Hypothesis Based on the Sum of Observations. *Annals of Mathematical Statistics* **23**, pp. 493–507.

Kaklamanis C., Krizanc D., Narayanan L. and Tsantilas Th. (1991) Randomized Sorting and Selection on Mesh Connected Processor Arrays. *Proc. ACM Symposium on Parallel Algorithms and Architectures.*

Krizanc D. (1991) A Note on Off-line Permutation Routing on a Mesh Connected Processor Array. *Proc. International Conference on Computing and Information*, pp. 418–421.

Krizanc D., Rajasekaran S. and Tsantilas T. (1988) Optimal Routing Algorithms for Mesh Connected Processor Arrays. *Proc. VLSI Algorithms and Architectures: AWOC*. Springer-Verlag Lecture Notes in Computer Science #319, pp. 411–22. To appear in *Algorithmica*.

Kumar M. and Hirschberg D. (1983) An Efficient Implementation of Batchers Odd-Even Merge Algorithm and its Application to Parallel Sorting Schemes. *IEEE Trans. Comp.*, **32**.

Kunde M. (1988) Routing and Sorting on Mesh Connected Processor Arrays. *Proc. VLSI Algorithms and Architectures: AWOC*. Springer-Verlag Lecture Notes in Computer Science #319, Springer-Verlag, pp. 423–33.

Kunde M. (1991) Concentrated Regular Data Streams on Grids: Sorting and Routing Near to the Bisection Bound. To be presented in the *IEEE Symposium on Foundations of Computer Science*.

Kunde M. and Tensi T. (1989) Multi-Packet Routing on Mesh Connected Arrays, *Proc. ACM Symposium on Parallel Algorithms and Architectures*, pp. 336–343.

Leighton T. (1990) Average Case Analysis of Greedy Routing Algorithms on Arrays. *Proc. ACM Symposium on Parallel Algorithms and Architectures*, pp. 2–10.

Leighton T. Makedon F. and Tollis I. G. (1989) A $2n - 2$ Step Algorithm for Routing in an $n \times n$ Array With Constant Size Queues. *Proc. ACM Symposium on Parallel Algorithms and Architectures*, pp. 328–35.

Ma Y., Sen S. and Scherson D. (1986) The Distance Bound for Sorting on Mesh Connected Processor Arrays is Tight. *Proc. 27th IEEE Symposium on Foundations of Computer Science*, pp. 255–263.

Makedon F. and Simvonis A. (1990) On bit Serial packet routing for the mesh and the torus. *Proc. 3rd Symposium on Frontiers of Massively Parallel Computation*, pp. 294–302.

Nassimi D. and Sahni S. (1980) An Optimal Routing Algorithm for Mesh Connected Parallel Computers. *J. ACM*, **27** pp. 6–29.

Raghavendra C. S. and Kumar V. K. P. (1986) Permutations on Illiac IV-Type Networks. *IEEE Trans. Comp.*, **35** pp. 662–669.

Rajasekaran S. (1988) *Randomized Parallel Computation*. PhD Thesis, Aiken Computing Lab., Harvard University.

Rajasekaran S. and Overholt R. (1990) Constant Queue Routing on a Mesh. *Proc. Symposium on Theoretical Aspects of Computer Science.* Springer-Verlag Lecture Notes in Computer Science #480, pp. 444–455. To appear in *J. of Parallel and Distributed Computing.*

Rajasekaran S. and Raghavachari M. (1991) Optimal Randomized Algorithms for Multipacket and Cut Through Routing on the Mesh. To be presented in the *IEEE Symposium on Parallel and Distributed Processing*, Dallas, Texas.

Rajasekaran S. and Tsantilas T. (1987) An Optimal Randomized Routing Algorithm for the Mesh and A Class of Efficient Mesh like Routing Networks. *Proc. 7th Conference on Foundations of Software Technology and Theoretical Computer Science.* Springer-Verlag Lecture Notes in Computer Science #287, pp. 226–241.

Schnorr C. P. and Shamir A. (1986) An Optimal Sorting Algorithm for Mesh Connected Computers. *Proc. 18th ACM Symposium on Theory of Computing*, pp. 255–263.

Thompson C. D. and Kung H. T. (1977) Sorting on a Mesh Connected Parallel Computer. *Comm. ACM*, **20**, pp. 263–270.

Valiant L. G. (1982) A Scheme for Fast Parallel Communication. *SIAM J. Comp.* **11**, pp. 350–361.

Valiant L. G. and Brebner G. J. (1981) Universal Schemes for Parallel Communication. *Proc. 13th ACM Symposium on Theory of Computing*, pp. 263–277.

CHAPTER 10

Asynchronous Iterative Algorithms: Models and Convergence

Aydın Üresin
York University, Canada
and
Michel Dubois
University of Southern California, USA

10.1   Introduction

A significant number of algorithms in the areas of both numerical and non-numerical computing fall into the class of *iterative algorithms*. These algorithms are characterised by an iteration operator F successively applied to some data x, starting with the initial value of the data $x(0)$, and are represented by the following recursive relationship:

$$x(k + 1) = F(x(k)), \quad k = 0, 1, 2, \ldots \tag{10.1}$$

where $x(k)$ is the value of x at the k-th iteration.

Some examples of numerical analysis problems that are solved by iterative algorithms are linear systems of equations, partial and ordinary differential equations and optimisation (Bertsekas and Tsitsiklis 1989a). Similarly, many graph problems such as network flow, shortest path, and transitive closure can be cast into the above formulation. Iterative algorithms and paradigms are also quite common in artificial intelligence and computer vision, including constraint satisfaction algorithms, execution of production systems and logic programs.

Typically, the iterated data x and the iteration operator F have multiple components, and this allows us to write (10.1) as

$$
\begin{aligned}
x_1(k + 1) &= F_1(x(k)) \\
x_2(k + 1) &= F_2(x(k)) \\
&\vdots \\
x_n(k + 1) &= F_n(x(k))
\end{aligned}
$$

for all $k = 0, 1, 2, \ldots$ It immediately follows that in each iteration, F_is can be computed in parallel in a multiprocessor system.

10.1.1 Multiprocessors

A multiprocessor is composed of several independent processors, each capable of executing a separate program (Hwang and Briggs 1984; Stone 1987). The processors are interconnected in a way which permits programs to exchange data and synchronise activities.

There are two major classes of multiprocessors, namely, *shared memory multiprocessors* and *message passing multiprocessors*. In shared memory multiprocessors, there is a globally shared memory which can be accessed by all the processors through an interconnection network. In message passing multiprocessors, each processor has its own local memory which cannot be accessed by other processors. Processors communicate by messages.

The computation of an F_i in (10.1) can be referred to as a *task*, and the assignment of tasks to available time slots of processors is called the *task allocation*. We will assume that the execution of a task is exclusively performed by a processor without interruption. If (10.1) is implemented on a multiprocessor in which there are as many processors available as components, then the computation of each F_i can be assigned to a different processor. Assuming that the processors in the system are identical, and ignoring the communication between the processors, the task allocation scheme in this case is unique. However, it is not uncommon for there to be fewer processors available than components. In this latter case, the allocation of F_is to processors is not unique and *static allocation* or *dynamic allocation* may be adopted. In the static allocation, the assignment of F_is to processors is made a priori before the start of the computations and remains fixed throughout. In the dynamic case, the decision as to which component to execute next is made dynamically at runtime. Typically, there is a global queue of tasks and when a processor becomes available, the current state of the task queue determines the task it executes next. Conceptually, the allocation scheme based on a unique global queue implies that the selection of the next task is *centralised*. In a scheme based on static allocation, the processors do not have to consult a global structure to select the next task: this decision is *decentralised*. It is also possible to design an allocation scheme which is a combination of the two extreme cases. For example, we can maintain multiple queues in the system; each queue schedules only a predetermined group of components and is accessed only by a predetermined group of processors. The advantage of the static allocation is that it suffers no overhead of maintaining and runtime accessing a global structure for scheduling purposes. However, in the case when it is not possible to predict the total workload (i.e., the total amount of computation to execute F), and not possible to distribute the load equally among the processors, the static allocation results in a poor load balance. Dynamic allocation improves load balance, but at the expense of the overhead caused by global queue accesses.

10.1.2 Synchronous Iterative Algorithms

If we desire to strictly enforce the sequence of events described by (10.1), regardless of the task allocation strategy we implement, we must ensure that all the computations for the k-th iteration are completed before the computations for the $(k+1)$-st iteration can start. This requires a *barrier synchronisation* after each iteration. A barrier synchronisation is a logical point in the control flow of an algorithm at which all the processors must arrive before any of them are allowed to proceed further. The iterative algorithms for which the state changes are exactly described by (10.1) are called *synchronous iterative algorithms* or *synchronous iterations*, and they use barrier primitives extensively. They suffer performance degradation caused by synchronisation barriers. Synchronisation barriers impose two kinds of performance penalties. The first one is due to the load imbalance, i.e., due to the fluctuations in the time taken by the processors to complete their work section before arriving at the barrier. The second kind of penalty results from the execution of the barrier.

The approximate overhead introduced by the load imbalance was formulated by Kruskal and Weiss (1984). They estimated the expected time T to complete the execution of Q independent tasks allocated by batches of size K on P processors. It was assumed that running times of tasks were independent identically distributed (i.i.d.) random variables with mean μ and variance σ^2. They require some additional restrictions. Many common distributions such as exponential and Gaussian distributions and large values of P satisfy the requirements. The result is as follows.

$$T \approx \frac{Q}{P}\mu + \sigma\sqrt{2K \cdot \log P}.$$

It should be clear that the second term corresponds to the penalty of load imbalance.

The second form of performance penalty is related to the implementation of the synchronisation barrier. In shared memory multiprocessors, a common implementation involves two stages (Axelrod 1986; Goodman *et al.* 1989). In the first stage, the processors check in at the barrier by simply incrementing a global counter which is initially reset. The last processor that checks in records this event, typically by setting a shared flag which is initially reset. Each processor, after checking in, polls the flag until it is set. The last processor will see the counter value as P (the number of processors), and will set the flag. Regarding the execution time of the barrier, the worst case occurs when the processors arrive at the barrier simultaneously. Usually, the counter can be accessed by no more than one processor at a time. Assume the worst case, and suppose that incrementing the counter takes t_c units of time. Then, the barrier execution takes at least $P t_c$ units. In general, the time complexity of this type of barrier is roughly $O(P)$.

The above described linear barrier is clearly damaging to the system performance for large numbers of processors. In a faster scheme, more than one barrier counters are organised as a tree structure (Axelrod 1986; Goodman *et al.* 1989). The processors are partitioned and each partition is assigned to a leaf of the tree. The maximum value of the counter on a leaf is the size of the corresponding partition, and the maximum value of an inner node counter is equal to the number of its children. Each node also has a flag associated with it and it is set when its counter reaches the maximum value. The leaf counters count the number of processors in each partition that have checked in, and the other counters count the number of set child flags. The barrier execution is complete when the root flag is set. If the tree is balanced, it has $O(\log P)$ levels. Since we can consider the whole process as a propagation from the leaves to the root, the complexity of this type of barrier is roughly $O(\log P)$. Although for a large P, logarithmic barriers should be preferred over linear barriers, for smaller numbers of processors they may not be worthwhile.

In message passing systems, the barrier at the end of each iteration k involves communicating all the components of $x(k)$ to all the processors. The time complexity of this operation (multinode broadcast) depends on the interconnection architecture, and it is given in Bertsekas and Tsitsiklis (1989a) for several architectures. In most cases it is $O(P)$.

So far we have assumed that the computation of each F_i in (10.1) depends on all the components of x. Sometimes there is only local dependency and it is not necessary to demand that all the computations in the k-th iteration are completed before the computations in the next iteration can start. It follows that if the computation of an F_i depends only on a subset of all the components, we do not need a barrier, and we can design synchronisation schemes in which the computation of F_i waits only for the components it depends on. Evidently, this type of relaxed synchronisation reduces the overhead caused by both load unbalance and execution of synchronisation. The disadvantage is that programming such a scheme is very complicated.

10.1.3 Asynchronous Iterative Algorithms

An *asynchronous iterative algorithm* (also called *asynchronous iteration*) can be defined as an algorithm which is obtained by simply removing the synchronisation points between the iterations in a synchronous iterative algorithm. Each processor participating in an asynchronous algorithm computes at full speed and efficiency. It uses component values as it needs them, even if they are not the most recent ones, and it releases new values as it produces them (Kung 1976). In its most general form, an asynchronous iteration allows any type of task scheduling: static, dynamic, centralised or decentralised. Since in a decentralised scheme the processors do not have information about the status of each other, it is possible that the same component is computed by more than one processor at the same time. As a result of this redundancy, the computation of a component does not necessarily use the most recent value of even the same component.

Let us make it clear that asynchronous iterations are not always free of all forms of synchronisation. In shared memory systems, the need for synchronisation arises in order to protect the integrity of data elements, i.e., to provide *atomic* access to data elements. For example, each component of x in (10.1) should be accessed atomically. This means that while a processor is updating a component x_i, no other processor should be allowed to read x_i, possibly having an intermediate and meaningless value. In practical systems, atomic access to a data component is usually guaranteed by hardware, if the size of the component is the same as the size of an addressable physical memory location. However, components of x may be more complex data, such as high precision floating point numbers, records, lists. They may also be smaller pieces of addressable memory locations, such as bits and characters. In these cases, components of x should be accessed in *critical sections*. A critical section with label L is a form of synchronisation, and can be defined as a code section of a parallel program such that only one critical section with label L can be executed at any given time. A common implementation of critical sections is to use the LOCK and UNLOCK operations (Hwang and Briggs 1984). These operations are atomic and have an argument of data type *lock*. A lock is defined to have two states: open and locked. LOCK(L) waits until the lock L is open and it locks L when it is open. UNLOCK(L) simply changes the state of L to open. Therefore, a critical section with label L is equivalent to the following code:

LOCK(L)
execute critical section
UNLOCK(L).

Critical sections are also required to access task queues when centralised allocation schemes are employed.

Asynchronous iterations were first introduced by Chazan and Miranker (1969). They were then called *chaotic iterations*. The assumptions of their computational model were twofold: all the components are updated infinitely often, and there exists an upper bound such that the input components for the updates cannot be more outdated than the bound. They showed that asynchronous iterations corresponding to a linear system of equations with non-negative coefficients converge to the desired solution if and only if the spectral radius of the underlying matrix is less than one. Kung (1976) discussed issues related to implementations of asynchronous iterations in multiprocessors. Robert, Chàrnay and Musy (1975) considered *serial-parallel iterations* for solving non-linear problems. Serial-parallel scheme of computations is a restricted model and it does not allow outdated input values. Miellou (1975), Robert (1976) and Miellou and Spiteri (1985) also study asynchronous iterations corresponding to non-linear operators. Baudet proved a convergence condition for non-linear problems, under the most general asynchronous computational model, where no bound exists for the recency of the input components. The general computational model described by Baudet was later called *totally asynchronous* by Bertsekas and Tsitsiklis (1989a). Spiteri (1986), Miellou (1986) and Mitra (1987) studied asynchronous iterations for some applications in numerical computing.

Bertsekas (1983) obtained a general convergence result that does not assume real data domains. Miellou (1974) and Bertsekas (1983) considered the convergence for monotone operators. Tseng *et al.* (1990), Tsitsiklis (1984), Tsitsiklis and Bertsekas (1986) and Tsitsiklis *et al.* (1986) discuss several applications for a model more restricted than total asynchronism but weaker than serial-parallel computations. This model is called *partial asynchronism.* The partially asynchronous computational model was also considered by Lubachevsky and Mitra (1986).

Robert derived a convergence condition in terms of dependency information, for serial-parallel computations, and when the data domain is binary (Robert 1986, 1987). Tsitsiklis (1987) considered a similar computational model and established general necessary and sufficient conditions for convergence.

Bertsekas and Tsitsiklis (1989b) studied some results related to convergence rate and termination of asynchronous iterations. The book by these authors contains most of the previous work in the area of asynchronous iterations (Bertsekas and Tsitsiklis 1989a).

10.2 Mathematical Background

Throughout this chapter, $X = X_1 \times X_2 \times \cdots \times X_n$ denotes the set from which the shared data x in (10.1) take values and the iteration operator $F : X \mapsto X$ in (10.1) can be decomposed as

$$F(x) = (F_1(x), F_2(x), \ldots, F_n(x)), \quad \forall x \in X,$$

where $F_i : X \mapsto X_i$, for all $i = 1, 2, \ldots, n$. X is said to be *finite* if it contains a finite number of elements. Furthermore, $\mathcal{N}$, $\mathcal{N}^+$ and $\Re$ denote the sets of non-negative integers, positive integers and real numbers, respectively.

We say that a sequence $\{z(k)\}$ elements of which take values from a set X *converges* to $z^* \in X$ and write $\lim_{k \to \infty} z(k) = z^*$ in either or both of the following cases.

- There exists an integer m such that $z(k) = z^*$ for all $k \geq m$.

- $X \subseteq \Re^n$ and for all $\epsilon > 0$, there exists an integer m such that $|z_i(k) - z_i^*| < \epsilon$ for all $k \geq m$ and $i = 1, 2, \ldots, n$. [1]

A sequence $\{Z(k)\}$ of subsets of X is said to *converge* to $z^* \in X$ ($\lim_{k \to \infty} Z(k) = z^*$) iff all sequences $\{z(k)\}$ such that $z(k) \in Z(k)$, for

---

[1] The reader should be cautioned here that this is not the most general definition of convergence, although it will serve our purpose. In particular, it may not be adequate when X is an infinite and non-numerical domain. In general, a topology on X should be given beforehand and convergence is defined in terms of the topology (Munkres 1975). This general approach is beyond our scope, but can be chosen in case of need. The results of this chapter would still be valid under the general convergence definition.

all k, converge to z^*. An element z^* of X is said to be a *limit point* of a sequence $\{z(k)\}$ if there exists a subsequence of $\{z(k)\}$ that converges to z^*.

A *fixed point* of F is defined to be an element of X satisfying $x^* = F(x^*)$. The set of fixed points of F will be denoted as X^*. Also, a function $F : X \mapsto X$ is said to be *continuous* on a set X if for all converging sequences $\{z(k)\}$ of elements of X

$$\lim_{k \to \infty} F(z(k)) = F(\lim_{k \to \infty} z(k)).$$

From this definition, it is clear that when X is finite, every function F under which X is closed is continuous.

Some of our results will be given in terms of an ordering relation. A relation $\preceq$ in a set X is called a *partial order on X* iff, for every $a, b, c \in X$

- (*Reflexivity*) $a \preceq a$,

- (*Antisymmetry*) $a \preceq b, b \preceq a \Rightarrow a = b$,

- (*Transitivity*) $a \preceq b, b \preceq c \Rightarrow a \preceq c$. □

The ordered pair $(X, \preceq)$ is called a *partially ordered set* or a *poset*. An element $a \in X$ is a *minimal element of X* if no element $b \in X$ exists such that $a \neq b$ and $b \preceq a$. Similarly, we can also define a *maximal element of X*. We write $a \prec b$ if $a \preceq b$, but $a \neq b$, for all $a, b \in X$.

10.3 Asynchronous Iterations

10.3.1 Formal Definition

The following definition formulates the general model of asynchronous iterative algorithms. It was first introduced by Chazan and Miranker (1969), and was then called *chaotic iterations*. We define it in terms of a *schedule* ($\mathcal{S}$), along with an iteration operator and initial data. Informally, a schedule is a timing of accesses to global data. For example, in a shared memory multiprocessor these accesses include both the fetching and the storing of iterate components. In message passing systems the schedule describes the timing of component updates in local memories and communication of these updates to remote processors.

Definition 10.3.1 (Asynchronous Iteration)
An asynchronous iteration with respect to the schedule $\mathcal{S} = (\{\alpha(t)\}, \{\tau(t)\})$ corresponding to F and starting with $x(0) \in X$ is a sequence $\{x(t)\}$ such that

$$x_i(t) = \begin{cases} x_i(t-1) & \text{if } i \notin \alpha(t) \\ F_i(x_1(\tau_1^i(t)), x_2(\tau_2^i(t)), \ldots, x_n(\tau_n^i(t))) & \text{if } i \in \alpha(t) \end{cases}$$

for all $t = 1, 2, \ldots$ and $i = 1, 2, \ldots, n$. It is denoted by $(F, x(0), \mathcal{S})$, where $\{\alpha(t)\}$ is a sequence of subsets of $\{1, 2, \ldots, n\}$ and $\{\tau(t)\}$ is a sequence of elements of $\mathcal{N}^{n \times n}$ satisfying $0 \leq \tau_j^i(t) < t$ for all $t = 1, 2, \ldots$ and $i, j = 1, 2, \ldots, n$. □

$\alpha(t)$ is the *update set at t* and we say that the i-th component is *updated at t* if $i \in \alpha(t)$. We assume that all of the components are updated at 0, i.e., $\alpha(0) = \{1, 2, \ldots, n\}$. The vector $u^i(t)$ with components $u_j^i(t) = x_j(\tau_j^i(t))$ is called the *input of the i-th component at t* and $F_i(u^i(t))$ is said to be *generated at t for the i-th component*. The definition implies that the computations start at $t = 0$.

In this formulation, synchronous iterations in fact belong to the class of asynchronous iteration. As an example, the schedule given below corresponds to the synchronous point-Jacobi iteration (Baudet 1978), where the inputs for the computations in each iteration are generated in the previous iteration.

$$\alpha(t) = \{1, 2, \ldots, n\}$$

$$\tau_j^i(t) = t - 1,$$

for all $t = 1, 2, \ldots$ and $i, j = 1, 2, \ldots, n$. Similarly, the schedule corresponding to the Gauss-Seidel iteration, where the components are updated sequentially and where each update uses the most recent iterate values, is shown in the following:

$$\alpha(t) = \{1 + [(t - 1) \bmod n]\},$$

$$\tau_j^i(t) = t - 1,$$

for all $t = 1, 2, \ldots$ and $i, j = 1, 2, \ldots, n$.

10.3.2 Interpretations of t

It should be clear that for a given execution of an asynchronous algorithm, the schedule S is not uniquely defined. A virtually unlimited number of physical interpretations of t as well as x are possible.

A simple interpretation of t would be the index of each update instance of the shared memory. Ordering of t conforms with the real time ordering, i.e., if the updates with index t and $t + 1$ occur at real times $r(t)$ and $r(t + 1)$, then $r(t) < r(t + 1)$ and there are no other updates between $r(t)$ and $r(t + 1)$, for all t. With this interpretation, $x(t)$ is the value of shared data right after the t-th update. If the i-th component is updated at this instance, then this update is the result of the computation of F_i using the value of the j-th component right after the update with index $\tau_j^i(t)$, i.e., it is fetched sometime no earlier than $r(\tau_j^i(t))$ and before $r(\tau_j^i(t) + 1)$.

Consider the parallel synchronised execution of the Jacobi iteration for three components and three processors. Suppose that the computation time of the first component is 10 real time units. Similarly, the computation times of the second and the third components are 20 and 30 real time units, respectively. With the interpretation described as above, the update instances of the first, the second and the third components in the first iteration are $t = 1, 2$, and 3, respectively. Then, at $t = 4$ the first component is updated, etc.

Although this interpretation of t is perfectly legitimate, it does not show the parallelism inherent in the computations. It describes the sequence of updates as if the same computations were implemented on a uniprocessor system. We wish to define the times of updates in a way that exhibits the parallelism in the computation. For example, in the above Jacobi iteration example, the exact order of updates in real time is not relevant to the convergence of the algorithm. We could also consider that, in each iteration, all the components are updated all at once.

In general, let us consider a sequence $\{\psi(t)\}$ of real time instances and the corresponding sequence $\{I(t)\} \equiv \{(\psi(t), \psi(t+1)]\}^2$ of the time intervals such that

- $I(0) \equiv (\psi_0, 0]$, where $\psi_0 < 0$ is an arbitrary time instance before the computation starts,[3]

- for all t, $I(t)$ covers at least one update instance of some component and at most one update instance of each component,

- for all $t = 1, 2, \ldots$ all the inputs of the updates in $I(t)$ are generated no later than in $I(t-1)$.

Given such $\{I(t)\}$, $x(t)$ can be interpreted as the value of the shared data right at the end of $I(t)$, i.e., at $\psi(t+1)$. If the i-th component is updated in $I(t)$, $\tau_j^i(t)$ can be interpreted as the index of the interval at which the j-th component of the input vector of this update is generated. Since there is an infinite number of sequences of the above form, for the same computation, an infinite number of interpretations of $\{x(t)\}$ is possible. If $I(t)$s contain single updates, then this interpretation is the same as the one described previously. In order to show maximum parallelism, we will define $I(t)$ such that it contains the maximum number of updates that satisfy the above two conditions. With this definition, $I(t)$ is unique for a given computation.

10.3.3 Redundancy

In the schedules above, as well as in most typical cases, the computation of a component x_i uses its most recent value as an input component. In other words, $\tau_i^i(t) = t - 1$, for all $i = 1, 2, \ldots, n$ and $t = 1, 2, \ldots$ In general, this condition holds when there are no redundancies in the computations. Redundancy means that the same component may be computed by multiple processors at the same time. Nonredundancy can be identified by the following cases.

1. Static allocation is adopted and the component sets assigned to processors are not overlapping, i.e., each component is assigned to a unique processor.

---

[2]$(a, b]$ denotes the set of real numbers $\{x | a < x \leq b\}$.
[3]It is assumed that the computations start at real time zero.

2. Dynamic allocation is adopted and the task scheduling scheme is centralised. For example, the system may keep a unique copy of the program code for each F_i so that x_i cannot be computed by more than one processor. Or, the global queue controlling the allocation of the tasks may be unique and does not allow the scheduling of a task which is presently executing.

Furthermore, these are the only cases that guarantee $\tau_i^i(t) = t - 1$, for all $i = 1, 2, \ldots, n$ and $t = 1, 2, \ldots$, because of the simple reason that in the computational systems we are assuming in this chapter, no assumption can be made about execution times of tasks. In the following, an example of a redundant computation is given.

Example 10.3.1 Assume two components in the shared data computed by two processors in a shared memory system. Each processor computes the components in order using whichever value it fetches from the shared store. Figure 10.1 shows such a schedule. For example, the arrows pointing to $t = 4$ can be interpreted as follows. The processor assigned to component 1 fetches the first component of the shared data at 1 and the second one at 2. Using these values it computes the second component and updates the shared data at 4. The other arrows can be interpreted similarly. The computation of the second component in this example is redundant in the following way. The update of the second component at 3 uses the initial value of that component without being aware of the fact that the most recent value of the second component is generated at 2. □

10.4 Totally Asynchronous Iterations

10.4.1 Definition

Obviously, Definition 10.3.1 would not be of any value if we did not have any tools to predict the convergence of an algorithm described by it. It should be clear that without any restriction, it is not possible to guarantee convergence, for all possible schedules. At the very least, we should require that all the components should be updated sufficiently many times, i.e., no component drops out of the computations forever (*progress of updates*). Furthermore, computations should not use an outdated component value as their input forever (*progress of inputs*), although they are allowed to do so for a while. These are very weak conditions which impose practically no restrictions on the system. Fortunately however, for many problems, they are the only requirements for convergence to the desired solutions. They were first introduced by Baudet (1978), and the computational models that satisfy these conditions were called *totally asynchronous* by Bertsekas and Tsitsiklis (1989a). It is formally defined as follows.

t	1	2	3	4	$\cdots$
$\alpha(t)$	$\{1\}$	$\{2\}$	$\{2\}$	$\{1\}$	$\cdots$
$\tau_1^1(t)$	0	–	–	1	$\cdots$
$\tau_2^1(t)$	0	–	–	2	$\cdots$
$\tau_1^2(t)$	–	0	1	–	$\cdots$
$\tau_2^2(t)$	–	0	0	–	$\cdots$

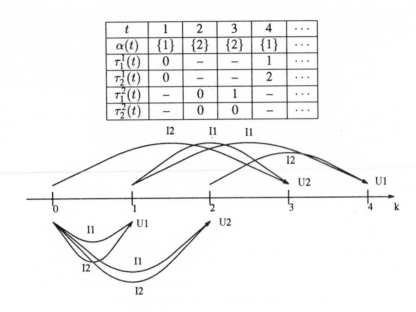

Ui: Update of the i-th component
Ii: The input component i

Figure 10.1: A redundant computation

Model 10.4.1 (Total Asynchronism)   • *(Progress of Updates) Each $i =$
$1, 2, \ldots, n$ occurs in an infinite number of $\alpha(t)$s, i.e., each component is
updated infinitely often.*

• *(Progress of Inputs) Given any $t_1 = 0, 1, \ldots$, there exists $t_2 > t_1$ such
that $\tau_j^i(t) \geq t_1$, for all i, j and $t \geq t_2$, i.e., after a component value is
generated, this value can be used as inputs of only a finite number of
updates.*   □

It is useful to state at this point the following property for convergent totally
asynchronous iterations.

Lemma 10.4.1 *Let F be a continuous function on $X = X_1 \times X_2 \times \cdots \times
X_n$. Then every point of convergence of all totally asynchronous iterations
$(F, x(0), S)$ with $x(0) \in X$, is also a fixed point of F in X.*

Proof. Consider the sequence $\{u^i(k)\}$ of input vectors for the i-th component.
From progress of inputs, $\{u^i(t)\}$ converges to the same point x^* as $\{x(t)\}$.
From the definition of continuity

$$x_i^* = \lim_{t \to \infty} u_i^i(t) = \lim_{t \to \infty} x_i(t) = \lim_{t \to \infty} F_i(u^i(t)) = F_i(x^*).$$

Since this holds for all $i = 1, 2, \ldots, n$, the point of convergence is a fixed point. □

In Section 10.4.3, we will prove a general condition on the iteration operator F, under which all totally asynchronous iterations converge to the fixed point of F. It will further be proven that this condition is also necessary for convergence if the domain X of the shared data contains finite number of elements. Before that, we will discuss how to interpret totally asynchronous iterations implemented in some shared memory architectures.

10.4.2 Interpretations of Total Asynchronism

In Section 10.3.2 we have already shown that the interpretation of the *time* of $x(t)$ is not unique. Another degree of freedom we may exploit is in the interpretation of the *place* of $x(t)$. In this section, we explain this in three example shared memory systems. The last two examples show that totally asynchronous iterations allow multiple inconsistent copies of x in the system.

1. Consider a shared memory multiprocessor, where the computations are controlled by a static allocation scheme. Each component is assigned to be computed by fixed processors throughout the execution of an asynchronous iteration. The requirement of total asynchronism on this system is trivial: no processor is allowed to drop out of the execution forever (progress of updates), and the computation time of each component is finite (progress of inputs). Because of the static allocation, this property satisfies both conditions of Model 10.4.1. Convergence is not affected when some of the processors occasionally drop out of the computations. However, common sense suggests that this increases the execution time.

2. Definition 10.3.1 by no means implies that x should be associated with a particular physical device, such as shared memory. Suppose that in the multiprocessor system, in addition to the shared memory, each processor has a local storage which can only be accessed by itself. The processors may generally be allocated dynamically to components. Assume, with no loss of generality, that each local storage is large enough to hold the whole x and that the initial value of x is loaded into each local storage before the computation starts. The processors work (read and write) with their local copies of x, but once in a while, at some unspecified times, some of the local updates (not necessarily all) propagate to the shared memory, and also once in a while, the processors load the most recent value of the shared memory to the local storages. An immediate suggestion for $x(t)$ may be to interpret it as the value of the shared data (in the global memory) right at t. However, this conflicts with Definition 10.3.1. Since the processors are working on the local copies, the input vectors may never be seen in the global memory. Even if they are data transfer to memory may take so long that the input vectors appear in the memory before they are used in the local updates which violates $\tau_j^i(t) < t$. Now

suppose that $x_i(t)$ is interpreted such that it is the value of the most recent local update of x_i no later than t (in any local storage). Then, if x_i is updated at t, $\tau_j^i(t)$ can be interpreted as the index of the interval in which the j-th component of the input data for this update is generated. Notice that it may have been generated in the local storage, or it may have been generated in another processor and may have been propagated to the local storage through the global memory. The implications of total asynchronism for this system are as follows:

- Each component should once in a while be updated.

- Local updates should once in a while be propagated to the global memory in a finite period of time.

- Local copies should once in a while be replaced with recent component values of x in the global memory in a finite period of time.

3. We now consider a simpler architecture on which asynchronous iterations execute with the same efficiency. It can easily be noticed that, for our point of view, the sole function of the shared memory in the above system is to provide communication between the processors. Therefore, this part of the system can be replaced with an interconnection network which assumes the same responsibility. In other words, the last two conditions of total asynchronism for the above system can be combined so that we have the following conditions for total asynchronism:

- Each component should be updated occasionally.

- Newer updates of each component of x should occasionally be communicated to each processor.

This leads to the architecture where processors maintain their own copies of the iterated data x in their local memories which interact through an interconnection network. Processors read their local copies. A write operation updates its local copy, and also notifies the interconnection network about this transaction. The interconnection network is responsible for broadcasting this update. The originating processor does not have to wait for the completion of this broadcast. As long as the updates are eventually propagated, the convergence is maintained. It should be noted that the ordering of the broadcasts does not affect the convergence. In other words, the interconnection network does not have to transmit writes in the same order as it receives them.

10.4.3   Main Convergence Result

We first give a different but equivalent formulation of Model 10.4.1 which is stated by the following lemma. This will be convenient in our later proofs.

Lemma 10.4.2 *A schedule* $S = (\{\alpha(t)\}, \{\tau(t)\})$ *is totally asynchronous if and only if there exists a sequence of integers* $\{\varphi(k)\}$ *satisfying the following conditions:*

- $\varphi(0) = t_0$ *and* $\varphi(1) = 0$, *where* $t_0 < 0$ *is some arbitrary integer.*

- *Each component* $i = 1, 2, \ldots, n$ *is updated at some* t *such that*

$$\varphi(k) < t \leq \varphi(k+1), \quad \forall k = 0, 1, 2, \ldots$$

- $t > \varphi(k) \Rightarrow \tau_j^i(t) \geq \varphi_j(k-1) > \varphi(k-1), \ \forall k = 1, 2, \ldots, \forall i = 1, 2, \ldots, n$ *where* $\varphi_j(k-1)$ *is defined as the first instance later than* $\varphi(k-1)$ *at which the j-th component is updated.*

Proof. The proof is by mathematical induction. The proof of basis clause (the case for $k = 1$) is trivial. Suppose that there exists $\varphi(0), \varphi(1), \ldots, \varphi(l-1)$ satisfying the conditions. From progress of updates, there exists a t' such that all the components are updated in the interval $(\varphi(l-1), t']$. On the other hand, from progress of inputs, given $\varphi_s(l-1)$ there exists a t'_s such that $\tau_s^i(t) \geq \varphi_s(l-1)$, for all $i, j = 1, 2, \ldots, n$ and $t \geq t'_s$. If $\varphi(l)$ is chosen to be the maximum of all t' and t'_ss, it satisfies the above conditions.

It is also easy to see that the schedule is totally asynchronous, given the conditions of the lemma. □

Following the terminology of Robert (1987), we call the set $\Phi(k) = \{j | \varphi(k) < j \leq \varphi(k+1)\}$ the k-th *pseudocycle* of S, and the sequence $\{\Phi(k)\}$ is said to be the *pseudocycle sequence* associated with S. There are an infinite number of $\{\varphi(k)\}$s and $\{\Phi(k)\}$s satisfying the above conditions, given a schedule. In the rest of the chapter $\{\varphi(k)\}$ and $\{\Phi(k)\}$ (when there is no ambiguity) will represent an arbitrary one of these, although we will not explicitly mention it every time.

Because of the above lemma, a totally asynchronous schedule is also called a *pseudoperiodic schedule*. As we shall show in Theorem 10.4.1, at the end of a pseudocycle $\Phi(k)$ (from $\varphi(k)$ to $\varphi(k+1)$), it is guaranteed that there is an "improvement" made towards the solution. Therefore, associating $\Phi(k)$s to actual periods of time suggests an obvious way of estimating the total computation time.

Now we define a class of operators that will later be shown to converge to the unique solution for all totally asynchronous schedules. In the next section, it is also proven that this is the largest such class for finite data domains.

Definition 10.4.1 An operator F is an *asynchronously contracting operator* (ACO) on a subset $X(0)$ of X iff there exists a sequence of sets $\{X(k)\}$ that satisfies the following conditions:

- (*Box Condition*) For all $k = 0, 1, 2, \ldots$, $X(k)$ is a Cartesian product of n sets; i.e.,

$$X(k) = X_1(k) \times X_2(k) \times \cdots \times X_n(k)$$

- For all $k = 0, 1, 2, \ldots$, $X(k+1) \subseteq X(k)$, and furthermore, $\{X(k)\}$ converges to a fixed point of F in $X(0)$;

- $x \in X(k) \Rightarrow F(x) \in X(k+1)$, for all $k = 0, 1, 2, \ldots$ $\qquad\square$

As will be shown in Theorem 10.4.1, $X(k)$ is the set in which the iterates stay after the k-th pseudocycle and therefore the above conditions constitute sufficient conditions for convergence of totally asynchronous iterations. The following lemma further states that the point of convergence, therefore the fixed point, is unique.

Lemma 10.4.3 *F has a unique fixed point in a set $X(0)$ if it is asynchronously contracting on $X(0)$.*

Proof. Existence of a fixed point is guaranteed by definition of ACO (second condition). Uniqueness can be shown by contradiction. Suppose that there exists two distinct fixed points $x^*, y^* \in X(0)$. Then, from the third condition, $x^*, y^* \in X(k)$, for all k. This violates the second condition. $\qquad\square$

The last two conditions in the above definition are intuitive. In fact, any definition of contraction imposes conditions of a similar nature. On the other hand, the box condition might seem counter-intuitive; yet, it is the requirement of an ACO that is due to the asynchronous nature of the computations which shows as *asynchronous distortions* in asynchronous algorithms. When asynchronous distortion occurs, an iterate value in $X(k)$ is generated, which is not possible to reach by successive applications of the iteration operator starting from the initial data. It will be shown later that there exists a schedule such that each point of $X(k)$ is visited in the k-th pseudo-cycle, when $X(0)$ contains a finite number of elements. In other words, all points in $X(k)$ are reachable by asynchronous distortion. Since we want convergence for *all* schedules, we need to guarantee the invariance of $X(k)$ under asynchronous distortion. Therefore, the box condition is *necessary* for convergence for finite $X(0)$ when no restriction is enforced on the schedule. Whether it is necessary for domains that contain an infinite number of elements, however, is an open question. The reason for the uncertainty is that the proof of the necessary condition for the finite case is based on the construction of an asynchronous iteration such that all of the points in $X(k)$ are visited in the k-th pseudo-cycle of this asynchronous iteration. If $X(k)$ contained an infinite number of elements, then the constructed schedule would not be pseudoperiodic.

Example 10.4.1 Figure 10.2 displays an iteration operator in which the arc from each element in the domain points to the value we obtain when we apply the operator F to that element. Consider the iteration that starts with $x(0) = (1, 2)$ and that converges synchronously as well as for many asynchronous schedules to $y = (2, 1)$, as long as the iterates do not leave the path from $x(0)$ to y. However, some schedules may cause distortion; i.e., the iterates may be "deflected" to the cycle in the lower left part of the domain. For instance, let there be two processors: P1 and P2. P1 first updates x_1 and then x_2. Right

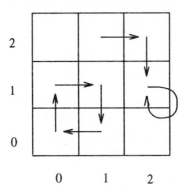

Figure 10.2: Asynchronous distortion

after this, P2 updates x_2, using the initial value of x_1 and the most recent value of x_2. At this point, the iterate value is $z = (1, 0)$, and starting with z, the algorithm cycles even if the remaining part of the schedule is synchronous. $\square$

In the example above, we observe that the cause of asynchronous distortion is the redundant computations of x_2. If we disallow this so that each update of a component uses the most recent value of that component, then the iterates always converge to y. Therefore, the box condition is a necessary condition only when the schedule is not restricted.

We now formally state and prove that if F is asynchronously contracting, then the corresponding asynchronous iteration converges to the fixed point of F, for all totally asynchronous schedules.

Theorem 10.4.1 (Sufficient Condition) *If F is an ACO on a set $X(0)$, then any asynchronous iteration $\{x(t)\} = (F, x(0), \mathcal{S})$ with $x(0) \in X(0)$ converges to the unique fixed point x^* of F in $X(0)$, for all totally asynchronous schedules $\mathcal{S}$.*

Proof. We will simply show by induction that starting from $\varphi_i(k)$, the i-th component of the iterates indefinitely enters $X_i(k)$ for all $i = 1, 2, \ldots, n$; i.e.,

$$t \geq \varphi_i(k) \;\Rightarrow\; x_i(t) \in X_i(k), \quad \forall i = 1, 2, \ldots, n, \; k = 0, 1, 2, \ldots \quad (10.2)$$

Suppose this is true for $k = l$ (induction hypothesis) and consider the update at $\varphi_{i'}(l+1)$. Since the i-th component of the input vector for this update is generated no earlier than $\varphi_i(l)$, the input vector is in $X(l)$, by the hypothesis. In other words, for $t = \varphi_{i'}(l+1)$

$$u^{i'}(t) \;=\; (x_1(\tau_1^{i'}(t)), \; x_2(\tau_2^{i'}(t)), \ldots) \;\in\; X(l).$$

Note that this would not always be true if $X(l)$ were not a Cartesian product. As a consequence, we have

$$x_{i'}(j) \;=\; F_{i'}(u^{i'}(t)) \;\in\; X_{i'}(l+1)$$

for $t = \varphi_{i'}(l+1)$, and by induction it can be generalised for $t \geq \varphi_{i'}(l+1)$. Since this is valid for all $i' = 1, 2, \ldots, n$, (10.2) holds for $k = l + 1$, and by induction for all $k = 0, 1, 2, \ldots$ Here, we are omitting the proof of the basis (that all the iterates are in $X(0)$), but the argument is the same.

(10.2) implies that after all the components are updated in the k-th pseudo-cycle the iterates indefinitely fall into $X(k)$, and therefore the iteration converges to the same point as the sequence $\{X(k)\}$. $\qquad\square$

Asynchronous contraction is an extremely simple concept which generalises and simplifies the convergence conditions given by other researchers. In the first paper on asynchronous algorithms (Chazan and Miranker 1969), linear iterations of the form $F(x) = Ax + b$, where A is an $n \times n$ matrix were considered. It was proven that for convergence of asynchronous iterations, it is necessary and sufficient that the spectral radius of $|A|$, $(\rho(|A|))$ is strictly less than 1. In Baudet, this result was generalised to cover non-linear iteration operators.

The convergence condition which is closest to asynchronous contraction was given by Bertsekas (1983). In fact, it can be shown to be equivalent to asynchronous contraction. In the formulation of Bertsekas (1983), the box condition and the inclusion property were not obvious. These were made clear later by Üresin and Dubois (1989a, 1989b) and Bertsekas and Tsitsiklis (1989a).

10.4.4 Necessary Condition

The following definitions and the subsequent lemmas establish the background required for the proof of Theorem 10.4.2, which states that asynchronous contraction is not only sufficient but also necessary for convergence under total asynchronism, for finite data domains.

Definition 10.4.2 Let $\{x(t)\} = (F, x(0), \mathcal{S})$ be an asynchronous iteration and $\{\Phi(k)\}$ be a pseudocycle sequence associated with it. The *outcome* of $\{x(t)\}$ in the k-th pseudocycle, $P(k) = P_1(k) \times P_2(k) \times \cdots \times P_n(k)$, is the set such that

$$P_i(k) = \{x_i(t) | i \in \alpha(t) \text{ and } t \in \Phi(k)\}, \ \forall i = 1, 2, \ldots, n, \ k = 0, 1, 2, \ldots\square$$

In the following discussions, without explicitly mentioning it we assume that $P(k)$, $P'(k)$, $P''(k)$, $\ldots$ correspond to $\{x(t)\}$, $\{x'(t)\}$, $\{x''(t)\}$, $\ldots$ respectively.

Lemma 10.4.4 is based on the concept of *merging* asynchronous iterations. Merging $\{x'(t)\}$ and $\{x''(t)\}$, each starting with the same initial vector $x(0)$, means the construction of another asynchronous iteration $\{x(t)\}$ in the following way. In the k-th pseudocycle of $\{x(t)\}$, first the outputs of the k-th pseudocycle of $\{x'(t)\}$ are obtained in the same order, then this procedure is repeated for the k-th pseudocycle of $\{x''(t)\}$, as shown in Example 10.4.2.

Example 10.4.2

$$
\begin{aligned}
\{x'(t)\} &= \underbrace{(1,1)}_{\Phi'(0)} \underbrace{(2,1)\,(2,2)}_{\Phi'(1)} \underbrace{(3,2)\,(3,3)}_{\Phi'(2)} \cdots \\[2pt]
\{x''(t)\} &= \underbrace{(1,1)}_{\Phi''(0)} \underbrace{(1,5)\,(5,5)}_{\Phi''(1)} \underbrace{(6,5)\,(6,6)}_{\Phi''(2)} \cdots \\[2pt]
\{x(t)\} &= \underbrace{(1,1)}_{\Phi(0)} \underbrace{(2,1)\,(2,2)\,(2,5)\,(5,5)}_{\Phi(1)} \underbrace{(3,5)\,(3,3)\,(6,3)\,(6,6)}_{\Phi(2)} \cdots
\end{aligned}
$$

Each of these sequences starts with the same initial vector $x(0) = (1,1)$, and for $k > 0$, $\Phi(k)$ has two phases. For example, in $\Phi(1)$ first x_1 then x_2 are updated, each time yielding the value 2, because this phase corresponds to $\Phi'(1)$. In the second phase of $\{\Phi(1)\}$, x_2 and x_1 are updated respectively, each time yielding 5, as in $\Phi''(1)$. □

Lemma 10.4.4 *Let* $\{x'(t)\} = (F, x(0), S')$ *and* $\{x''(t)\} = (F, x(0), S'')$ *be totally asynchronous iterations w.r.t.* S' *and* S'', *respectively. There exists an asynchronous iteration* $\{x(t)\} = (F, x(0), S)$ *such that* $P(k) = P'(k) \cup P''(k)$ *for all* $k = 0, 1, 2, \ldots$

Proof. The desired $\{x(t)\}$ is constructed by merging $\{x'(t)\}$ and $\{x''(t)\}$. The formal definition of the schedule for $\{x(t)\}$ is given by the series of formulas below.

$$
\varphi(k) = \begin{cases} -1 & \text{if } k = 0 \\ 0 & \text{if } k = 1 \\ \varphi'(k) + \varphi''(k) & \text{otherwise.} \end{cases}
$$

Literally, the length of the k-th pseudocycle of $\{x(t)\}$ is the sum of the lengths of the k-th pseudocycles of $\{x'(t)\}$ and $\{x''(t)\}$, for $k > 0$.

$$
\delta'(t) = \begin{cases} 0 & \text{if } t = 0 \\ t + \varphi''(k) & \text{if } t > 0 \text{ and } t \in \Phi'(k) \end{cases}
$$

$$
\delta''(t) = \begin{cases} 0 & \text{if } t = 0 \\ t + \varphi'(k+1) & \text{if } t > 0 \text{ and } t \in \Phi''(k). \end{cases}
$$

δ' (δ'') simply maps the indices of the sequence $\{x'(t)\}$ ($\{x''(t)\}$) to the corresponding indices of $\{x(t)\}$. The inverse mapping is defined by

$$
\delta^{-1}(t) = \begin{cases} 0 & \text{if } t = 0 \\ t - \varphi''(k) & \text{if } t > 0 \text{ and } 0 < t - \varphi(k) \le |\Phi'(k)| \\ t - \varphi'(k+1) & \text{if } |\Phi'(k)| < t - \varphi(k) \le |\Phi'(k)| + |\Phi''(k)|. \end{cases}
$$

δ^{-1} maps the indices of $\{x(t)\}$ to the corresponding indices of $\{x'(t)\}$ or $\{x''(t)\}$. In this definition, the second (third) line corresponds to the elements of the first (second) phase of $\Phi(k)$. If t is in the first (second) phase of a pseudocycle, the value of $\alpha(t)$ is the same as the value of α' (α'') for the corresponding element of $\{x'(t)\}$ ($\{x''(t)\}$); i.e.

$$\alpha(t) = \begin{cases} \alpha'(\delta^{-1}(t)) & 0 < t - \varphi(k) \leq |\Phi'(k)| \\ \alpha''(\delta^{-1}(t)) & \text{if } |\Phi'(k)| < t - \varphi(k) \leq |\Phi'(k)| + |\Phi''(k)|. \end{cases}$$

τ can be defined in a similar manner, as follows:

$$\tau_j^i(t) = \begin{cases} \delta'(\tau'^i_j(\delta^{-1}(t))) & 0 < t - \varphi(k) \leq |\Phi'(k)| \\ \delta''(\tau''^i_j(\delta^{-1}(t))) & \text{if } |\Phi'(k)| < t - \varphi(k) \leq |\Phi'(k)| + |\Phi''(k)|. \end{cases}$$

It is obvious that the resulting sequence indeed satisfies the assumptions of asynchronous iteration and the requirement of the lemma. $\square$

Lemma 10.4.5 *Let* $\{x^{(1)}(t)\} = (F, x(0), \mathcal{S}^{(1)})$, $\{x^{(2)}(t)\} = (F, x(0), \mathcal{S}^{(2)})$, $\cdots$ *and* $\{x^{(m)}(t)\} = (F, x(0), \mathcal{S}^{(m)})$ *be asynchronous iterations corresponding to* F *and starting with* $x(0)$. *Then, there exists* $\{x(t)\} = (F, x(0), \mathcal{S})$, *such that*

$$P(k) = P^{(1)}(k) \cup P^{(2)}(k) \cup \cdots \cup P^{(m)}(k), \quad \forall k = 0, 1, 2, \ldots$$

Proof. A straightforward generalisation of the previous lemma. $\square$

Lemma 10.4.5 states that the pseudocycles of different asynchronous iterations on the same F and $x(0)$ can be merged to form a new asynchronous iteration.

Lemma 10.4.6 *Let* $\{x'(t)\} = (F, x(0), \mathcal{S}')$ *be an asynchronous iteration. If* $a_{i_1} \in P'_{i_1}(k)$ *and* $a_{i_2} \in P'_{i_2}(k)$, *then there exists an asynchronous iteration* $\{x(t)\} = (F, x(0), \mathcal{S})$ *such that for some* $p \in \Phi(k)$, $x_{i_1}(p) = a_{i_1}$, *and* $x_{i_2}(p) = a_{i_2}$ *for all* $k > 0$, *and* $i_1 \neq i_2$.

Proof. Let a_{i_1} and a_{i_2} be the results of the updates at t_1 and $t_2 \in \Phi'(k)$, respectively; i.e.,

$$i_1 \in \alpha'(t_1), \quad x'_{i_1}(t_1) = a_{i_1},$$

and

$$i_2 \in \alpha'(t_2), \quad x'_{i_2}(t_2) = a_{i_2}.$$

Now, construct the first k pseudocycles of $\{x(t)\}$ by inserting an iterate at the end of the k-th pseudocycle of $\{x'(t)\}$ such that a_{i_1} and a_{i_2} are the outputs of this update. More formally, define

- $\alpha(t) = \alpha'(t)$, $\tau(k) = \tau'(k)$; therefore, $x(t) = x'(t)$, for $t \leq \varphi'(k + 1)$;

- $\alpha(p) = \{i_1, i_2\}$, $\tau^{i_1}(p) = \tau'^{i_1}(j_1)$, $\tau^{i_2}(p) = \tau'^{i_2}(j_2)$;

- $\varphi(l) = \varphi'(l)$ for $l \leq k$;

- $\varphi(k+1) = \varphi'(k+1) + 1$,

where $p = \varphi'(k+1) + 1$. Also, for $t > p+1$, $\alpha(t)$ and $\tau(t)$ are chosen freely to satisfy the conditions of total asynchronism. Furthermore, it can easily be seen that the first k pseudocycles of $\{x(t)\}$ satisfy these conditions. □

The generalisation of the above proof gives the following lemma.

Lemma 10.4.7 *Let* $\{x'(t)\} = (F, x(0), S')$ *be a totally asynchronous iteration. There exists* $\{x(t)\} = (F, x(0), S)$ *such that for any n-dimensional vector* $a \in P'(k)$, *there exists a t in the k-th cycle of* $\{x(t)\}$ *such that* $x(t) = a$. □

This lemma simply states that given a vector a, each component of which is the output of an update in the k-th pseudocycle of a particular asynchronous iteration, we can construct another asynchronous iteration such that the value of the global data is a, at some time instance in the k-th pseudocycle.

Now, we can prove Theorem 10.4.2, which says that the sufficient condition of convergence given in Theorem 10.4.1 (i.e., F being asynchronously contracting) is also necessary for convergence for finite data domains.

Theorem 10.4.2 (Necessary Condition) *Let* F *be an operator defined in a domain* X *that contains a finite number of elements. If all totally asynchronous iterations* $\{x(t)\} = (F, x(0), S)$, *starting with* $x(0) \in X$ *and corresponding to* F, *converge to the same fixed point* x^*, *then* F *is asynchronously contracting on a set* $X(0) \subseteq X$, *where* $x(0) \in X(0)$.

Proof. Define $X(k) = X_1(k) \times X_2(k) \times \cdots \times X_n(k)$ such that

$$X_i(k) = \{a_i | a_i \in P_i(k) \text{ for some } S\}, \quad \forall i = 1, 2, \ldots, n, \ \forall k = 0, 1, 2, \ldots$$

We shall show by mathematical induction that there exists an asynchronous iteration $\{x'(t)\} = (F, x(0), S')$ such that

$$P'(k) = X(k), \quad \forall k = 0, 1, 2, \ldots$$

We will show only the induction part, since the argument for $k = 0$ is very similar. Assume that the above statement is true for $k < l$. Since the domain of F is finite, the iterates of all asynchronous iterations can take a finite number of values; therefore, $X(l)$ contains a finite number of elements. Consequently, there are $m - 1$ schedules that satisfy

$$X(l) = P^{(1)}(L) \cup P^{(2)}(l) \cup \cdots \cup P^{(m-1)}(l),$$

and from the hypothesis, there exists $S^{(m)}$, such that,

$$P^{(m)}(k) = X(k), \quad k < l.$$

For all $j = 1, 2, \ldots, m$ and for all $k = 0, 1, 2, \ldots$, $P^{(j)}(k) \subseteq X(k)$.

From Lemma 10.4.5 there exists an asynchronous iteration, $\{x'(t)\} = (F, x(0), S')$ such that

$$P'(k) = X(k), \quad \forall k \leq l,$$

and by induction, $P'(k) = X(k)$ for all $k = 0, 1, 2, \ldots$ Then, from Lemma 10.4.7 there exists an asynchronous iteration $\{x(t)\} = (F, x(0), S)$ such that

$$X(k) \subseteq \{x(j)|j \in \Phi(k)\}, \quad \forall k = 0, 1, 2, \ldots \qquad (10.3)$$

We can now show that the sequence $\{X(k)\}$ satisfies the conditions of Definition 10.4.1 :

- Box condition is satisfied by definition.

- The definition of pseudocycle permits us to coalesce two consecutive pseudocycles into one. In other words, given a pseudocycle sequence, $\{\Phi(k)\}$ corresponding to $\{x(t)\}$, there exists $\{\overline{\Phi}(k)\}$ such that $\overline{\Phi}(k) = \Phi(k) \cup \Phi(k + 1)$, which is also a pseudocycle sequence. Therefore,

$$X(k + 1) \subseteq X(k) \cup X(k + 1) \subseteq \overline{P}(k) \subseteq X(k).$$

  From (10.3) $\{X(k)\}$ converges to x^* and the second condition of Definition 10.4.1 is satisfied.

- There exists $\{x'(t)\} = (F, x(0), S')$ such that $P'(k) = X(k)$ for $k = 0, 1, 2, \ldots$ Consequently, for all $x \in P'(k)$ and $i = 1, 2, \ldots, n$, there exists $\{x''(t)\} = (F, x(0), S'')$ such that

  - $P''(k) = P'(k)$, and
  - $x_i(t_i) = F_i(x)$ for some $t_i \in \Phi''(k + 1)$.

Therefore, $F_i(x) \in P_i''(k + 1)$ for all $i = 1, 2, \ldots, n$, and hence $F(x) \in X(k + 1)$, which completes the proof.   □

Finally, for the sake of completeness, we give the following theorem, which is the restatement of Theorem 10.4.1 and Theorem 10.4.2 combined.

Theorem 10.4.3 *Let F be an operator defined in a domain X that contains a finite number of elements. For all totally asynchronous schedules, any asynchronous iteration $(F, x(0), S)$ corresponding to F and starting with any initial guess $x(0)$ converges to a fixed point of F if and only if F is asynchronously contracting in a set $X(0)$ that contains $x(0)$.*   □

10.4.5   Other Sufficient Conditions

Although conceptually very simple, asynchronous contraction may be difficult to verify in practical problems. In this section, we will give simpler conditions on the iteration operator F each of which will be shown to be a special case of asynchronous contraction. This means that in each case, every totally asynchronous iteration corresponding to F and starting with the initial vector $x(0) \in X$ is guaranteed to converge to the unique fixed point x^*.

Monotonic Operators

Monotonic operators were first shown to converge under asynchronism, by Bertsekas (1982, 1983). In the following proposition, $\{y(k)\}$ refers to the sequence such that

- $y(0) = x(0)$,

- $y(k) = F(y(k-1))$, $\forall k \in \mathcal{N}^+$.

In other words, $\{y(k)\}$ is the synchronous iteration corresponding to F and starting with $x(0)$. Furthermore, $\preceq_i$ denotes a partial order on the elements of X_i, for all $i = 1, 2, \ldots, n$. We write $a \preceq b$ if $a_i \preceq_i b_i$, for all $a, b \in X$ and $i = 1, 2, \ldots, n$.

Condition 10.4.1

- *F is continuous in X.*

- $F(x(0)) \preceq x(0)$.

- $\{y(k)\}$ *converges to an element of X.*

- *F is monotone in X; i.e., for all $a, b \in X$,* $a \preceq b \Rightarrow F(a) \preceq F(b)$.

Proof. From the first condition, X is closed under F, and therefore the elements of $\{y(k)\}$ take values from the set X. The continuity also implies that $\{y(k)\}$ converges to a fixed point x^* of F.

From the second condition and from the monotonicity

$$y(0) \succeq y(1) \succeq \ldots y(k) \succeq y(k+1) \succeq \ldots \succeq x^*,$$

define $\{X(k)\}$ such that

$$X(k) = \{a | x^* \preceq a \preceq y(k) \text{ and } a \in X\}.$$

Obviously, $X(k)$ satisfies the first two conditions of Definition 10.4.1 (asynchronous contraction). On the other hand, from monotonicity (last condition), for all $a \in X(k)$, $x^* \preceq F(a)$, and

$$F(a) \preceq F(y(k)) = y(k+1).$$

Also, since $F(a) \in X$ (from closure property), $F(a) \in X(k+1)$. Hence, all the conditions of Definition 10.4.1 are satisfied. $\qquad\square$

When X is a finite set, Condition 10.4.1 can be simplified as follows.

Condition 10.4.2

- X *is a finite set.*

- X *is closed under F; i.e., for all $a \in X$,* $F(a) \in X$.

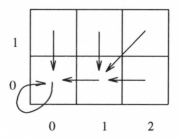

Figure 10.3: A monotone operator

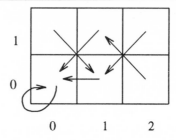

Figure 10.4: An asynchronously contracting operator which is not monotone

- F is nonexpansive in X; i.e., for all $a \in X$, $F(a) \preceq a$.

- F is monotone in X; i.e., for all $a, b \in X$, $a \preceq b \Rightarrow F(a) \preceq F(b)$.

Proof. For finite domains the closure property implies continuity. Therefore, we only need to show the third condition of Condition 10.4.1. Since F is nonexpansive and monotone,

$$y(0) \succeq y(1) \succeq \ldots y(k) \succeq y(k+1) \succeq \ldots$$

From the antisymmetry property of $\preceq$, no two elements of the sequence $\{y(k)\}$ are the same, except when they are equal to the fixed point. Consequently, $\{y(k)\}$ converges and Condition 10.4.1 applies. ☐

Example 10.4.3 The operator given by Figure 10.3 satisfies the conditions of the above propositions when $\preceq$ is selected as the regular ordering relation defined on numbers. The notation of the figure is the same as the one given in Example 10.4.1. ☐

The question may arise of whether the conditions of Proposition 10.4.1 and 10.4.2 are necessary for convergence. The following counter-example shows that they are not.

Example 10.4.4 F is an operator defined on the elements of $X = \{0, 1, 2\} \times \{0, 1\}$, as shown in Figure 10.4. F is asynchronously contracting on X, because if we take

$$X(1) = \{0, 1\} \times \{0, 1\}, X(2) = \{0, 1\} \times \{0\},$$
$$X(3) = X(4) = \cdots = \{(0, 0)\},$$

then for all k,

$$x \in X(k) \Rightarrow F(x) \in X(k + 1).$$

Suppose that F also satisfies the conditions of the above propositions. Since F is nonexpansive,

$$F(0, 1) \preceq (0, 1) \Rightarrow 1 \preceq_1 0,$$

and

$$F(1, 0) \preceq (1, 0) \Rightarrow 0 \preceq_1 1,$$

which is a contradiction, and therefore, F is *not* nonexpansive but still asynchronously contracting. □

We can redefine $\preceq$ such that Example 10.4.4 shows a monotone operator with respect to the new $\preceq$. We can state this in the following way.

Condition 10.4.3 *F is a contracting operator and $\{X(k)\}$ is the corresponding nested sequence of domains as defined in Definition 10.4.1. Furthermore, the following restriction is imposed on F.*

$$x \in R(k) \Rightarrow F(x) \in R(k + 1), \quad \forall k = 0, 1, 2, \ldots,$$

where $R(k)$ is defined as

$$R(k) = X(k) - X(k + 1), \quad \forall k = 0, 1, 2, \ldots$$

□

The above restriction simply states that the application of the operator to some data in $R(k)$ moves the data to $X(k + 1)$, but not any further, i.e., to $X(k+2)$. It is obvious that the restriction does not contradict with the definition of asynchronous contraction. Therefore, it constitutes a sufficient condition for convergence. Furthermore, it is implied by Condition 10.4.1. If we define $\preceq$ such that we say $x \preceq y$ if $k_x \leq k_y$, where k_x and k_y are the integers satisfying $x \in R(k_x)$ and $y \in R(k_y)$, for all $x, y \in X$. Then, Condition 10.4.1 holds, but $\preceq$ is not a partial order any longer since it violates the second condition (antisymmetry) of the definition of partial order.

Componentwise Contraction

In the following case, each application of the iteration operator makes each component smaller with respect to a partial order.

Condition 10.4.4

- X *is a finite set.*

- X *is closed under* F; *i.e., for all* $a \in X$, $F(a) \in X$.

- *There exists a fixed point* $x^* \in X$ *such that for all* $a \in X$ *and* $i = 1, 2, \ldots, n$, $F_i(a) \prec_i a_i$ *if* $a_i \neq x_i^*$, *and* $F_i(a) = x_i^*$ *otherwise.*

Proof. x^* is the unique fixed point in X. This can easily be seen by contradiction, i.e., by assuming the existence of another fixed point $y^* \neq x^*$. Then, the last condition is not satisfied for $a = y^*$. Now, define $\{X(k)\}$ such that

$$X_i(k) = X_i(k-1) - R_i(k-1), \quad \forall k > 0, \forall i,$$

where $R_i(k)$ is the set of maximal elements of $X_i(k)$, except x_i^*; i.e.,

$$R_i(k) = \{a_i | (a_i \in X_i(k)) \text{ and } (\forall b_i \in X_i(k), \ a_i \preceq_i b_i \Rightarrow a_i = b_i)\} - \{x_i^*\}.$$

It is obvious that every $X_i(k)$ has a maximal element, because of the fact that $X_i(k)$ is finite. Therefore, $R_i(k)$s are non-empty, except when $X_i(k) = \{x_i^*\}$. Thus, $X_i(k+1) \subset X_i(k)$ except when $X_i(k) = \{x_i^*\}$, from which the first two conditions of Definition 10.4.1 are satisfied.

To prove the last part, we first note that $x^* \in X$, and since for all $i = 1, 2, \ldots, n$, $x_i^* \notin R_i(k)$, then $x^* \in X(k)$, for all $k = 0, 1, 2, \ldots$ Now, for a given $a \in X(k)$, there are two possibilities, for all k:

- If $a = x^*$, then $F(a) = x^* \in X(k+1)$;

- if $a \neq x^*$, then $F_i(a) \prec_i a_i$ or $a_i = x_i^*$, and therefore, $F_i(a) \notin R_i(k)$, which implies that $F_i(a) \in X_i(k+1)$, for all $i = 1, 2, \ldots, n$.

In both cases, the conditions of asynchronous contraction are satisfied. $\square$

Like Condition 10.4.1, Condition 10.4.4 is not necessary for asynchronous contraction. Example 10.4.4 can be used to support this. The major difference between Condition 10.4.2 and Condition 10.4.4 is that the latter does not impose monotonicity.

Example 10.4.5 Figure 10.5 describes an operator satisfying Condition 10.4.4 but not Condition 10.4.2. In this example, the only ordering relation that would satisfy the second part of Condition 10.4.2 is the one defined for numbers in the usual sense, but for this ordering relation, monotonicity does not hold: although $(2, 0) \preceq (2, 1)$, $F(2, 0) \succ F(2, 1)$. $\square$

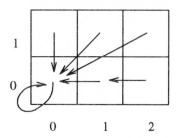

Figure 10.5: A componentwise contracting operator which is not monotone

A Condition Based on Dependencies

The following condition is expressed in terms of the dependency matrix $D(F)$ of the iteration operator, where D_{ij} is 0 if F_i does not depend on x_j and it is 1 otherwise. It is a generalisation of the result given by Robert (1986, 1987) and proven for a special class of asynchronous iterations in which each update of the shared data always uses the most recent value of the data as input.

Condition 10.4.5 *The transitive closure of the dependency matrix D(F) is null.*

Sketch of Proof. Let,

$$
D(F) = \begin{bmatrix}
0 & 0 & 0 & \cdots & 0 & 0 & 0 \\
0 & 0 & 0 & \cdots & 0 & 0 & * \\
0 & 0 & 0 & \cdots & 0 & * & * \\
& & & \cdots & & & \\
0 & * & * & \cdots & * & * & *
\end{bmatrix},
$$

without loss of generality. Otherwise, we can obtain the above matrix by permuting on the rows and the columns (Robert 1986). For any x, $F_1(x)$ is constant, say x_1^* (from the first row). Given, $x_1 = x_1^*$, $F_2(x)$ is constant, say x_2^* (from the second row). Similarly,

$$
x_1 = x_1^*, \ldots, x_i = x_i^* \;\Rightarrow\; F_{i+1}(x) = x_{i+1}^*.
$$

Define,

$$
\begin{aligned}
X(0) &= X_1(0) \times X_2(0) \times \cdots X_n(0) \\
X(1) &= \{x_1^*\} \times X_2(0) \times \cdots \times X_n(0) \\
X(2) &= \{x_1^*\} \times \{x_2^*\} \times X_3(0) \times \cdots \times X_n(0)
\end{aligned}
$$

$$\vdots$$

Then, F is asynchronously contracting due to Definition 10.4.1. □

This condition is also not sufficient for asynchronous contraction, e.g., it is not satisfied for the operator given in Example 10.4.4, because each component of F depend on each component of x.

Numerical Problems

The result of this section is due to Baudet (1978). The following notations are adopted. If z is a vector, $|z|$ denotes the vector with components $|z_i|$. Also, for any $x, y \in \Re^n$, $x \leq y$ implies $x_i \leq y_i$, for all i. An operator $F : \Re^n \mapsto \Re^n$ is said to be a *P-contracting operator* on a subset $X = X_1 \times X_2 \times \cdots \times X_n$ of $\Re^n$ if there exists a nonnegative $n \times n$ matrix A such that $\rho(A) < 1$ (the spectral radius of A) and,

$$|F(x) - F(y)| \leq A\,|x - y|, \quad \forall\, x, y \in X$$

(Ortega and Rheinboldt 1970).

Condition 10.4.6 $F : \Re^n \mapsto \Re^n$ *is P-contracting on a subset* $X = X_1 \times X_2 \times \cdots \times X_n$ *of* $\Re^n$, *where* X *is closed under* F.

Sketch of Proof. A complete proof can be found in Baudet (1978). Here, our goal is to relate it to the concept of asynchronous contraction. The starting point of the proof in Baudet (1978) is a lemma that can be stated as follows. Let B be a nonnegative square matrix. Then $\rho(B) < 1$ if and only if there exists a positive scalar ω and a positive vector v such that

$$B\,v \leq \omega\,v \quad \text{and} \quad \omega < 1.$$

Without loss of generality, let us assume that the solution vector x^* is null. Otherwise, the origin of the coordinate system can be translated to the solution vector. Then, there exists a nonnegative $n \times n$ matrix A, a positive scalar $\omega < 1$ and a positive vector v such that

- $|F(x)| \leq A\,|x|$, for all $x \in X$, and

- $A\,v \leq \omega\,v$.

We can choose a scalar c such that $c\,v$ is larger than the absolute value of $x(0)$, and $X(k)$s defined as follows satisfy the definition of asynchronous contraction (Definition 10.4.1):

$$X(k) = \{z \mid |z| \leq c\,\omega^k\,v\}.$$

$\square$

Obviously, the result of the original paper on asynchronous iterations is a straightforward consequence of the above condition (Chazan and Miranker 1969). It states that for a linear operator given by $F(x) = A\,x + b$, $\rho(|A|) < 1$ is a sufficient condition for convergence. Chazan and Miranker (1969) further proved that this condition is also necessary.

10.5 Nonredundant Asynchronous Computations

10.5.1 Definition

In most computations, the computation of a component x_i always uses the most recent value of x_i as an input component. As we discussed in Section 10.3 this can be identified by the fact that the computation of a component by multiple processors at the same time is not possible. We call this property *nonredundancy* and formulate it as follows.

Model 10.5.1 (Nonredundant Asynchronism)
Besides the requirements of Model 10.4.1 (total asynchronism), there holds:

$$\tau_i^i(t) = t - 1, \quad \forall\, i = 1, 2, \ldots, n, \ t = 1, 2, \ldots$$

$\square$

10.5.2 Convergence

Lemma 10.5.1 *A nonredundant asynchronous iteration $\{x(t)\} = (F, x(0), \mathcal{S})$ converges to a fixed point if every limit point of the input sequence $\{u^i(t)\}$ is a fixed point, for all $i = 1, 2, \ldots, n$, where F is continuous in a domain that contains $x(0)$.*

Proof. Let us fix i and define $\{u^i(t_l)\}$ as the subsequence of $\{u^i(t)\}$ obtained by eliminating the instances at which the i-th component is not updated. Now consider a limit point x^* of $\{u^i(t)\}$. There exists a subsequence $\{u^i(t_k)\}$ of $\{u^i(t_l)\}$ that converge to x^*. From continuity of F and since x^* is a fixed point

$$\lim_{t_k \to \infty} x_i(t_k - 1) = \lim_{t_k \to \infty} u_i^i(t_k) = F_i(\lim_{t_k \to \infty} u^i(t_k))$$
$$= \lim_{t_k \to \infty} (F_i(u^i(t_k))) = \lim_{t_k \to \infty} x_i(t_k),$$

which means that $\{x_i(t_k - 1)\}$ and $\{x_i(t_k)\}$ converge to the same point. On the other hand, we can partition $\{u^i(t_l)\}$ into subsequences such that each subsequence converges to a fixed point. We can repeat the above arguments for each subsequence and as a result, we conclude that $\{x_i(t)\}$ converges. Therefore, $\{x(t)\}$ converges and from Lemma 10.4.1 it converges to a fixed point. $\square$

The immediate consequence of the above lemma can be stated as follows.

Proposition 10.5.1 *Let $F : X \mapsto X$ be an operator, $\{X(k)\}$ be a sequence of subsets of X and $X^* = X_1^* \times X_2^* \times \cdots \times X_n^* \subseteq X$ be a set of fixed points of F. Then, all nonredundant asynchronous iterations with $x(0) \in X(0)$ converge to a fixed point in X^* if the following conditions hold.*

- *(Box Condition) For all $k = 0, 1, 2, \ldots, n$, $X(k)$ is the Cartesian product of n sets, i.e.,*

$$X(k) = X_1(k) \times X_2(k) \times \cdots \times X_n(k).$$

- *For all $k = 0, 1, 2, \ldots,$ $X^* \subseteq X(k+1) \subseteq X(k)$, and furthermore, every limit point of all the sequences $\{z(k)\}$ such that $z(k) \in X(k)$ is a member of X^*.*

- *$x \in X(k) \Rightarrow F(x) \in X(k+1), \quad k = 0, 1, 2, \ldots$*

Proof. Let $\{\varphi(k)\}$ be the pseudocycle sequence. Then, from the proof of Theorem 10.4.1

$$u^i(t) \in X(k), \quad \forall t > \varphi(k+1), \quad \forall i = 1, 2, \ldots, n.$$

Therefore, for all i, every limit point of $\{u^i(t)\}$ converges and the previous lemma applies. $\qquad\square$

Proposition 10.5.2 *Given posets $(X_i, \preceq_i)$, for all $i = 1, 2, \ldots, n$, let $(X, \preceq)$ be a poset such that $x \preceq y$ if $x_i \preceq y_i$, for all $i = 1, 2, \ldots, n$ and $x, y \in X$. Then, all nonredundant asynchronous iterations $\{x(t)\} = (F, x(0), \mathcal{S})$ with $x(0) \in X = X_1 \times X_2 \times \cdots \times X_n$ satisfy the property that $x(t) \preceq x(t+1)$, for all $t = 0, 1, 2, \ldots,$ if*

- *X is closed under F, and*

- *for all $i = 1, 2, \ldots,$*

$$F_i(x) \preceq x_i, \quad \forall x \in X$$

Proof. For all $t = 1, 2, \ldots$ and $i = 1, 2, \ldots, n$, there are two cases possible:

- $i \in \alpha(t)$ (x_i is updated at t). In this case,

$$x_i(t) = F_i(u^i(t)),$$

where,

$$u_i^i(t) = x_i(\tau_i^i(t)) = x_i(t-1).$$

From the second condition,

$$x_i(t) \preceq_i x_i(t-1).$$

- $i \notin \alpha(t)$ (x_i is not updated at t). By definition, $x_i(t) = x_i(t-1)$. $\qquad\square$

It immediately follows that if the conditions of the above lemma are satisfied, then all nonredundant asynchronous iterations converge to a fixed point in the following cases:

- X is a finite set.

- $F : \Re^n \mapsto \Re^n$ is a real and continuous function and $\preceq_i$ is defined as the order relation $\leq$ in the usual sense.

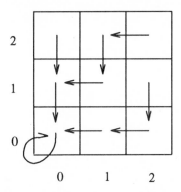

Figure 10.6: An operator convergent under nonredundancy but not under total asynchronism

In both cases, a decreasing sequence implies convergence and from Lemma 10.4.1, convergence to a fixed point is guaranteed.

Example 10.5.1 The conditions of Proposition 10.5.2 do not imply asynchronous contraction. The operator given in Figure 10.6 supports this: it satisfies the conditions of the proposition, although it is not asynchronously contracting. □

10.6 Partial Asynchronism

10.6.1 Definition

Partial asynchronism can simply be defined as a model which satisfies the conditions of both nonredundancy and bounded delay. Results which are specific to this model were given by Lubachevsky and Mitra (1986), Tsitsiklis (1984), Tsitsiklis and Bertsekas (1986), Tsitsiklis *et al.* (1986), Tseng *et al.* (1990), Bertsekas and Tsitsiklis (1989a) and the term partial was introduced by Bertsekas and Tsitsiklis (1989a) and Tseng *et al.* (1990). Partial Asynchronism is formally defined as follows.

Model 10.6.1 (Partial Asynchronism) *There exists a positive integer B such that:*

- *For every $i = 1, 2, \ldots, n$ and for every $t \geq 0$ at least one element t' of the set $\{t, t+1, \ldots, t+B-1\}$ satisfies $i \in \alpha(t')$.*

- *$t - B \leq \tau_j^i(t) < t$, for all i, j and t such that $i \in \alpha(t)$.*

- *$\tau_i^i(t) = t - 1$ for all i and t such that $i \in \alpha(t)$.* □

By definition, convergence under asynchronism with no redundancy implies convergence under partial asynchronism. This means that the results of the previous section can be used for partial asynchronism. In the following, we further prove that if the domain X of data is finite, then convergence under partial asynchronism for all B implies convergence under nonredundant asynchronism.

Lemma 10.6.1 *Let F be a closed function in a finite set X and let $x(0) \in X$. If all asynchronous iterations $(F, x(0), S)$ converge under partial asynchronism, then all asynchronous iterations $(F, x(0), S)$ also converge under nonredundancy.*

Proof. In this proof, given a schedule, $\{\varphi(k)\}$ will denote a particular pseudocycle sequence, for example the minimum pseudocycle sequence. Now, given $x(0)$ and F, let us fix a partially asynchronous schedule. Let $P(k)$ be the set of values generated in the k-th pseudo-cycle. Also let $v(k)$ be the last value of the data in the k-th pseudocycle, i.e., $v(k) = x(\varphi(k + 1))$. We observe that there exists no two k' and k'' such that $P(k') \equiv P(k'')$ and $v(k') = v(k'')$, unless all the elements of $P(k')$ and $v(k')$ are fixed points. Otherwise, there would exist a partially asynchronous schedule, for which some points are repeated indefinitely, which are not fixed points. Since the domain of data is finite, it follows that there are finite choices of $(P(k), v(k))$ pairs. As a result, there exists an integer m such that all partially asynchronous iterations converge in at most m pseudocycles. This means that all nonredundant asynchronous iterations also converge in at most m pseudocycles. This can be seen by contradiction. Suppose that there exists a nonredundant asynchronous iteration that does not converge in m pseudocycles, then there would be a partially asynchronous schedule which does not converge in m pseudocycles.   □

10.6.2   Convergence Conditions

Maximum Contraction

This section is due to Tseng *et al.* (1990) and Bertsekas and Tsitsiklis (1989a). The results can also be used to obtain the results of Lubachevsky and Mitra (1986).

In this section, $F : \Re^n \mapsto \Re^n$ is a real function. Let X^* be the set of fixed points of F and for each $x \in \Re^n$ let $||x|| = \max_{i=1,2,...,n} |x_i|$ denote the maximum norm of x. For any $x \in \Re^n$, we denote by $\rho(x)$ the distance of x from X^*, defined by

$$\rho(x) = \inf_{y \in X^*} ||x - y||.$$

Finally, given any $x \in \Re^n$ and $x^* \in X^*$, we let $I(x; x^*)$ be the set of indices of coordinates of x that are farthest away from x^*, that is,

$$I(x; x^*) = \{i \mid |x_i - x_i^*| = ||x - x^*||\},$$

and we also denote

$$U(x; x^*) \quad = \quad \{y \in \Re^n \mid y_i = x_i \text{ for all } i \in I(x; x^*),$$

$$\text{and} |y_i - x_i^*| < ||x - x^*|| \text{ for all } i \notin I(x; x^*)\}.$$

In other words, $U(x; x^*)$ is the set of all vectors y with $||y - x^*|| = ||x - x^*||$ that agree with x in the components that are farthest away from x^*.

The function F is assumed to satisfy the following assumption:

Assumption 10.6.1

(a) F is continuous.

(b) The set of fixed points X^* is convex and nonempty.

(c) $||F(x) - x^*|| \leq ||x - x^*||, \ \forall x \in \Re^n, \forall x^* \in X^*$.

(d) For every $x \in \Re^n$ and $x^* \in X^*$ such that $||x - x^*|| = \rho(x) > 0$, there exists some $i \in I(x; x^*)$ such that $F_i(y) \neq y_i$, for all $y \in U(x; x^*)$.

(e) For any $i, x \in \Re^n$, and $x^* \in X^*$, if $F_i(x) \neq x_i$, then $|F_i(x) - x_i^*| < ||x - x^*||$. □

Proposition 10.6.1 *Let F be a function that satisfies Assumption 10.6.1. Then all partially asynchronous iterations $(F, x(0), S)$ converge to some element of X^*.*

Sketch of proof. For $t = 0, 1, \ldots,$ define

$$z(t) = (x(t - B + 1), \ldots, x(t)),$$

$$d(z(t)) = \min_{x^* \in X^*} \{\max\{||x(t - B + 1) - x^*||, \ldots, ||x(t) - x^*||\}\}.$$

From Assumption 10.6.1(c), we first observe that

$$d(z(t + 1)) \leq d(z(t)), \ \forall t \geq 0.$$

Let us now fix t. Let $d(z(t)) = \beta > 0$. The first goal is to show that there exists an integer m such that

$$d(z(t + m)) < d(z(t)). \tag{10.4}$$

For this purpose, we observe that if for some $t' \geq t$, and i, $|x_i(t') - x_i^*| < \beta$, then $|x_i(t'') - x_i^*| < \beta$, for all $t'' \geq t'$. This results from the last condition of partial asynchronism, and Assumption 10.6.1(e).

Consider $2n$ consecutive periods of time instances after t, such that the length of each period is B, i.e., $[t + 1, t + B], [t + B + 1, t + 2B], \ldots$ Assume that $B - 1$ units after the end of these periods, d value of z is still β, i.e.,

$$z(t) = z(t + 1) = \cdots = z(t + 2nB + B - 1) = \beta.$$

Otherwise, (10.4) is obtained.

From assumptions (c) and (d), it can be concluded that for any two consecutive periods, there exists a component i and an update instance t_i of this component in these cycles, such that the distance from x_i to x_i^* ($|x_i - x_i^*|$) is β before t_i and it is less than β, after t_i. This means that after $2n$ periods, the distances from all the components are less than β. Then,

$$d(z(t + 2nB + B - 1)) < d(z(t)). \tag{10.5}$$

As a result, $\{d(z(t)\}$ converges to a limit d^*. If $d^* = 0$, $\{x(t)\}$ converges to a fixed point and the proof is complete. Assuming $d^* > 0$ will lead to a contradiction as follows. Since $\{z(t)\}$ is bounded, there exists a subsequence of it that converges to some z^* and since d is continuous $d(z^*) = d^*$. Now let $\Delta t = 2nB + B - 1$. We can express $z(t + \Delta t)$ as a function of $z(t)$:

$$z(t + \Delta t) = g(z(t); \theta(t)),$$

where $\theta(t)$ is some composition of F determined by the part of the schedule between t and $t + \Delta t$. Because of the finite delay assumption, there are only a finite number of $\theta(t)$s and therefore there exists a subsequence $\{z'(t)\}$ of $\{z(t)\}$ such that $\theta(t)$s are constant and equal to θ and $\{z'(t)\}$ converges to z^*. This means for another subsequence $\{z''(t)\}$

$$z''(t) = g(z'(t), \theta).$$

Since F is continuous, θ and g is a continuous function of $z'(t)$. Therefore, $\{z''(t)\}$ converges to $g(z^*, \theta)$.

From (10.5) $d(g(z^*, \theta)) < d^*$, which is a contradiction. $\square$

Gradient-Like Optimisation Algorithms

The problem looked at in this section is the minimisation of a cost function $G : \Re^n \longmapsto \Re$. The iteration operator F for the solution of this problem is $F(x) = x - \gamma s(x)$, where $s(\cdot)$ is some function related to $G(\cdot)$. The synchronous iteration corresponding to F converges to the solution if γ is small enough (Bertsekas and Tsitsiklis 1989a). The following result states that the convergence is preserved under partial asynchronism if γ is smaller than a threshold that depends on B. It was originally obtained by Tsitsiklis (1984), Tsitsiklis and Bertsekas (1986) and Tsitsiklis *et al.* (1986).

The following conditions are assumed.

Assumption 10.6.2

(a) *There holds $G(x) \geq 0$, for every $x \in \Re^n$.*

(b) *(Lipschitz Continuity of ∇G) The function G is continuously differentiable and there exists a constant K_1 such that*

$$||\nabla G(x) - \nabla G(y)|| \leq K_1 ||x - y||, \quad \forall x, y, \in \Re^n.$$

(c) (Descent Property Along Each Coordinate) *For every i and $u \in \Re^n$, we have*

$$s_i(u)\nabla_i G(u) \le 0.$$

(d) *There exists positive constants K_1 and K_2 such that*

$$K_2 |\nabla_i G(u)| \le |s_i(u)| \le K_3 |\nabla_i G(u)|, \quad \forall u \in \Re^n, \quad \forall i.$$

□

Proposition 10.6.2 *Under Assumption 10.6.2, a partially asynchronous iteration*
$(F, x(0), S)$ converge if $0 < \gamma < \gamma_0$, where

$$\gamma_0 = \frac{1}{[(1 + B + n B) K_1 K_3]}.$$

□

The complete proof requires the manipulation of the given assumptions and the descent lemma and is involved. We will not give a complete proof, but the key idea is to write the sequence generated by the partially asynchronous schedule as

$$x(t) = x(t - 1) + \gamma\, r(t),$$

where

$$r_i(t) = s_i(u^i(t)).$$

The only difference with the synchronous algorithm is that in the synchronous case

$$r_i(t) = s_i(x(t - 1))$$

and the difference between $x(t - 1)$ and $u^i(t)$ is bounded, because

$$|u^i_j(t) - x_j(t)| = \gamma\Big| \sum_{\tau = r^i_j(t)}^{t} r_j(\tau)\Big| \le \gamma \sum_{\tau = t - B}^{t} |r_j(\tau)|.$$

This suggests that the analysis follows the same steps as the one for the synchronous case. However, partial asynchronism introduces some additional error terms which depend on B and γ. It follows that given B, the error terms do not affect the convergence when γ is small enough.

10.7 Dynamic Iteration Operators

Up to this point we have considered only fixed point iterations for which the iteration operator remains fixed throughout the computations. There are cases, however, in which F changes dynamically as the computation progresses. For example, a common form of synchronous iterations looks like the following.

$$\textbf{for } k := 1 \textbf{ to } m \textbf{ do}$$
$$\textbf{forall } 1 \leq i \leq n \textbf{ do}$$
$$x_i := F_i(x, k)$$

The keyword **forall** indicates that x_is are computed in parallel, for the current value of k. **For** has the same meaning as in most declarative languages, and it implies that all the computations for k should be completed before the computations for $k + 1$ can start. In other words, there is a barrier between each iteration.

In the above loop, we immediately observe that the iteration operator takes k as a parameter in the k-th iteration. This suggests an asynchronous implementation, for which the iteration operator takes k as a parameter in the k-th pseudocycle. For such asynchronous implementations the straightforward generalisation of the convergence condition given by Definition 10.4.1 (asynchronous contraction) can be given as follows.

Condition 10.7.1 *There exists a sequence of sets* $\{X(k)\}$ *that satisfies the following conditions.*

- *(Box Condition) For all* $k = 0, 1, 2, \ldots, X(k)$ *is a Cartesian product of* n *sets; i.e.,*

$$X(k) = X_1(k) \times X_2(k) \times \cdots \times X_n(k).$$

- *For all* $k = 0, 1, 2, \ldots, X(k+1) \subseteq X(k)$, *and furthermore,* $\{X(k)\}$ *converges to a fixed point of* F ;

- $x \in X(k) \Rightarrow F(x, k) \in X(k+1)$, *for all* $k = 0, 1, 2, \ldots$ □

Here we omit the proof that the condition above is sufficient for convergence of totally asynchronous iterations, in which, for all k, the iteration operator takes k as a parameter, in the k-th pseudo-cycle, because this is almost identical to the proof of Theorem 10.4.1.

How to implement the asynchronous version of the above synchronous loop still needs some consideration. The main issue in this respect is how to keep track of the pseudocycles. Simply maintaining a single, global variable k that contains the current pseudocycle at all times is not sufficient, because by using only this information, it is not possible to update k, i.e., to determine when the next pseudocycle starts. However, if we maintain a variable P_CYCLE_i which contains the pseudocycle of the i-th component, for all i, then from this information it is possible to determine k and also the next pseudocycle of the i-th component. In other words, for all i, P_CYCLE_i is the variable such that $x_i \in X_i(P\_CYCLE_i)$, at all times. Therefore, at all times, $x \in X(\min_i\{P\_CYCLE_i\})$, i.e., $k = \min_i\{P\_CYCLE_i\}$. Furthermore, after each update of the i-th component, the new value of P_CYCLE_i can be updated by

$$P\_CYCLE_i := \min_j\{P\_CYCLE_j\} + 1.$$

These ideas amount to the following asynchronous loop.

$task(i)$;
    begin $\{task\}$
        $k := \min_j \{P\_CYCLE_j\}$;
        $x_i := F_i(x, k)$;
        $P\_CYCLE_i := k + 1$;
        if $(P\_CYCLE_i = m)$ **then terminate else return**;
    end $\{task\}$

In this notation, a task is an indivisible unit of execution that is held by the same processor from its start until the statement **return** or **terminate** is reached. There is a pool of tasks in the system: initially n of them, one for each component. There are also a number of processors, and each available processor selects a task from the pool and executes it until the **return** statement. At this point it returns the task to the pool and makes another task selection, or possibly continues with the same task. The execution of the **terminate** statement in $task(i)$ discards this task from the system for good, because x_i has reached the solution. It should also be mentioned that P_CYCLE_is are initially assigned to 0. We also assume a computational system that supports atomic accesses (e.g. by hardware) to individual components of P_CYCLE and x. If not, accesses to P_CYCLE and x should be performed in critical sections.

A final remark that has to be made is that the asynchronous implementation described above is not always correct when the underlying schedule is totally asynchronous. As an example, suppose that at some instance of the computations two processors start executing the same task (say i) at exactly the same time. They execute the first two statements in parallel, generating the same update of x_i. Since both processors do the same work on x_i the next pseudocycle of the i-th component should be obtained by incrementing k once. However, execution of the third statement by two processors one after another, in our example, increments P_CYCLE_i more than once, which is incorrect. To avoid this type of race condition, the schedule should be nonredundant. This result seems to contradict the above generalisation of Theorem 10.4.1 which states that the totally asynchronous implementation is correct. However, the general result does not consider implementation details and assumes an abstract mechanism that automatically makes the pseudocycle information available to the processors, as the computation proceeds, and since this abstraction is not available to the real-life implementation, it needs the additional restriction (nonredundancy) for correct operation.

10.8   Other Models of Asynchronism

10.8.1   Temporally Ordered Schedules

Another reasonable restriction we can impose is to assume that the information in the system is received in the order it is generated by the processors. In other words, after a processor uses, as input, the value of component x_i which is generated at some time instance t, it does not use values of x_i older than t. Another way to state this is that τ_j^is have the monotonicity property as functions of t. Bertsekas and Tsitsiklis (1989b), made this assumption for convergence rate comparison of asynchronous and synchronous iterations. However, we are not aware of any convergence condition which is specific to this model. The formal statement of the model is as follows.

Model 10.8.1 (Temporally Ordered Schedule) *For all $i, j = 1, 2, \ldots, n$ and $t = 0, 1, 2, \ldots$,*

$$\tau_j^i(t) \le \tau_j^i(t + 1).$$

□

We call this "temporal" ordering, because the ordering of τ_j^is conforms with the ordering of time. Bertsekas and Tsitsiklis (1989b), proved the following result as an intermediate step to other results concerning the convergence rate.

Proposition 10.8.1 *Suppose that Condition 10.4.1 (monotonicity) holds and let $\{x(t)\} = (F, x(0), S)$ be an asynchronous iteration, where S is a horizontally ordered schedule. Then, there holds $x(t + 1) \preceq x(t)$, for all t.*

Sketch of Proof. The proof is by induction. Here we omit the basis clause and assume that

$$x(t) \preceq x(t - 1) \preceq \ldots \preceq x(1) \preceq x(0).$$

Let us fix a component i. There are three cases for i.

- It is not updated at $t + 1$.

- It is updated at $t + 1$ and its most recent update before $t + 1$ occurs at some $t' > 0$.

- It is updated at $t + 1$ and not updated previously.

It is trivial that the first case yields $x_i(t + 1) \preceq x_i(t)$. In the second case, from the requirement of the schedule and from the induction hypothesis,

$$u^i(t + 1) \preceq u^i(t').$$

From monotonicity,

$$F_i(u^i(t + 1)) = x_i(t + 1) \preceq F_i(u^i(t')) = x_i(t).$$

The third case can be handled in a similar manner.   □

10.8.2 Spatially Ordered Schedules

In other cases, the ordering conforms with j. In other words, for the computation of some F_i, the input components are received (or fetched) in some predetermined order. For example, some component x_r is always fetched before the other component x_s. We call this property *spatial ordering of the schedule* and formally state it as follows.

Model 10.8.2 (Spatially Ordered Schedule) *For all $i = 1, 2, \ldots, n$, there exists a permutation function $\pi_i : \{1, 2, \ldots, n\} \mapsto \{1, 2, \ldots, n\}$ such that*

$$\tau_r^i(t) \leq \tau_s^i(t)$$

if

$$\pi_i(r) \leq \pi_i(s),$$

for all $t = 0, 1, \ldots$ □

This is called "spatial" ordering, because the ordering of τ_j^is conforms with the ordering of the components. It may be an interesting research problem to derive convergence conditions specific to these models.

10.8.3 Serial Computations

Another mode of operation is *serial* which assumes negligible computation time. In other words, $\tau_j^i(t) = t - 1$, for all i, j and t. This has been studied by Robert (1987), and Tsitsiklis (1987) derived necessary and sufficient conditions for convergence.

10.9 Conclusion

In recent years, we have witnessed significant advances in parallel processing technology. Today, multiprocessor systems are very popular and support parallel processing. Given this trend towards more parallelism, it is not unrealistic to anticipate systems, in the near future, with large numbers of processors, for example a few hundreds, maybe thousands. On the other hand, research in parallel algorithms and programming, in the past, concentrated either on 'simulating' known sequential algorithms on parallel systems, or on designing new parallel algorithms, exploiting the parallel nature of the applications, but with a synchronous computational model in mind. Consequently, multiprocessor implementation of such algorithms involves extensive synchronisations. While the value of such research cannot be denied, it is also important to understand and exploit the additional dimension of parallelism introduced by multiprocessors: asynchronism. This is even more critical for multiprocessing systems with large number of processors, since it is more difficult to support synchronisation in such systems.

In this chapter, we have explored an aspect of asynchronous computing which does not appear in sequential or synchronous parallel computing, that is, convergence of an iterative computation even when the sequence of interactions among the participating processors shows a chaotic behaviour. For this purpose, we have defined models of asynchronous computations, and have derived convergence results for most of these models. The idea of asynchronous iterative algorithms with chaotic processor interactions is not new. Previous work emphasised numerical computing; in this chapter, more general results applicable also to symbolic computing have been derived. We have concentrated on underlying fundamental issues rather than on application driven approaches.

One of the major criticisms of asynchronous algorithms is that they require mathematical knowledge too sophisticated for average algorithm designers. This chapter demonstrates that significant results can be obtained by relying on very little mathematical knowledge and that the fundamental concepts are in fact very simple.

In the future, we may expect many applications of asynchronous algorithms in various areas, especially in artificial intelligence. This would be facilitated if asynchronism was integrated in the semantics of languages. Several parallel languages have been designed by integrating parallelism in existing sequential languages. A similar approach should be taken for asynchronism. It seems natural for example to extend production system and logic programming languages to allow asynchronous computations which are not derived from uniprocessor computations. Another possible area of future research is the design of architectures for asynchronous computations. In order to achieve maximum performance from asynchronous algorithms, systems must be designed with fully asynchronous data accesses and transfers. Overlapping of data transfer and computation will become more critical as the gap between the speed of interprocessor communication and the speed of processors increases.

References

Axelrod T. S. (1986) Effects of synchronization barriers on multiprocessor performance. *Parallel Computing*, **3**, pp. 129–40.

Baudet G. M. (1978) Asynchronous iterative methods for multiprocessors. *J. ACM*, **25**(2), pp. 226–44.

Bertsekas D. P. (1982) Distributed dynamic programming. *IEEE Trans. on Automatic Control*, **27**, pp. 610–6.

Bertsekas D. P. (1983) Distributed asynchronous computation of fixed points. *Mathematical Programming*, **27**, pp. 107–20.

Bertsekas D. P. and Tsitsiklis J. N. (1989a). *Parallel and Distributed Computation*. Prentice Hall.

Bertsekas D. P. and Tsitsiklis J. N. (1989b). Convergence Rate and Termination of Asynchronous Iterative Algorithms. In *Proc. International Conference on Supercomputing*, Crete, Greece.

Chazan D. and Miranker W. (1969) Chaotic Relaxation. *Linear Algebra and Its Applications*, **2**, pp. 199–222.

Goodman J. R., Vernon M. K. and Woest P. J. (1989) Efficient Synchronization Primitives for Large-Scale Cache-Coherent Multiprocessors.

Hwang K. and Briggs F. A. (1984) *Computer Architecture and Parallel Processing*. McGraw-Hill.

Kruskal C. P. and Weiss A. (1984) Allocating Independent Subtasks on Parallel Processors. In *ICPP*, pp. 236–40.

Kung H. T. (1976) Synchronized and Asynchronous Parallel Algorithms for Multiprocessors. In Traub J. F., editor, *Algorithms and Complexity : New Directions and Recent Results*. Academic Press, New York.

Lubachevsky B. D. and Mitra D. (1986) A Chaotic Asynchronous Algorithm for Computing the Fixed Point of a Nonnegative Matrix of Unit Spectral Radius. *J. ACM*, **33**(1), pp. 130–50.

Miellou J.-C. (1974) Itérations chaotiques à retards, études de la convergence dans le cas d'espaces partilellement ordonneés. *CRAS*, (278), pp. 957–60.

Miellou J.-C. (1975) Algorithmes de relaxation à retards. *Revue d'Automatique, Informatique et Recherche Opérationnelle*, **9**(R–1), pp. 55–82.

Miellou J.-C. (1986) Asynchronous iterations and order intervals. In Michel Cosnard *et al.*, (Ed.) *Parallel Algorithms and Architectures*. North-Holland.

Miellou J.-C. and Spiteri M. (1985) Un critère de convergence pour des méthodes générales de point fixe. *Math. Modelling and Numer. Analy.*, **19**, pp. 645–69.

Mitra D. (1987) Asynchronous relaxations for the numerical solution of differential equations by parallel processors. *SIAM J. Sci. Stat. Comput.*, **8**, pp. 43–58.

Munkres J. R. (1975) *Topology: A First Course*. Prentice Hall, Englewood Cliffs, NJ.

Ortega J. M. and Rheinboldt W. C. (1970) *Iterative Solution of Nonlinear Equations in Several Variables*. Academic Press.

Robert F. (1976) Contraction en norme vectorielle: Convergence d'itérations chaotiques pour des équations non linéaires de point fixe à plusieurs variables. *Lin. Algeb. and Appl.*, **13**, pp. 19–36.

Robert F. (1986) *Discrete Iterations*. Springer-Verlag, Berlin.

Robert F. (1987) Itérations Discrètes Asynchrones. Technical Report R.R. 671-M, Informatique et Mathématiques Appliquées de Grenoble.

Robert F., Charnay M. and Musy F. (1975) Itérations chaotiques série-parallèle pour des équations non-linéaires de point fixe. *Aplikace Mat.*, **20**, pp. 1–38.

Spiteri P. (1986) Parallel asynchronous algorithms for solving boundary value problems. In Michel Cosnard *et al.* (Ed.) *Parallel Algorithms and Architectures*. North-Holland.

Stone H. S. (1987) *High-performance computer architecture*. Addison-Wesley.

Tseng P., Bertsekas D. P. and Tsitsiklis J. N. (1990) Partially Asynchronous, Parallel Algorithms for Network Flow and Other Problems. *SIAM J. Cont. and Opt.*, **28**(3), pp. 678–710.

Tsitsiklis J. N. (1984) *Problems in Decentralized Decision Making and Computation*. PhD thesis, Dept. of Electrical Engineering and Computer Science, Massachusetts Institute of Technology, Cambridge.

Tsitsiklis J. N. (1987) On the stability of asynchronous iterative processes. *Mathematical Systems Theory*, **20**, pp. 137–53.

Tsitsiklis J. N. and Bertsekas D. P. (1986) Distributed asynchronous optimal routing in data networks. *IEEE Trans. on Automatic Control*, **31**, pp. 325–32.

Tsitsiklis J. N., Bertsekas D. P. and Athans M. (1986) Distributed asynchronous deterministic and stochastic gradient optimization algorithms. *IEEE Trans. on Automatic Control*, **31**, pp. 803–12.

Üresin A. and Dubois M. (1989a). Parallel asynchronous algorithms for discrete data. Technical Report CENG 89–15, Computer Engineering Division, Department of Electrical Engineering-Systems, University of Southern California. Also to appear in *J. ACM*.

Üresin A. and Dubois M. (1989b). Sufficient conditions for the convergence of asynchronous iterations. *Parallel Computing*, **10**, pp. 83–92.

CHAPTER 11

Parallel Dynamic Programming

Tom Archibald

University of Edinburgh, UK

11.1   Introduction

Dynamic programming was first proposed by Bellman (1957) as a technique for solving sequential decision processes. In these problems a number of decisions have to be made in sequence. The decision taken is based only on the state of the process. Each decision involves a cost and affects the future state. To minimise total cost, a balance must be struck between incurring low immediate cost and keeping out of states in which high costs are unavoidable. All dynamic programming formulations are based on the principle of optimality which states that 'an optimal policy has the property that whatever the initial state and initial decision are, the remaining decisions must constitute an optimal policy with regard to the state resulting from the first decision' (Bellman 1957 p 83). In a deterministic formulation the decision taken dictates the next state or states (see Sections 11.2 and 11.3). In other cases the decision taken may only influence the possible outcomes of a chance event. Such formulations are called stochastic (see Section 11.4) and the objective is to minimise the expected cost.

A very general dynamic programming formulation is given in equation 11.1, where n is the number of states, v_i is the minimum cost with initial state i, K_i is the set of all decisions permissible in state i, $f_i^k : \Re^n \to \Re$ is some function and v is the vector with components v_j for $j = 1, 2, \ldots, n$.

$$v_i = \min_{k \in K_i} f_i^k(v) \text{ for } i = 1, 2, \ldots, n \qquad (11.1)$$

The applications of sequential decision processes are extremely diverse, as can be seen from the host of examples described in Bellman and Dreyfus (1962) and Howard (1960). While there are many practical problems which can be easily solved using dynamic programming (for instance TeX, the widely used text processing package, uses a finite stage deterministic dynamic programming formulation to decide where to place line breaks in order to achieve the most aesthetically pleasing output (Knuth 1981)), real life problems often have such large data sets that the computation of their solution is beyond the scope of serial machines due to memory and processing time constraints. This has restricted the extent to which dynamic programming has influenced decision making.

As a result the efficient implementation of dynamic programming algorithms is an area which has been at the centre of a considerable research effort for the last three decades. Bellman and Dreyfus (1962) suggest techniques which simplify the computational requirements of the algorithm for some applications, while in the early 1980s a number of comparisons of the computational performance of solution algorithms for the important class of problems known as Markov decision processes were carried out (Porteus 1981; Thomas *et al.* 1983; Hendrikx *et al.* 1984).

The application of parallel processing has long been advocated as a potential solution to the computational intractability of the dynamic programming algorithm on serial machines. Bellman and Dreyfus (1962) describe a parallel method for the solution of one-dimensional allocation processes. Tabak (1968) identifies dynamic programming as suitable for parallel processing and assesses the likely impact of the evolving technology on this area. Casti *et al.* (1973) describe a number of algorithms which exploit the parallelism inherent in the dynamic programming formulation, but since these algorithms require a vast number of processors they were never implemented. Al-Dabass (1980) extends these ideas, performs extensive theoretical comparisons and implements the algorithms developed on a two processor parallel computer. Of course since then parallel computers have grown in size and become more widely available. As a result the number of articles on the subject of parallel dynamic programming has increased dramatically. The remainder of this chapter will be concerned with these developments.

In Section 11.2, the parallel solution of shortest path problems by dynamic programming is considered. It is interesting to note that one application of this class of problems is in the design of optimal routing algorithms for computer networks. Section 11.3 deals with parallel methods for optimal problem subdivision. Recently this problem has received more attention from parallel processing practitioners than any other dynamic programming formulation, probably because of its importance in combinatorial optimisation. Section 11.4 describes a number of approaches to the parallel solution of Markov decision processes.

11.2 Shortest Path Problems

The problem of finding the shortest path from one vertex (vertex 1) to all other vertices in a digraph can be formulated as a deterministic dynamic programming problem as follows:

$$
\begin{aligned}
v_1 &= 0 \\
v_i &= \min_{k \in E_i} \{c_{ki} + v_k\} \text{ for } i = 2, \ldots, n
\end{aligned}
\tag{11.2}
$$

where n is the number of vertices in the digraph, v_i is the length of the shortest path from 1 to i, E_i is the set of vertices which lie at the root of an edge into i and c_{ij} is the length of the edge from i to j. The cardinality of E_i is equal to the in-degree of i.

If every cycle has positive length and there exists a path from 1 to all other vertices, then equation 11.2 has a unique solution.

Bertsekas (1987) gives more details of this formulation as well as a number of applications. An example with relevance to parallel processing is the problem of routing data in a computer network. The time required to send a message of length m along a link joining node i to node j in a computer network can be modelled as $\sigma_{ij} + m/\rho_{ij}$, where σ_{ij} is the set-up time for the link and ρ_{ij} is the rate of transmission for the link. With $c_{ij} = \sigma_{ij} + m/\rho_{ij}$ in equation 11.2, v_j can be interpreted as the time required to send a message of length m from node 1 to node j using optimal routing.

One solution method for shortest path problems is Ford's algorithm (Ford and Fulkerson 1962). In this algorithm each value is initialised so that v_j is an upper bound on the length of the shortest path from 1 to j and $v_1 = 0$. At an iteration each edge in the digraph is considered in turn. When an edge from i to j is encountered, v_j is updated according to $v_j = \min\{v_j, c_{ij} + v_i\}$. The algorithm terminates when none of the values change during an iteration and at this point v_j is equal to the length of the shortest path from 1 to j.

If D is the maximum number of edges in a shortest path between 1 and any other vertex, convergence will be detected in at most $D+1$ iterations. However there are many instances of Ford's algorithm, each one corresponding to a different order of the edges, and some require far fewer iterations than others. Most other shortest path methods can be viewed as a version of Ford's algorithm with a specific order for updating the edges.

Bertsekas and Tsitsiklis (1989) suggest using a parallel iterative solution method to solve equation 11.2. One such method, a distributed version of Ford's algorithm, involves partitioning the vertex set of the digraph into $\{S_t\}_{t=1}^P$, where P is the number of processors used, and using the following recurrence relation.

$$v_j^0 = \infty \text{ for } j = 2, \ldots, n$$
$$v_1^r = 0 \text{ for } r \geq 0$$
$$v_j^r = \min_{i \in E_j} \left\{ c_{ij} + \left\{ \begin{array}{l} v_i^r \text{ if } i, j \in S_t \text{ for any } t \text{ and } i < j \\ v_i^{r-1} \text{ otherwise} \end{array} \right\} \right\} \text{ for } r \geq 1.$$

At the rth iteration processor t can calculate the values $\{v_j^r : j \in S_t\}$ independently of all other processors. Synchronisation is required at the end of each iteration at which point a check for convergence is performed.

The method terminates when the values at two successive iterations are identical. Although the number of iterations is bounded above by the maximum number of edges in any shortest path, generally as the number of processors used increases, so does the number of iterations required. Therefore in an efficient parallel implementation a balance must be struck between reduced calculation time per iteration, due to the parallel execution, and increased number of iterations.

The method is analogous to a Gauss-Seidel method for solving linear equations and as the number of processors increases, the convergence of such a method approaches that of a Jacobi method, see Section 11.4. Bertsekas and

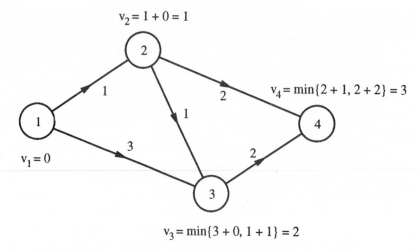

$$v_2 = 1 + 0 = 1$$

$$v_4 = \min\{2 + 1, 2 + 2\} = 3$$

$$v_1 = 0$$

$$v_3 = \min\{3 + 0, 1 + 1\} = 2$$

Figure 11.1: A simple shortest path problem in an acyclic digraph

Tsitsiklis (1989) also consider a Jacobi method and discuss in more depth the importance of the choice of starting values.

If the digraph is acyclic, then the vertices can be ordered so that an edge from i to j satisfies $i < j$ and the solution of the problem is greatly simplified. With such an order E_j is a subset of $\{1, 2, \ldots, j - 1\}$.

In some cases an order which achieves this is apparent from the application. For example consider a replacement problem in which the system is to be in operation for n time periods. Let c_{ij} equal the cost of replacing the incumbent machine at period i with one which will last until period j, then v_j can be interpreted as the minimum cost of maintaining the system until period j and the order of the vertices clearly has the above property. Shortest path problems in acyclic digraphs can also arise in scheduling problems and resource allocation problems.

If the structure of the problem does not yield a suitable order then it is possible to calculate one, the time required being linear in the number of edges in the digraph. Since this significantly reduces the solution time for the problem, it is worthwhile if a solution is to be calculated for several sets of edge lengths on the same digraph.

For this special case of the shortest path problem, the version of Ford's algorithm described above converges after only one iteration when one processor is used. In Figure 11.1, the solution procedure is illustrated for a simple example with $n = 4$.

Nicol (1989) describes three methods of distributing Ford's algorithm when the digraph is acyclic and compares their performance on a class of randomly generated problems using a shared memory multiprocessor. While two of the algorithms are based on commonly used techniques, the third is a novel approach which Nicol (1989) calls the block window algorithm. It is suggested that this algorithm will be particularly effective for sparse digraphs and the results obtained appear to bear this out. The remainder of this section describes the three algorithms considered by Nicol (1989) and examines some of the points brought to light by the comparison.

For the rest of this section, it is assumed that the digraph is acyclic and the vertices are ordered so that all edges from i to j satisfy $i < j$.

The calculation of v_i in equation 11.2 only requires the values in $\{v_k : k = 1, \ldots, i - 1\}$ and so the values $\{v_i : i = 2, \ldots, n\}$ can be calculated in sequence. The first parallel algorithm considered follows this approach using several processors in the computation of each value. This can be achieved by partitioning the sets E_i into $\{E_{it}\}_{t=1}^P$, where P is the number of processors used, and calculating $\min_{k \in E_{it}} \{c_{ki} + v_k\}$ on processor t. The calculation of v_i is then completed by the coordinated use of the processors in a global minimisation. This final step will typically require a lot of synchronisation; however in the first step processors can function independently.

The second approach uses the general purpose iterative method described earlier.

In the block window algorithm, the vertex set of the digraph is partitioned into blocks consisting of P vertices denoted by

$$\{W_t = \{(t - 1)P + 1, \ldots, tP\}\}_{t=1}^{n/P}$$

where P is the number of processors used. Each block is considered in sequence and the values for vertices within a block are calculated in parallel using a two-step procedure. In the first step it is assumed that the values for vertices within the block being considered do not depend upon each other and so only edges with roots outwith the block need be considered. The second step checks the validity of this assumption and repeats any calculation that is necessary. Explicitly the algorithm can be written as follows:

Calculation procedure for block r

Step 1

On processor t, calculate

$$u_j = \min_{i \in E_j \setminus W_r} \{c_{ij} + v_i\} \text{ where } j = (r - 1)P + t.$$

Step 2

On processor t, if

$$z_j \equiv \min_{i \in E_j \cap W_r} \{c_{ij} + u_i\} < u_j,$$

where again $j = (r - 1)P + t$, then set a flag.

If the flag is not set on any processor, $v_j = u_j \; \forall j \in W_r$.

Otherwise determine $\hat{t}$, the lowest processor number for which the flag is set, and then, with $\hat{k} = (r - 1)P + \hat{t}$, $v_j = u_j$ for $j \in W_r$, $j < \hat{k}$ and the values $\{v_j : j \in W_r, j \geq \hat{k}\}$ have to be recalculated sequentially using

$$v_j = \min \left\{ u_j, z_j, \min_{i \in E_j, i \geq k} \{c_{ij} + v_i\} \right\}.$$

The calculation of the values for the first block will have to be performed sequentially. The algorithm requires synchronisation before the start of the second step and also before the value of $\hat{t}$ can be determined.

The validity of step 2 hinges on proposition 1 of Nicol (1989) which states that

$$\text{if } u_j \leq \min_{i \in E_j \cap W_r} \{c_{ij} + u_i\} \text{ for all } j \in W_r \text{ such that } j < k,$$

$$\text{then } v_\ell = u_\ell \text{ for all } \ell \in W_r \text{ such that } \ell < k.$$

If $E_j \cap W_r = \emptyset$ then the condition is taken as being satisfied. The proof of this proposition is straightforward using induction.

Since the calculation of every edge sum involves the same computation, it is always possible to achieve a good load balance for the first algorithm by, as far as possible, assigning the same number of edges to each set E_{jt}. The load imbalance can be no worse than one addition and one comparison.

The first algorithm is effective when the digraph is large and dense, because in this situation the in-degree of the vertices is large compared to the number of processors, and consequently the time saved by the parallel evaluation of the edge sums (i.e. $c_{ki} + v_k$) is far greater than the overheads of the synchronisation required by the global minimisation. For highly sparse digraphs this will not be the case.

The other two algorithms partition the problem on the vertex set and so do not rely on the high density of the digraph to provide scope for parallelism. However for a fixed number of vertices, the overheads associated with the synchronisation required in these algorithms will become less significant as the density of the digraph increases.

If the vertex set is large compared to the number of processors, it will usually be possible to find a partition for the iterative method which results in good load balancing. For digraphs with shortest paths consisting of few edges, the number of iterations required will remain small as the number of processors used increases and hence the efficiency of the iterative method will be high. This is certainly not a factor which affects the efficiency of the other two algorithms.

As presented the block window algorithm only achieves good load balancing during step 1 if the cardinality of the sets $E_j \backslash W_r$ is similar for all $j \in W_r$. If this is not the case, the vertex set of the digraph can be partitioned into blocks consisting of more than P vertices. Each block can then be partitioned into P sets in a manner which will achieve good load balancing during step 1.

The efficiency of the block window algorithm also depends upon the number of vertices recalculated sequentially during step 2. If the proportion of edges which join vertices in the same block is low, this number is likely to be low and any load imbalance during the test in step 2 will be insignificant. This will be the case if, for example, the in-degree of the vertices is large compared to the size of the blocks.

Nicol (1989) implements the three algorithms on the Flex/32 (Matelan 1985), which is a shared memory MIMD parallel computer. The test problems have 1024 vertices and are generated in a way which ensures that the expected in-degree of the vertices is the same for almost all vertices. Edges joining i to j where $j - i > \omega$ are not considered by the generation procedure. In the test problems ω is chosen to be 64 and 128. The sparsity of the problems varies from 1% to 12.5%.

The theoretical analysis determines conditions under which the assumption in the first step of the block window algorithm (i.e. that values for vertices within the same block do not influence each other) has a high probability of being valid. Even though not all the test problems meet these conditions, the results show that the block window algorithm consistently out performs the other two on 2, 4, 8, and 16 processors. On 16 processors the block window algorithm achieves an efficiency of just over 50% when the sparsity is 12.5%, while the other two algorithms achieve efficiencies of under 20%.

11.3   Optimal Problem Subdivision

One dynamic programming formulation which arises in many combinatorial optimisation problems has the following functional equation:

$$v_{ij} = \min_{i \le k < j} \left\{ c_{ikj} + v_{ik} + v_{(k+1)j} \right\} \text{ for } 1 \le i < j \le n. \qquad (11.3)$$

For example, this approach can be used to determine the optimal parenthesization of a matrix product $M_1 M_2 \ldots M_n$. Let r_i be the number of rows in M_i and r_{n+1} be the number of columns in M_n. For the product to be well defined, M_i must have r_{i+1} columns. With $v_{ii} = 0$ and $c_{ikj} = r_i(2r_{k+1} - 1)r_{j+1}$ (the number of operations performed when calculating the product of one matrix with order $r_i \times r_{k+1}$ and another with order $r_{k+1} \times r_{j+1}$), v_{ij} can be interpreted as the minimum number of operations required in the calculation of the matrix product $M_i M_{i+1} \ldots M_j$.

To appreciate the value of an optimal parenthesization consider the following trivial case involving three small matrices. Calculate $M_1 M_2 M_3$ where M_1, M_2 and M_3 have orders (3×2), (2×4) and (4×1) respectively. Computing $M_1(M_2 M_3)$ requires 23 operations, while computing $(M_1 M_2)M_3$ requires 47 operations. The fact that the range should be so great in such a small example, illustrates the importance of parenthesization to the efficient calculation of large matrix products.

Another application of this dynamic programming formulation arises in the design of chemical plant to separate chemical mixtures into their components (Hendry and Hughes 1972). Since on serial machines the problem cannot be solved quickly enough to be of use to design engineers, parallel solution methods have been sought (Fraga and McKinnon 1991).

Since the solution of equation 11.3 is computationally intensive (requiring $\frac{n^3}{2} - \frac{n^2}{2}$ operations even when c_{ikj} requires no calculation) and the formulation is widely applicable, the development of efficient parallel algorithms for this problem has received a lot of attention recently.

Rytter (1988) shows that on a shared memory multiprocessor which allows concurrent reads, but does not permit concurrent writes (the CREW P-RAM model), equation 11.3 can be solved in $\log^2 n$ time using $\frac{n^6}{\log n}$ processors. Further if concurrent writes are permitted (the CRCW P-RAM model) the solution time can be reduced to $\log n$, because with this model the minimum of a set of values can be determined in a time which is independent of the size of the set (Kucera 1982).

Louka and Tchuente (1988) propose an algorithm for a distributed memory multiprocessor which solves equation 11.3 in $O(n)$ time using a systolic array consisting of $\frac{n^2}{8} + O(n)$ processors. A number of other methods which achieve $O(n)$ time were known at the time, but the best of these requires twice this number of processors (Gachet *et al.* 1986).

The performance achieved by the algorithms presented in these two papers must be considered as a theoretical bound on the performance which can be obtained, since these algorithms are not practical for solving realistic examples of equation 11.3 on the parallel computers currently available because of the large number of processors they require.

Miguet and Robert (1989) and Edmonds *et al.* (1990) consider the case in which the number of processors P is much smaller than n. The former uses a distributed memory multiprocessor, the latter a shared memory multiprocessor. Although designing algorithms for different architectures, the two papers use a similar strategy to distribute the calculation among the processors. This is loosely based on the following observations.

The values in the set $S_d = \{v_{ij} : j - i = d\}$ can be calculated independently given a knowledge of the values in the set $\bigcup_{t=0}^{d-1} S_t$. More precisely, calculating v_{ij} where $j - i = d$ requires $\{v_{ik} : k - i < d\}$ and $\{v_{kj} : j - k < d\}$. Thus in a parallel algorithm which solves equation 11.3, the sets S_d could be evaluated in order of increasing d, with the calculation of the values in each set being distributed. The set S_d corresponds to a diagonal of the cost matrix $\{v_{ij}\}_{i<j}$, which is upper triangular (see Figure 11.2).

The number of values in S_d is $n - d$ and calculating any value in the set requires $3d - 1$ operations. Therefore calculating the same number of values from S_d on each processor will achieve perfect load balancing, and when this is not possible the imbalance need be no more than $3d - 1$ operations.

If this imbalance is significant (as it might be since S_{n-1} contains only one element which requires $3n - 4$ operations to update), then, at the expense of some additional communication or synchronisation, the calculation of individual values could be distributed or the calculation of values in S_d could be

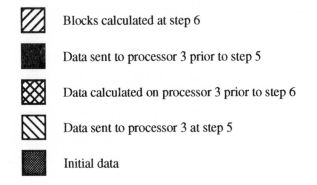

Blocks calculated at step 6

Data sent to processor 3 prior to step 5

Data calculated on processor 3 prior to step 6

Data sent to processor 3 at step 5

Initial data

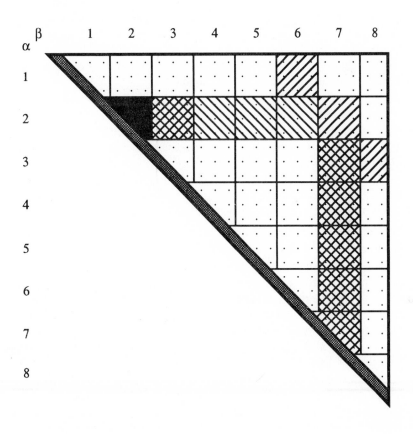

Figure 11.2: The data used by processor 3 at step 6 of the algorithm in a simple example with $n = 24$, $P = 4$ and $\ell = 3$

started as soon as the evaluation of S_e, where $e = \left\lceil \frac{d-1}{2} \right\rceil$, is complete.

Neither approach attempts this kind of fine load balancing. Edmonds *et al.* (1990) switch to a serial algorithm near the end of the solution, when the scope for parallel execution is not sufficient to keep all processors occupied. This prevents the synchronisation cost becoming dominant. Miguet and Robert (1989) assume that the load imbalance in the later stages of the algorithm is insignificant.

In both cases, the aim is to find an allocation of the values of the cost matrix to processors for calculation which achieves a good load balance throughout the solution, but does not introduce high communication or synchronisation overheads.

One method is to allocate the values in $V_t = \{v_{ij} : 1 \leq i < j, j \equiv (t-1) \bmod P\}$ to processor t and evaluate the sets S_d in sequence. This approach allocates columns of the cost matrix to processors in a wrap around fashion and achieves as good a load balance as possible, since the number of values in V_t satisfying $j - i = d$ is less than or equal to $\left\lceil \frac{n-d}{P} \right\rceil$. This is easy to see because $S_d = \{v_{ij} : d+1 \leq j \leq n, i = j - d\}$ and out of g consecutive numbers, at most $\left\lceil \frac{g}{h} \right\rceil$ are equal to f modulo h.

Both studies consider such an allocation, but find that the communication or synchronisation overheads are high and this limits the speed-ups which can be obtained. It is possible to evaluate a number of the sets S_d concurrently, and in so doing the cost of communication or synchronisation can be reduced at the expense of a poorer load balance.

This is the basis of the narrow-band algorithm proposed by Edmonds *et al.* (1990). The number of sets evaluated in parallel falls as the algorithm proceeds and is chosen to keep the load imbalance to a minimum. It is shown that the arithmetic cost of the algorithm is $O\left(\frac{n^3}{6P}\right)$, so that the dominant term is the same as for the method above, but the synchronisation cost is much less, $O(|\log_c n|)$ where $c = \frac{2P-1}{2P+1}$ provided P is much smaller than n.

Applied to a distributed memory parallel computer, this method of allocating values to processors results in a complex distribution of the data. The approach adopted by Miguet and Robert (1989) allocates blocks of ℓ consecutive columns of the cost matrix to processors, the rth block being allocated to processor t if $r \equiv (t-1) \bmod P$. This results in a straightforward distribution of the data. Note that if ℓ does not divide $n-1$, the final block will have less than ℓ columns.

The rows are partitioned in a similar way so that the cost matrix is partitioned into blocks $\{B_{\alpha\beta}\}$ where $1 \leq \alpha \leq \beta \leq \left\lceil \frac{n-1}{\ell} \right\rceil$. At the rth step of the algorithm, processor t calculates the values in blocks $B_{(\beta-r+1)\beta}$ where $\beta \equiv (t-1) \bmod P$ and then sends these values to processor $t+1$ along with the values in the previous $P - 2$ (or $\max\{0, r-2\}$ if $r < P$) blocks from row $\beta - r + 1$. For the final block ($\beta = \left\lceil \frac{n-1}{\ell} \right\rceil$) no values need be passed on, although doing so will not increase communication overheads and may simplify the code. This procedure is illustrated in Figure 11.2.

The communication protocol described is optimal in the sense that no processor receives any value more than once and all values received by a processor are required in its calculations. Since processor t communicates with processors $t-1$ and $t+1$ only, a ring of processors is the optimal topology for this protocol. In this case the communication cost is $O(\frac{n^2}{2})$. The communication protocol used by Miguet and Robert (1989) involves some redundancy and so is less efficient, it has a cost of $O(\frac{n^3}{P\ell})$. The arithmetic cost of the algorithm is $O\left(\frac{n^3}{6P}\right)$. Miguet and Robert (1989) construct an estimator for the choice of ℓ which maximizes the efficiency for a given n and P.

Both papers present numerical results. Edmonds *et al.* (1990) use a Sequent S27 parallel computer with 10 processors and 32 Mbytes of global shared memory. Using 9 processors the narrow-band algorithm achieves an efficiency of 92% for one application with $n = 1000$. Miguet and Robert (1989) use a ring of 16 T414 transputers, each transputer having its own local memory. Using all 16 processors and a block size of 6, an efficiency of 81% is achieved for a different application with $n = 1080$.

11.4 Markov Decision Processes

11.4.1 Introduction

Markov decision processes comprise an important class of dynamic programming problems. Applications arise in maintenance and replacement, inventory control, queuing models, harvesting policies, resource allocation and many other areas (White 1985). The calculation of the minimum expected cost for an infinite horizon time invariant discounted Markov decision process involves solving the following equation:

$$v_i = \min_{k \in K_i} \left\{ r_i^k + \beta \sum_{j \in S} p_{ij}^k v_j \right\} \forall\, i \in S. \tag{11.4}$$

It is known that equation 11.4 has a unique solution v^* and the rule which assigns to state i the action which minimises the right hand side of equation 11.4 for state i is an optimal policy (Bertsekas 1987).

An infinite horizon time invariant discounted Markov decision process can be described as follows.

The set of all possible states (or the state space) S is finite. Actions are chosen at regular time intervals or stages. For each state $i \in S$ an action is chosen from the action space K_i, which is finite.

When action k is chosen in state i an immediate cost r_i^k is incurred and the probability that the system will be in state j at the next stage is given by the transition probability p_{ij}^k. Transition probabilities satisfy

$$\sum_{j \in S} p_{ij}^k = 1 \,\forall\, i \in S,\, \forall\, k \in K_i$$

and $\quad 0 \le p_{ij}^k \le 1 \,\forall\, i, j \in S,\, \forall\, k \in K_i.$

Future returns are discounted by a factor β, where $0 \leq \beta < 1$.

The single item inventory problem is an easily understood example which helps to define the various components of a Markov decision process.

Depending on the circumstances stock levels may be reviewed daily, weekly or whatever. The state of the system is the number of units currently in stock and the state space is limited by the storage capacity.

The action is the number of units to be ordered. Any units ordered will arrive in stock after a delay referred to as the lead time. The action space in state i is finite and depends upon i since $i + k$ cannot exceed the storage capacity.

The immediate cost will typically consist of a fixed cost for placing an order, a cost per unit ordered, a cost per unit of stock held and a cost per unit of demand which cannot be met. The transition probabilities are determined by the demand for the item. For example if the lead time is one stage, then

$$
p_{ij}^{k} = \begin{cases} Pr(\text{demand in 1 stage} \geq i) \text{ if } j = k \\ Pr(\text{demand in 1 stage} = i - j + k) \text{ if } j \geq k. \\ 0 \text{ otherwise} \end{cases}
$$

A problem which is of much more practical interest is the multiple item inventory problem. In this case the state is a vector where each component is equal to the current stock level of one of the items and the action is the number of units of each item to be ordered. The transition probabilities will have a more complex form as the demand for each item need not be independent. This problem aptly illustrates the curse of dimensionality which plagues many real applications. Even a modest example with 10 items and a storage capacity for 6 of each has a million states.

Unlike in Sections 11.2 and 11.3 where the methods used find the exact solution in a finite number of steps, methods for solving equation 11.4 only find a good approximation to the optimal solution.

11.4.2   Value Iteration Methods

Value iteration methods are iterative schemes for solving Markov decision processes and are similar to iterative methods for solving systems of linear equations. These methods are good candidates for solution methods for Markov decision processes because transition matrices are usually large and sparse, the conditions under which iterative methods are often preferred for systems of linear equations, and they find an approximation which is guaranteed to be within a given tolerance of the optimal solution.

Previous studies (Thomas *et al.* 1983; Archibald *et al.* 1990) suggest that Pre-Jacobi and Gauss-Seidel are the two best algorithms. These are defined as follows.

Pre-Jacobi

$$
v^0 = 0
$$

$$v_i^n = \min_{k \in K_i} \left\{ r_i^k + \beta \sum_{j \in S} p_{ij}^k v_j^{n-1} \right\}.$$

Gauss-Seidel

$$v^0 = 0$$

$$v_i^n = \min_{k \in K_i} \left\{ \frac{r_i^k + \beta \sum_{j < i} p_{ij}^k v_j^n + \beta \sum_{j > i} p_{ij}^k v_j^{n-1}}{1 - \beta p_{ii}^k} \right\}.$$

Jacobi is another value iteration method which is useful for comparison with parallel Gauss-Seidel.

Jacobi

$$v^0 = 0$$

$$v_i^n = \min_{k \in K_i} \left\{ \frac{r_i^k + \beta \sum_{j \neq i} p_{ij}^k v_j^{n-1}}{1 - \beta p_{ii}^k} \right\}.$$

Further details of these schemes can be found in Bertsekas (1987).

Since these two schemes have linear update rules v^n can be expressed in terms of v^{n-1} by $v^n = M^n v^{n-1}$, although in the case of Gauss-Seidel M^n is never calculated explicitly. M^n is called the iteration matrix for the policy selected at the nth iteration.

The iteration matrices for Pre-Jacobi as applied to Markov decision processes have a simple form, β times a stochastic matrix. The largest eigenvalue of the iteration matrices is therefore β and the corresponding eigenvector is $(1, 1, \ldots, 1)^T$. A good estimate of v^* can be obtained from v^n by an acceleration along the direction of this eigenvector (Porteus 1971). As a result, the convergence of Pre-Jacobi is governed by the 2nd eigenvalues of the iteration matrices used (Morton and Wecker 1977).

Although the iteration matrices for Gauss-Seidel do not have this stochastic property, since the method always uses the most recent estimates available, good convergence is often obtained.

11.4.3 A Synchronous Approach

Archibald *et al.* (1990) implement Pre-Jacobi and Gauss-Seidel on a distributed memory MIMD machine using the following techniques.

For a network of P processors, partition the state space S into P sets denoted by $\{S_t\}_{t=1}^P$. During an iteration let each processor r update the states in S_r and send the new estimates to all other processors (in a procedure known as a broadcast) before the start of the next iteration.

It is straightforward to apply this principle to Pre-Jacobi and the resulting parallel algorithm has exactly the same convergence properties as serial Pre-Jacobi. Figure 11.3(a) illustrates how this algorithm would be scheduled on a

(a) Synchronous case with no overlap

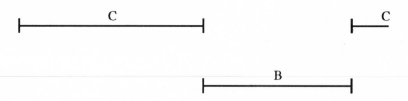

(b) Asynchronous case with overlap

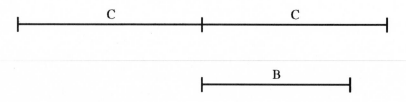

(c) Phased pipeline algorithm

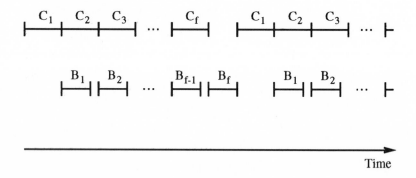

Time

Figure 11.3: A comparison of the three approaches

processor. Updating the states is represented by the event C and this must be completed before the broadcast, represented by B, can commence. Subsequent calculation cannot take place until the broadcast is completed.

With Gauss-Seidel it is not generally possible to update any of the states in parallel. However two modifications of the Gauss-Seidel algorithm which allow calculations to be executed in parallel and use more up-to-date estimates than Pre-Jacobi are suggested.

In the first, GS1, a processor uses the most recent estimates for states local to that processor and estimates from the previous iteration for all other states. This algorithm has exactly the same communication requirements as parallel Pre-Jacobi.

For both parallel Pre-Jacobi and GS1 the partition of the state space is chosen so that updating the states in each set involves approximately the same number of operations. In this way a good load balance is achieved. Communication overheads would be minimised if each set in the partition contained the same number of states; however it is assumed that the resulting gain would be insignificant when compared with the load imbalance introduced.

In the second parallel Gauss-Seidel method, GS2, every processor is kept informed of the most recent estimates for all states. Each processor updates one of its states and then passes this new estimate to all other processors and receives a new estimate from every other processor before continuing with its calculation. For this algorithm each processor is allocated the same number of states. Although this allocation minimises the communication cost, load balance will be poor if states which are updated concurrently take different times to update. Again Figure 11.3(a) applies, but in this case C refers to the updating of one state and B to the broadcast of one estimate from each processor.

For good convergence of GS1 and GS2, the transition matrices corresponding to the policies selected during the solution should have considerable weight in the shaded areas of Figure 11.4 and Figure 11.5 respectively. These areas correspond to transitions to states for which up-to-date estimates are available when the probabilities of these transitions are used in the calculation. Since the areas become smaller as the number of processors increases, the convergence of the two parallel Gauss-Seidel methods becomes slower as the number of processors increases. Eventually the convergence is equivalent to that of the serial Jacobi algorithm which is known to be very poor (Thomas *et al.* 1983). As a consequence of this in cases where Gauss-Seidel is the best serial algorithm, Pre-Jacobi is often the best parallel algorithm.

The convergence of GS2 is generally faster than that of GS1, but this must be offset against considerably higher communication overheads and generally poorer load balancing.

All these parallel algorithms allow the problem data to be distributed which is an important consideration since, as noted in the introduction, practical examples of dynamic programming problems have huge data sets. The only data which is duplicated on all processors is the current estimates. These are accessed

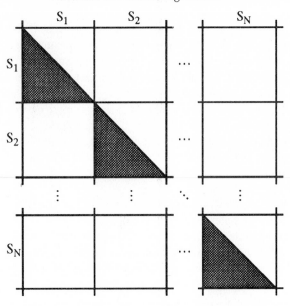

Figure 11.4: The critical area for GS1

more often than they are changed, so duplication reduces communication time.

The results presented by Archibald *et al.* (1990) were obtained using a Meiko computing surface (Trew and Wilson 1991) and a number of test problems each of which has a different set of properties. Using 16 processors an average efficiency of around 85% was obtained by parallel Pre-Jacobi. However because of the additional iterations required by GS1 and GS2, these algorithms only achieved an efficiency of around 55% and 45% respectively.

In many structured problems, the states can be ordered so that most of the weight in the transition matrix for the optimal policy lies just below the diagonal. This is good for the convergence of both GS1 and GS2, but because of the additional communication overheads of GS2, GS1 performs better.

In problems where the transition structure is random, GS2 exhibits better convergence than GS1 (because the shaded area in Figure 11.5 is greater than that of Figure 11.4) and consequently out performs GS1.

The results show that there is no algorithm which is best for all problems and suggest that the algorithm used should be matched to the properties of the problem to be solved.

Archibald *et al.* (1990) conclude that for very large sparse problems communication overheads will be the major factor limiting the speed-up obtainable. This effect might be reduced by following an asynchronous approach with less frequent communication.

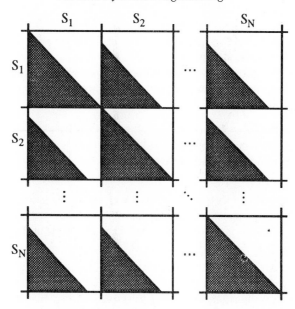

Figure 11.5: The critical area for GS2

11.4.4 An Asynchronous Approach

Bertsekas (1982) describes an asynchronous parallel algorithm for dynamic programming and proves that it converges under certain assumptions. The algorithm applies to a wide class of formulations including shortest path problems and finite and infinite horizon Markov decision processes. A general iterative algorithm is considered which when applied to Markov decision processes encompasses value iteration.

As before, the state space is partitioned into P sets denoted by $\{S_t\}_{t=1}^{P}$ and processor r is responsible for updating the states in S_r.

Processor r is called a neighbour of processor t if the estimates for states in S_r influence the estimates for states in S_t. For Markov decision processes this happens if and only if under some action there exists a transition with non-zero probability from a state in S_t to a state in S_r. The estimates for states updated on a processor depend only on the estimates for states which are updated on one of its neighbours. Therefore a processor need only receive estimates from its neighbours and send its estimates to processors of which it is a neighbour. In some circumstances this modification alone will lead to reduced communication overheads, for instance if P is very large or the problem has special structure.

At any time a processor can be in one of three phases. It can be updating the states it has been allocated or transmitting its latest estimates to one or more processors of which it is a neighbour or idling. It is assumed that a processor can update and receive concurrently, transmit and receive concurrently and transmit to more than one processor at a time. However processors are not able to update and transmit in parallel.

Under the following two assumptions, it is shown that an asynchronous algorithm organised in this way converges to the optimal solution of the problem.

Assumption 1

There exists $T > 0$ such that for all processors r and for all time intervals of length T, processor r performs at least one update of all states in S_r and at least one transmission to all processors of which it is a neighbour.

Assumption 2

There exists vectors $\underline{v}$ and $\overline{v}$ such that $\underline{v}$ converges monotonically increasing to v^*, $\overline{v}$ converges monotonically decreasing to v^* and $\underline{v} \leq v^{0,r} \leq \overline{v}$ for all processors r, where v^* is the optimal solution and $v^{0,r}$ is the initial approximation on processor r.

Further it is shown that for shortest path problems and finite horizon Markov decision problems, convergence occurs in finite time.

Since this algorithm uses more selective and less frequent communication, it will have lower communication overheads than the synchronous algorithm described in the previous section. A further reduction in communication overheads could be obtained if transmission and updating is allowed to occur concurrently.

Figure 11.3(b) illustrates such an asynchronous scheme for the case in which calculation and communication can be overlapped without penalty. Each time a processor updates all its states (represented by C), the new estimates are sent to all other processors in a broadcast (represented by B), but the processor does not wait for this broadcast to be completed before starting the next update. A processor uses the most recent estimates available in its calculation. Comparing this with Figure 11.3(a), it is apparent that the average time per update is far less, due to the reduced communication overheads.

Unfortunately, there are no bounds which relate the latest estimates with the optimal solution for an asynchronous algorithm, although it would be possible to perform a synchronous iteration periodically to test for convergence.

The good convergence properties of Pre-Jacobi depend upon the iteration matrices being stochastic. An asynchronous version of Pre-Jacobi will not have this property and consequently the rate of convergence will be slower. It is likely that this effect will outweigh the benefits of reduced communication overheads. However using the parallel Pre-Jacobi algorithm described in the next section, it is possible to reduce communication overheads by overlapping updating and broadcasting and still preserve the convergence properties of serial Pre-Jacobi.

When an asynchronous algorithm is applied to Gauss-Seidel, the effect on convergence will not be as bad as for Pre-Jacobi. However the algorithm described in the next section is still attractive as it offers lower communication overheads and better convergence than GS1.

11.4.5 The Phased Pipeline Algorithm

Archibald *et al.* (1991) develop a synchronous parallel algorithm based on the value iteration methods described in Section 11.4.2 which splits the work performed on each processor into phases so that updates and broadcasts can take place in parallel. The resulting algorithm is called the phased pipeline algorithm and it has the same convergence properties as the algorithm described in Section 11.4.3, but has significantly lower communication overheads.

As in the two preceding sections the state space is partitioned into P sets denoted by $\{S_t\}_{t=1}^{P}$, however unlike before each set S_t in this partition is itself partitioned into f sets denoted by $\{S_{t\ell}\}_{\ell=1}^{f}$. Each iteration of the algorithm involves $f + 1$ steps.

The first of these steps involves only calculation, with processor r updating the states in S_{r1}. For the next $f - 1$ steps, calculation and communication is overlapped on each processor. At step ℓ, where $2 \le \ell \le f$, three tasks are performed concurrently on processor r. The states in $S_{r\ell}$ are updated, the latest estimates for states in $S_{r(\ell-1)}$ are broadcast and the latest estimates for states in $S_{t(\ell-1)}$, where $t \ne r$, are received. All three tasks are completed before the processor proceeds to the next step. The final step involves only communication, with processor r broadcasting the latest estimates for states in S_{rf} and receiving the latest estimates for states in S_{tf}, where $t \ne r$.

A check for convergence can be performed at the end of each iteration.

When applying this method to Pre-Jacobi, in order to preserve the convergence properties of serial Pre-Jacobi, the new estimates received by a processor during an iteration are not used in that processor's calculation until the following iteration. Therefore implementing this algorithm requires some additional storage.

Figure 11.3(c) shows how this scheme might be scheduled on a processor for the case in which calculation and communication can be overlapped without penalty. For processor r, C_ℓ represents updating the states in $S_{r\ell}$ and B_ℓ represents broadcasting the latest estimates for states in $S_{r\ell}$ and receiving the latest estimates for states in $S_{t\ell}$, where $t \ne r$. In the time required to update all the states assigned to a processor, all but a few of the new estimates have been broadcast. In terms of time per update, this algorithm is far superior to the synchronous algorithm in Figure 11.3(a) and approaches the time per update achieved by the asynchronous algorithm in Figure 11.3(b) as the number of phases increases.

If this method is applied to Gauss-Seidel and a processor uses the most recent estimates for states local to that processor and estimates from the previous iteration for all other states, then the resulting algorithm has the same convergence properties as GS1, but has greatly reduced communication overheads.

However since the convergence of a parallel Gauss-Seidel algorithm is generally improved by the use of a greater number of up-to-date estimates, it is sensible for a processor to use the new estimates in its calculation as soon as they are received. This also eliminates the need for extra storage. The

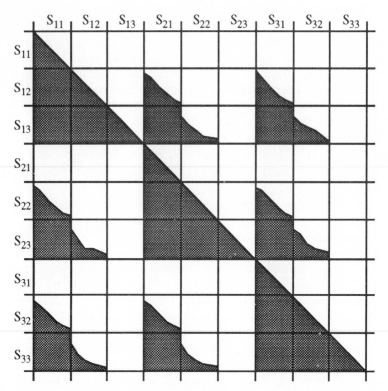

Figure 11.6: An example of the critical area for GS3 with $N = 3$ and $f = 3$

behaviour of the resulting algorithm, GS3, cannot be predetermined, but will depend upon the manner in which communication is scheduled.

For good convergence of GS3, the transition matrices corresponding to the policies selected during the solution should have considerable weight in some subset of the shaded area in Figure 11.5. Figure 11.6 shows one possible candidate for this area for a small example. The critical area in off-diagonal partially filled squares is determined by the scheduling of the communication and may vary from iteration to iteration. It is apparent from the figures that in terms of convergence GS3 is somewhere between GS1 and GS2. As the number of phases increases the convergence of GS3 approaches that of GS2, while the time per update approaches that of the asynchronous scheme in Figure 11.3(b).

The time per update can be further reduced by modifying the method so that the last step of one iteration is overlapped with the first step of the next iteration. This approach is not attractive for Pre-Jacobi, as it is not possible to retain the convergence properties of serial Pre-Jacobi, but may work well for Gauss-Seidel even though to test for convergence it is necessary to perform a synchronous iteration from time to time.

Archibald *et al.* (1991) report that when calculation and communication are overlapped on a transputer, communication interferes with calculation in two ways. Firstly the set-up of a link prior to transmission or reception requires the cpu and so blocks the calculation. Secondly during transmission or reception, the conflict between link buffers and the cpu for memory access slows the rate of calculation.

Archibald *et al.* (1991) incorporate these points into a model for an implementation of the phased pipeline algorithm on a Meiko computing surface (Trew and Wilson 1991) and use this model to estimate the number of processors and number of phases which give maximum speed-up for the problem to be solved. The results presented show a good agreement between the model's predictions and the actual solution times.

In one test problem with 60,000 states and an average of 3 actions per state, a speed-up of 60 times was obtained using 129 processors and 4 phases. The model predicts that to obtain a similar speed-up with the synchronous algorithm from Section 11.4.3 would require more than twice this number of processors. Due to the overheads involved in overlapping calculation and communication on a transputer, it is not worth using the phased pipeline algorithm for small problems (around 1000 states).

Archibald *et al.* (1991) conclude that the phased pipeline algorithm is an efficient solution method for large scale Markov decision processes even when there is significant overhead involved in overlapping calculation and communication.

11.4.6   Other Solution Methods

Policy iteration (Hartley *et al.* 1986) is another solution method for infinite horizon time invariant discounted Markov decision processes. Like value iteration, this is an iterative scheme and there is a close relationship with solution algorithms for systems of linear equations.

Each iteration involves two steps, policy evaluation and policy improvement. At the policy evaluation step the value of the current policy δ is evaluated by solving the following system of equations:

$$v_i = r_i^{\delta(i)} + \beta \sum_{j \in S} p_{ij}^{\delta(i)} v_j.$$

At the policy improvement step, one iteration of a value iteration method is performed to determine if there exists a policy with lower value. If no such policy can be found then the algorithm terminates, otherwise the policy found is evaluated at the next iteration. Any policy can be chosen to start the procedure.

Generally Pre-Jacobi or Gauss-Seidel is used in the policy evaluation step, so the techniques used to distribute the value iteration methods can be applied. However this is not a requirement of the algorithm and any suitable parallel method for solving systems of linear equations can be employed in the policy evaluation step. At the policy improvement step any parallel synchronous value iteration method can be used.

Another alternative would be to formulate the Markov decision process as a linear programming problem (Bertsekas 1987) as follows:

Maximize $\sum_{i \in S} v_i$

Subject to $v_i - \beta \sum_{j \in S} p_{ij}^k v_j \le r_i^k \quad \forall k \in K_i, \forall i \in S$

and apply a parallel linear programming algorithm.

11.5   Conclusions

The three dynamic programming formulations which have been discussed appear to be very different. The first two are deterministic, while the third is probabilistic. In the first formulation a decision results in a transition to exactly one state, while in the others transitions can be made to more than one state. However despite these differences, the parallel solution methods have much in common.

In a parallel algorithm for a dynamic programming problem the states can be considered in parallel and the actions for each state can be considered in parallel. The algorithms described by Louka and Tchuente (1988) and Rytter (1988) use both of these approaches, exploiting all the parallelism inherent in the problems and the theoretical analysis of calculation and communication time shows that a fast algorithm can be obtained. However most implementations only consider the states in parallel, exploiting a coarser grain parallelism. Even when this results in more iterations than the serial algorithm, the overall solution time can still be reduced.

The same approach can be employed on different architectures. For example although the algorithms developed by Miguet and Robert (1989) and Edmonds *et al.* (1990) to find optimal problem subdivisions, Section 11.3, are targeted at distributed memory multiprocessors and shared memory multiprocessors respectively, they use similar techniques and achieve similar efficiencies.

The aim of applying parallel processing techniques to dynamic programming problems is to reduce solution time and to increase the size of problem which can be solved. All the algorithms described achieve a significant speed-up over their serial counterparts and the size of problem now being routinely solved is far greater than in earlier studies. For example the performance comparison carried out by Porteus (1981) used test problems with up to 200 states, while in a recent paper on parallel dynamic programming (Archibald *et al.* 1991) problems with as many as 60,000 states are solved.

With larger faster parallel computers becoming available all the time, this work is only a beginning, but it indicates that parallel processing has an important role to play in the future of dynamic programming as a tool for decision makers.

References

Al-Dabass D. (1980) Two methods for the solution of the dynamic programming algorithm on a multiprocessor cluster. *Optimal Control Applications and Methods*, **1**, pp. 227–38.

Archibald T. W., McKinnon K. I. M. and Thomas L. C. (1990) Serial and parallel value iteration algorithms for discounted Markov decision processes. Working Paper **90/14**, Department of Business Studies, University of Edinburgh, Scotland. To appear in *European Journal of Operational Research*.

Archibald T. W., McKinnon K. I. M. and Thomas L. C. (1991) Performance issues for the iterative solution of Markov decision processes on parallel computers. *Working Paper*, Department of Business Studies, University of Edinburgh, Scotland. To appear in *Proc. of the 1991 Edinburgh Workshop on Parallel Numerical Analysis*.

Bellman R. (1957) *Dynamic Programming*, Princeton University Press, Princeton, New Jersey.

Bellman R. and Dreyfus S. (1962) *Applied Dynamic Programming*, Princeton University Press, Princeton, New Jersey.

Bertsekas D. P. (1982) Distributed dynamic programming. *IEEE Transactions on Automatic Control*, **AC–27**, pp. 610–16.

Bertsekas D. P. (1987) *Dynamic Programming: Deterministic and Stochastic Models*, Prentice-Hall, Englewood Cliffs, New Jersey.

Bertsekas D. P. and Tsitsiklis J. N. (1989) *Parallel and Distributed Computation: Numerical Methods*, Prentice-Hall, Englewood Cliffs, New Jersey.

Casti J., Richardson M. and Larson R. (1973) Dynamic programming and parallel computers. *Journal of Optimization Theory and Applications*, **12**, pp. 423–438.

Edmonds P., Chu E. and George A. (1990) Dynamic programming on a shared-memory multiprocessor. *Report* **CS–90–49**, Department of Computer Science, University of Waterloo, Canada.

Ford L. R. and Fulkerson D. R. (1962) *Flows in Networks*, Princeton University Press, Princeton, New Jersey.

Fraga E. S. and McKinnon K. I. M. (1991) Parallel optimization techniques for optimal heat integrated separation sequence generation. *Technical Report* **1991–2**, Department of Chemical Engineering, University of Edinburgh, Scotland. To appear in *Proc. of the 1991 Edinburgh Workshop on Parallel Numerical Analysis*.

Gachet P., Joinnault B. and Quinton P. (1986) Synthesizing systolic arrays using DIASTOL. In *Proc. of the International Workshop on Systolic Arrays*, (Ed. by W. Moore, A. McCabe and R. Urquhart), pp. 25–36. Adam Hilger, Bristol, UK.

Hartley R., Lavercombe A. C. and Thomas L. C. (1986) Computational comparison of policy iteration algorithms for discounted Markov decision processes. *Computers and Operations Research*, **13**, pp. 411–20.

Hendrikx M., Nunen J. van and Wessels J. (1984) On iterative optimization of structural Markov decision processes. *Mathematische Operationsforschung und Statistik*, **15**, pp. 439–60.

Hendry J. E. and Hughes R. R. (1972) Generating separation process flowsheets. *Chemical Engineering Progress*, **68(6)**, pp. 71–6.

Howard R. A. (1960) *Dynamic Programming and Markov Processes*, John Wiley, New York.

Knuth D. E. (1981) Breaking paragraphs into lines. *Software Practice and Experience*, **11**, pp. 1119–84.

Kucera L. (1982) Parallel computation and conflicts in memory access. *Information Processing Letters*, **14**, pp. 93–6.

Louka B. and Tchuente M. (1988) Dynamic programming on two dimensional systolic arrays. *Information Processing Letters*, **29**, pp. 97–104.

Matelan N. (1985) The Flex/32 multicomputer. In *Proc. of the 12th International Symposium on Computer Architecture*, pp. 209–13, Computer Society Press.

Miguet S. and Robert Y. (1989) Dynamic programming on a ring of processors. *Report* **89–01**, Laboratoire LIP, Ecole Normale Supérieure de Lyon, France.

Morton T. E. and Wecker W. R. (1977) Discounting, ergodicity and convergence for Markov decision processes. *Management Science*, **23**, pp. 890–900.

Nicol D. M. (1989) Parallel solution of sparse one-dimensional dynamic programming problems. *ICASE report* **89–17**, NASA, Langley Research Center, Hampton, Virginia.

Porteus E. L. (1971) Some bounds for discounted sequential decision processes. *Management Science*, **18**, pp. 7–11.

Porteus E. L. (1981) Computing the discounted return in Markov and semi-Markov chains. *Naval Research Quarterly*, **28**, pp. 567–77.

Rytter W. (1988) On efficient parallel computations for some dynamic programming problems. *Theoretical Computer Science*, **59**, pp. 297–301.

Tabak D. (1968) Computational improvement of dynamic programming solutions by multiprocessing techniques. *IEEE Transactions on Automatic Control*, **AC–13**, p. 596.

Thomas L. C., Hartley R. and Lavercombe A. C. (1983) Computational comparison of value iteration algorithms for discounted Markov decision processes. *Operations Research Letters*, **2**, pp. 72–6.

Trew A. and Wilson G. (Eds.) (1991) *Past, Present, Parallel: A survey of available parallel computing systems*, Springer-Verlag.

White D. J. (1985) Real applications of Markov decision processes. *Interfaces*, **15**, pp. 73–8.

CHAPTER 12

Parallel Derandomisation Techniques

Yijie Han

University of Kentucky, USA

12.1 Introduction

With the advance of computer technology it is now possible to build parallel computers with thousands even millions of processors. These parallel computers are capable of executing many computer instructions simultaneously. Thus the study of parallelism of problems to be solved by parallel computers and the design of parallel algorithms become an important research topic. Problems for which trivial sequential algorithms exist are sometimes intriguing problems to be solved in parallel. Many techniques have been invented to exploit parallelism. In this chapter we will outline a powerful technique for the design of parallel algorithms, the derandomisation technique.

The basic idea of derandomisation is to start with a randomised algorithm and then obtain a deterministic algorithm by applying the technique of derandomisation.

Derandomisation is a powerful technique because with the aid of randomisation the design of algorithms, especially the design of parallel algorithms, for many difficult problems becomes manageable. The technique of derandomisation offers us the chance of obtaining a deterministic algorithm which would be difficult to obtain otherwise. For some problems the derandomisation technique enables us to obtain better algorithms than those obtained through other techniques.

Although the derandomisation technique has been applied to the design of sequential algorithms (Raghavan 1988; Spencer 1987) these applications are sequential in nature and cannot be used directly to derandomise parallel algorithms. To apply derandomisation techniques to the design of parallel algorithms we have to study how to preserve or exploit parallelism in the process of derandomisation. Thus we put emphasis on derandomisation techniques which allow us to obtain fast and efficient parallel algorithms.

Every technique has its limit, so does the derandomisation technique. In order to apply derandomisation techniques, a randomised algorithm must be first designed or be available. Although every randomised algorithm with a finite sample space can be derandomised, it does not imply that the derandomisation approach is always the right approach to take. Other algorithm design techniques might yield much better algorithms than those obtained through derandomisation. Thus it is important to classify situations where derandomisation techniques have a large potential to succeed. In the design of parallel algorithms we need to identify situations where derandomisation techniques could yield good parallel algorithms.

Since derandomisation techniques are applied to randomised algorithms to yield deterministic algorithms, the deterministic algorithms are usually derived at the expense of a loss of efficiency (time and processor complexity) from the original randomised algorithms. Thus we have to study how to obtain randomised algorithms that are easy to derandomise and have small time and processor complexities.

12.2 The Scheme of Derandomisation

The basic idea of derandomisation is to obtain a deterministic algorithm by first designing a randomised algorithm and then removing the randomness from the randomised algorithm. Derandomisation is a powerful tool in the design of algorithms because it is usually easier to design a randomised algorithm than to design a deterministic algorithm. If we know how to design a randomised algorithm and how to derandomise the algorithm we will arrive at a deterministic algorithm.

A straightforward way of derandomisation is to examine all the sample points to find a good point. Suppose we know that the expectation of a random variable r is good through a probabilistic analysis. Then there is a sample point p on which the value of r is good. Here r could be a random variable indicating the failure rate, the size of an independent set, the execution time of a procedure, etc. Good is interpreted as the value being greater than or equal to (less than or equal to) a desired quantity. A sample point on which the value of r is good is called a good point. Since there exists a good point, by examining all sample points we are guaranteed to find a good sample point p. Suppose that a randomised algorithm outputs a value of r, then by using point p we have a deterministic algorithm which outputs a value no worse than $E[r]$.

The last paragraph gives the basic idea of derandomisation. This idea, when applied, often yields deterministic algorithms with an exponential number of operations. That is, in the case of a sequential algorithm it gives an exponential time algorithm, and in the case of a parallel algorithm it requires an exponential number of processors to achieve polylogarithmic time. This is caused by the size of the sample space which, in many situations, has an exponential number of points. Two approaches are usually assumed to obtain efficient deterministic algorithms. One is to use a small sample space, the other is to avoid exhaustive search of the sample space. The first approach requires a good design of the sample space while the second approach requires an efficient search strategy.

Both small sample space and fast search techniques are usually required when the derandomisation method is used to obtain efficient NC (Cook 1979, 1985) algorithms. This is because in many situations the search strategy can examine only a constant number of subspaces in polylogarithmic time. There are exceptions where exponential sized sample space is used to achieve efficient parallel algorithms.

Thus to obtain an efficient algorithm through derandomisation we have to design a sample space easy to search, to perform a probabilistic analysis showing that the expectation of a desired random variable is no less than demanded and to provide an efficient search technique which ultimately returns a good sample point.

12.2.1 The Design of Sample Space

Two considerations are usually given to the design of a sample space. One is the size of the sample space, *i.e.* the number of sample points in the space. The other is the independence of the random variables in the sample space.

It usually takes less resource to search a small sample space than it does to search a large one. This is particularly so when exhaustive search technique is used to locate a good sample point. In the case of the design of a parallel algorithm, a smaller sample space implies that fewer processors are needed in order to examine all the points in the sample space.

The independence of random variables in the sample space is also an important factor. When binary search technique is used to search the sample space, the random variables can be fixed one by one. In this case we usually want the random variables to be fixed independent of other unfixed random variables. The independence condition usually helps the probabilistic analysis showing that after fixing the random variables the remaining smaller sample space contains a good point. In many situations the use of independence also helps the computation of the expectation of the subspace and thus facilitates the design of an efficient algorithm. In the design of parallel algorithms the independence condition sometimes helps to fix several random variables simultaneously and independently, thus speeding up the derandomisation process. In these situations we usually require large degree independence among random variables.

Mutually independent random variables are used in the design of sequential algorithms (Raghavan 1988; Spencer 1987). When n 0/1-valued uniformly distributed mutually independent random variables are needed in the design of a randomised algorithm, the sample space $S = \{0, 1\}^n$ can be used. Such a sample space contains an exponential number of sample points. Efficient search technique is needed to locate a good point.

In the design of parallel algorithms we usually want a small sample space. This can be achieved by using limited independence among random variables. When n 0/1-valued uniformly distributed pairwise independent random variables are needed, a sample space containing $O(n)$ points can be used. Let $k = \lceil \log n \rceil$. The sample space is $\Omega = \{0, 1\}^{k+1}$. For each $a = a_0 a_1 ... a_k \in \Omega$, $Pr(a) = 2^{-(k+1)}$. The value of random variables x_i, $0 \le i < n$, on point a is $x_i(a) = (\sum_{j=0}^{k-1}(i_j a_j) + a_k) \bmod 2$, where i_j is the j-th bit of the binary expansion of i. This design is given by Luby (1988).

To obtain n 0/1-valued uniformly distributed k-wise independent random variables, a set of n vectors $S = \{i | i \in Z_2^l\}$ in which every k vectors are linearly independent can be used, where l is a function of n and k. The random variable x_i corresponding to $i \in S$ has value $x_i(a) = \sum_{j=0}^{l-1}(i_j a_j)$ mod 2. It can be verified that x_is are k-wise independent (Berger and Rompel 1989). When NC algorithms are to be designed using the binary search technique to locate a good sample point the sample space cannot be larger than $n^{\log^c n}$, therefore we can use n random variables with at most $(\log^c n)$-wise independence (Berger and Rompel 1989), where c is a constant. In this case l is bounded by $O(\log^{c+1} n)$.

The above mentioned designs only give random variables distributed uniformly on $\{0, 1\}$. When random variables with range greater than 2 or random variables distributed nonuniformly on $\{0, 1\}$ are needed, the sample space can be designed as follows. Let $q \geq n$ be a prime. n random variables which are d-wise independent and uniformly distributed on $\{0, 1, ..., q-1\}$ can be designed on a sample space S containing q^d points. Each point in S is assigned probability $1/q^d$. Let $x = < x_0, x_1, ..., x_{d-1} >$ be a point in S. The value of random variable r_i on x is $r_i(x) = (\sum_{j=0}^{d-1} x_j i^j)$ mod q. This design is given by Joffe (1974) and by Luby (1986). By imposing a function $f : \{0, 1, ..., q-1\} \mapsto \{0, 1, ..., p\}$ on random variable r_i we obtain a random variable taken value from set $\{0, 1, ..., p\}$. Luby (1986) used a table to implement function f and therefore obtained random variables nonuniformly distributed on $\{0, 1\}$. Although the designed sample space is rather compact in that it contains only q^d points, it is not easy to evaluate conditional expectation on a subspace and not easy to search the sample space using binary search method. Thus this space is a good design when exhaustive search method is used to convert a randomised algorithm to a deterministic algorithm. In particular, Luby used exhaustive search on such a sample space to obtain a fast parallel algorithm for the maximal independent set problem (Luby 1986).

Another design given by Luby (1988) is to use a set of 0/1-valued uniformly distributed pairwise independent random variables to construct random variables which are not uniformly distributed. Let $R_i = \{r_{i0}, r_{i1}, ..., r_{i,n-1}\}$ be a set of 0/1-valued uniformly distributed pairwise independent random variables. Take k sets, $R_0, R_1, ... R_{k-1}$. Let the random variables in R_is be mutually independent. Then the random variables $x_j = \sum_{t=0}^{k-1} r_{tj} 2^t, 0 \leq j < n$, are pairwise independent random variables uniformly distributed on $\{0, 1, ..., 2^k - 1\}$. A function can then be imposed on this range to obtain nonuniformly distributed random variables. This sample space contains $O(n^k)$ points. When k is not a constant the sample space contains more than a polynomial number of points. Thus exhaustive search cannot be used on this sample space when k is not a constant. However, this sample space is well suited for binary search. Luby showed how to search this sample space to obtain efficient parallel algorithms for the $\Delta + 1$ vertex colouring problem, the maximal independent set problem and the maximal matching problem (Luby 1989).

Recently Han and Igarashi gave a new design of a sample space on which n 0/1-valued uniformly distributed mutually independent random variables can be built (Han and Igarashi 1990). n 0/1-valued uniformly distributed mutually independent random variables r_i, $0 \leq i < n$, are used. Without loss of generality assume that n is a power of 2. A tree T which is a complete binary tree with n leaves plus a node which is the parent of the root of the complete binary tree (thus there are n interior nodes in T and the root of T has only one child). n variables $x_0, x_1, ..., x_{n-1}$ are associated with n leaves of T and the n random variable r_is are associated with the interior nodes of T. The n leaves of T are numbered from 0 to $n - 1$. Variable x_i is associated with leaf i. Variables x_i, $0 \leq i < n$, are randomised as follows. Let r_{i_0}, r_{i_1}, ..., r_{i_k} be the random variables on the path from leaf i to the root of T, where $k = \log n$. Random variable x_i is defined to be $x_i = (\sum_{j=0}^{k-1} i_j r_{i_j} + r_{i_k})$ mod 2. It can be verified that random variables x_i, $0 \leq i < n$, are uniformly distributed mutually independent random variables. Tree T is called a random variable tree. This design is particularly suited to the derandomisation of certain random variables (Han and Igarashi 1990).

Combining Luby's design (Luby 1988) and Han and Igarashi's design (Han and Igarashi 1990) we can obtain mutually independent random variables uniformly distributed in $\{0, 1, ..., 2^k - 1\}$ as follows. Let $R_i = \{r_{i0}, r_{i1}, ..., r_{i,n-1}\}$ be a set of 0/1-valued uniformly distributed mutually independent random variables. Take k sets, $R_0, R_1, ...R_{k-1}$. Let the random variables in R_is be mutually independent. Then the random variables $x_j = \sum_{t=0}^{k-1} r_{tj} 2^t, 0 \leq j < n$, are mutually independent random variables uniformly distributed on $\{0, 1, ..., 2^k - 1\}$. We can then apply a function on this range to obtain nonuniformly distributed random variables.

It is far from clear what kind of sample space is the best for the design of parallel algorithms through derandomisation. The choice of the sample space is related to the probabilistic analysis, the search technique, and therefore the problem to be solved. The works by Berger and Rompel (1989), Luby (1988), Han and Igarashi (1990), and Han (1991a, 1991b) began to demonstrate certain design paradigms for the construction of sample spaces good for achieving efficient parallel algorithms.

12.2.2 The Probabilistic Analysis

The probabilistic analysis is needed first to show that the expectation of a desired random variable is no worse than demanded. Since the expectation of the random variable is good, there must exist a good point. In the case when the space partition method is used to locate a good sample point, conditional expectation has to be analyzed. Luby's analysis for the vertex partition problem (Luby 1988) provides an excellent example here to showcase the probabilistic analysis of a problem.

The vertex partitioning problem is to label vertices of a graph $G = (V, E)$ with 0s and 1s, $l : V \mapsto \{0, 1\}$, such that the size of the crossing set

$|\{(i,j)|(i,j) \in E, l(i) \neq l(j)\}| \geq |E|/2$. $|V|$ 0/1-valued uniformly distributed mutually independent random variables x_i, $0 \leq i < n$, can be used (Luby 1988), one for each vertex. For $(i,j) \in E$, $f(x_i, x_j) = x_i \oplus x_j$ is 1 iff (i,j) is in the crossing set, where x_i (x_j) is the random variable associated with vertex i (j) and $\oplus$ is the exclusive-or operation. The total number of edges in the crossing set can be expressed as $F = \sum_{(i,j) \in E} f(x_i, x_j)$. The expectation of F is $E[F] = |E|/2$. Thus there exists a sample point p such that $F(p) \geq |E|/2$.

Since n mutually independent random variables are used, the sample space contains an exponential number of points. The sample space designed by Han and Igarashi (1990) can be used to obtain n mutually independent random variable. When a random variable tree is used, we can view the random variable x_i at the leaf of the random variable tree as a function of random variable r_is in the interior nodes of the tree. Thus we transform the original problem of locating a good point $(x_0, x_1, ..., x_{n-1})$ to the problem of locating a good point $(r_0, r_1, ..., r_{n-1})$. The conditional expectation after setting a random variable r at the lowest level in the random variable tree can be easily obtained. Let r be such a random variable and let x_i and $x_{i\#0}$ be the two children of r in the random variable tree, where $i\#0$ is the value obtained by complementing the 0-th bit of i. The conditional expectation when r is set is $E[F|r = 0] = E[F|x_i = x_{i\#0}]$ and $E[F|r = 1] = E[F|x_i = 1 - x_{i\#0}]$. The nice feature of Han and Igarashi's design is that all random variables at the lowest level of the random variable tree can be set simultaneously. Note that there are $n/2$ random variables at that level.

Luby (1988) designed a sample space which has only $O(n)$ points. Each point is assigned equal probability. The value of random variable x_i on point r is $(\sum_{k=0}^{\log n - 1} i_k r_k + r_{\log n})$ mod 2, where i_k (r_k) is the k-th bit of i (r). Thus Luby's design transforms the problem of locating a good point $(x_0, x_1, ..., x_{n-1})$ to the problem of locating a good point $(r_0, r_1, ..., r_{\log n})$. For function $f(x_i, x_j) = x_i \oplus x_j$, let $t = \min\{u|i_v = j_v$ for $v \geq u\}$, when r_k, $k < t$, is fixed, $E[f(x_i, x_j)]$ does not change. When r_t is fixed $E[f] = (f(0,0) + f(1,1))/2$ or $E[f] = (f(0,1) + f(1,0))/2$ depends on whether r_t is set to 0 or 1. Thus the sample space designed by Luby also has the feature that conditional expectations can be evaluated easily.

Probabilistic analysis is essential for establishing a randomised algorithm in the first place. It is also vital to the feasibility of an efficient derandomisation process. When space partition methods are used to search for a good point the conditional expectations need to be evaluated. Probabilistic analysis usually provides ways for evaluating these conditional expectations.

12.2.3 The Search Technique

The known search techniques can be put into three categories, namely exhaustive search, binary search and multiple space partition search.

When the exhaustive search method is used all points in the sample space are examined and the best point is chosen as the output. Exhaustive method can only be used when the size of the sample space is small. In the design of sequential algorithms the size of the sample space cannot be larger than a polynomial if polynomial time algorithm is demanded. In the design of parallel algorithms the size of the sample space cannot be larger than a polynomial if NC algorithms are demanded. Although exhaustive search method is sometimes considered as a brute force method, in many situations it yields very fast algorithms. In situations where the sample space and the probabilistic analysis are complicated the exhaustive search may be the only feasible method to search the sample space.

Binary search is usually considered the "standard" way to search the sample space efficiently. When the sample space contains n points and binary search is used to partition the space evenly then $\log n$ steps are enough to locate a good point. In contrast, the exhaustive method requires n operations. When function f has expectation $E[f|S_i]$ on space S_i, $i = 1, 2, 3$, and $S_1 = S_2 \cup S_3$, $S_2 \cap S_3 = \phi$, then either $E[f|S_2] \geq E[f|S_1]$ or $E[f|S_3] \geq E[f|S_1]$. Thus if the expectation of f on S_1 is good then the expectation of f is good on at least one of S_2 and S_3. Thus if the expectation of f on S_i, $1 \leq i \leq 3$, can be computed efficiently the search can continue on either S_2 or S_3. Binary search methods have been used by Raghavan (1988), Spencer (1987) in obtaining sequential algorithms through derandomisation. Luby (1988) and Berger and Rompel (1989) used binary search to obtain efficient parallel algorithms through derandomisation.

Multiple space partition method is a method used in the design of parallel algorithms. It partitions the sample space into several subspaces and evaluates the conditional expectation on each of the subspaces. A good subspace is chosen for the continuing search until a good point is located. The method is typically used in the design of parallel algorithms to speed up the computation when extra processor power is available to evaluate conditional expectations of subspaces. Luby (1988) used this method to speed up the derandomisation process of the PROFIT/COST problem. Not much is known on how to perform multiple space partition without resorting to extra processor power. Han and Igarashi (1990) showed a case where a sample space of size n is partitioned into $\sqrt{n}$ subspaces and a good subspace is found using only a linear number of processors.

Luby (1988) gave a technique for searching a sample space containing nonuniformly distributed 0/1-valued random variables. Let $\vec{y} = <y_i \in \{0, 1\}^p : i = 0, 1, ..., n - 1>$. Let $\vec{x_u}$, $p \leq u < q$, be totally independent random bit strings, each of length n. Let $B(\vec{x_{q-1}} \cdots \vec{x_1}\vec{x_0})$ be the function we are to search for a good point. Let $\vec{z}$ be a vector of n bits. Define $TB(\vec{y}) = E[B(\vec{x_{q-1}} \cdots \vec{x_{p+1}}\vec{x_p}\vec{y})]$. Then $E[TB(\vec{x_p}\vec{y})] = E[E[B(\vec{x_{q-1}} \cdots \vec{x_{p+1}}\vec{z}\vec{y})| \ \vec{z} = \vec{x_p}]] = E[B(\vec{x_{q-1}} \cdots \vec{x_p}\vec{y})] = TB(\vec{y})$. Luby's method is to find a $\vec{z}$ satisfying $TB(\vec{z}\vec{y}) \geq E[TB(\vec{x_p}\vec{y})] = TB(\vec{y})$, thus fixing the random bits in $\vec{x}_p$. Luby's method (Luby 1988, 1989) can be interpreted as follows. Solving the big problem by solving q small problems, one for each $\vec{x}_u$. These small problems are solved sequentially. After the small problems for $\vec{x}_u$, $0 \leq u < v$, are solved. Conditional expectations are evaluated based on the setting of bits x_{i_u}, $0 \leq u < v, 0 \leq i < n$, and the small problem for fixing $\vec{x}_v$ is ready to be solved.

12.2.4 Approximation

Approximation method is a method to simulate the derandomisation process by substituting the function to be search on with an approximation function. This method is typically used in situations where conditional expectation is hard to evaluate. With the use of an approximation function, the evaluation of conditional expectations becomes the evaluation of function values under certain restrictions. The approximation methods allow us to extend the applicability of derandomisation. Raghavan (1988) used this method in the design of a sequential algorithm. Luby (1988) used this method in the design of a parallel algorithm. The main problem involved here is to find a suitable approximation function. We will not discuss the approximation method in detail because the method is very much case-dependent and it usually involves complicated analysis.

12.3 Derandomisation Algorithms

In this section we will present derandomisation algorithms for a class of functions. These functions model several important problems for which fast and efficient deterministic parallel algorithms can be obtained through derandomisation. The derandomisation algorithms presented in this section are given by Berger and Rompel (1989), Han and Igarashi (1990), Han (1991a, 1991b) and Luby (1988). The parallel machine model we use is the Parallel Random Access Machine (PRAM, Fortune and Wyllie 1978). In a PRAM the memory cells are shared among processors. Each memory cell can be accessed by any processor in a step. PRAMs are usually classified into the Exclusive Read Exclusive Write (EREW) PRAM in which a memory cell cannot be read or written by more than one processor in a step, the Concurrent Read Exclusive Write (CREW) PRAM in which a memory cell can be read by several processors in a step, but simultaneous write to the same memory cell by several processors is not allowed, and the Concurrent Read Concurrent Write (CRCW) PRAM in which a memory cell can be read or written by several processors in a step.

12.3.1 Pairwise Independent Random Variables

We are to derandomise for functions of the form

$$B(\vec{x}) = \sum_i f_i(x_i) + \sum_{(i,j)} f_{i,j}(x_i, x_j).$$

Such functions arise in many cases. Luby (1989) formulated the vertex partitioning problem, the $\Delta + 1$ vertex colouring problem, the maximal independent set problem and the maximal matching problem by functions of such form. Functions f_i and $f_{i,j}$ are called PROFIT/COST functions and function B is called a BENEFIT function (Luby 1989). The problem of derandomising such a function is called the PROFIT/COST problem (Luby 1988). The formal definition is given below.

Let $\vec{x} = < x_i \in \{0,1\}^q : i = 0, ..., n-1 >$. Each point $\vec{x}$ out of the 2^{nq} points is assigned probability $1/2^{nq}$. Given function $B(\vec{x}) = \sum_i f_i(x_i) + \sum_{i,j} f_{i,j}(x_i, x_j)$, where f_i is defined as a function $\{0,1\}^q \to \mathcal{R}$ and $f_{i,j}$ is defined as a function $\{0,1\}^q \times \{0,1\}^q \to \mathcal{R}$. The general pairs PROFIT/COST (GPC for short) problem is to find a good point $\vec{y}$ such that $B(\vec{y}) \geq E[B(\vec{x})]$. B is called the general pairs BENEFIT function and $f_{i,j}$s are called the general pairs PROFIT/COST functions.

A special case of the GPC problem, called the bit pairs PROFIT/COST (BPC) problem, is a GPC problem in which $q = 1$. In a BPC problem function B is called the BPC BENEFIT function and f_is and $f_{i,j}$s are called the BPC PROFIT/COST functions.

For the simplicity of discussion we only consider BENEFIT functions of the form $B(\vec{x}) = \sum_{i,j} f_{i,j}(x_i, x_j)$.

The Bit Pairs PROFIT/COST Problem

For the bit pairs PROFIT/COST problem we obtain n mutually independent random variables using the following sample space (Han 1991a; Han and Igarashi 1990).

n 0/1-valued uniformly distributed mutually independent random variables r_i, $0 \leq i < n$, are used, without loss of generality assuming n is a power of 2. We build a tree T which is a complete binary tree with n leaves plus a node which is the parent of the root of the complete binary tree (thus there are n interior nodes in T and the root of T has only one child). The n random variables $x_0, x_1, ..., x_{n-1}$ of B are associated with n leaves of T and the n random variables $r_0, r_1, ..., r_{n-1}$ are associated with the interior nodes of T. The n leaves of T are numbered from 0 to $n-1$. Variable x_i is associated with leaf i.

We now randomise the variables x_i, $0 \leq i < n$. Let $r_{i_0}, r_{i_1}, ..., r_{i_k}$ be the random variables on the path from leaf i to the root of T, where $k = \log n$. Random variable x_i is defined to be $x_i = (\sum_{j=0}^{k-1} i_j r_{i_j} + r_{i_k}) \bmod 2$.

Lemma 12.3.1 *Random variables x_i, $0 \leq i < n$, are uniformly distributed mutually independent random variables.*

Proof: By flipping the random bit at the root of the random variable tree we see that each x_i is uniformly distributed in $\{0, 1\}$. To show the mutual independence we note that the mapping $m(\vec{r}) \mapsto \vec{x}$, where $x_i = (\sum_{j=0}^{\log n - 1} i_j r_{i_j} + r_{i_{\log n}})$ mod 2, is a one to one mapping. Thus $Pr(x_{i_1} = a_1, x_{i_2} = a_2, ..., x_{i_k} = a_k) = 2^{n-k}/2^n = 1/2^k = Pr(x_{i_1} = a_1)Pr(x_{i_2} = a_2) \cdots Pr(x_{i_k} = a_k)$. □

Tree T is called a random variable tree.

We are to find a sample point $\vec{r} = (r_0, r_1, ..., r_{n-1})$ such that $B(\vec{r}) \geq E[B] = \frac{1}{4}\sum_{i,j}(f_{i,j}(0, 0) + f_{i,j}(0, 1) + f_{i,j}(1, 0) + f_{i,j}(1, 1))$.

We fix random variables r_i (setting their values to 0s and 1s) one level in a step starting from the level next to the leaves (we shall call this level 0) and going upward on the tree T until level k. Since there are $k + 1$ interior levels in T all random variables will be fixed in $k + 1$ steps.

Now consider fixing random variables at level 0. Since there are only two random variables x_j, $x_{j\#0}$ which are functions of random variable r_i (node r_i is the parent of the nodes x_j and $x_{j\#0}$) and x_j, $x_{j\#0}$ are not related to other random variables at level 0, and since random variables at level 0 are mutually independent, they can be fixed independently.

Consider in detail the fixing of r_i which is only related to x_j and $x_{j\#0}$. We simply compute $f_0 = f_{j,j\#0}(0, 0) + f_{j,j\#0}(1, 1) + f_{j\#0,j}(0, 0) + f_{j\#0,j}(1, 1)$ and $f_1 = f_{j,j\#0}(0, 1) + f_{j,j\#0}(1, 0) + f_{j\#0,j}(0, 1) + f_{j\#0,j}(1, 0)$. If $f_0 \geq f_1$ then set r_i to 0 else set r_i to 1. This scheme allows all random variables at level 0 be set in parallel in constant time.

If r_i is set to 0 then $x_i = x_{i\#0}$, if r_i is set to 1 then $x_i = 1 - x_{i\#0}$. Therefore after r_i is fixed, x_i and $x_{i\#0}$ can be combined. The n random variables x_i, $0 \leq i < n$, can be reduced to $n/2$ random variables. BPC functions $f_{i,j}$, $f_{i\#0,j}$, $f_{i,j\#0}$, and $f_{i\#0,j\#0}$ can also be combined into one function. It can be checked that the combining can be done in constant time using a linear number of processors.

During the combining process variables x_i and $x_{i\#0}$ are combined into a new variable $x^{(1)}_{\lfloor i/2 \rfloor}$, functions $f_{i,j}$, $f_{i\#0,j}$, $f_{i,j\#0}$, and $f_{i\#0,j\#0}$ are combined into a new function $f^{(1)}_{\lfloor i/2 \rfloor, \lfloor j/2 \rfloor}$. After combining a new function $B^{(1)}$ is formed which has the same form of B but has only $n/2$ variables. As we stated above, $E[B^{(1)}] \geq E[B]$.

What we have explained above is the first step of the algorithm (Han and Igarashi 1990). This step takes constant time using a linear number of processors. After k steps the random variables at levels 0 to $k - 1$ in the random variable tree are fixed, the n random variables $\{x_0, x_1, ..., x_{n-1}\}$ are reduced to $n/2^k$ random variables $\{x^{(k)}_0, x^{(k)}_1, ..., x^{(k)}_{n/2^k-1}\}$, functions $f_{i,j}$, $i, j \in \{0, 1, ..., n-1\}$, have been combined into $f^{(k)}_{i,j}$, $i, j \in \{0, 1, ..., n/2^k-1\}$.

After $\log n$ steps $B^{(\log n)} = f^{(\log n)}_{0,0}(x^{(\log n)}_0, x^{(\log n)}_0)$. The bit at the root of the random variable tree is now set to 0 if $f^{(\log n)}_{0,0}(0,0) \geq f^{(\log n)}_{0,0}(1,1)$, and 1 otherwise. Thus Han and Igarashi's algorithm (Han and Igarashi 1990) solves the BPC problem in $O(\log n)$ time with a linear number of processors.

0	1	4	5
2	3	6	7
8	9	12	13
10	11	14	15

Figure 12.1: File-major indexing for the two dimensional array

Theorem 12.3.1 (Han and Igarashi 1990) *A sample point* $\vec{r} = (r_0, r_1, ..., r_{n-1})$ *satisfying* $B(\vec{r}) \geq E[B]$ *can be found in* $O(\log n)$ *time using a linear number of processors and* $O(n^2)$ *space.* □

A close examination of the process of derandomisation of the algorithm shows that functions $f_{i,j}$ are combined according to the so-called file-major indexing for the two dimensional array, as shown in Figure 12.1. In the file-major indexing the $n \times n$ array A is divided into four subfiles $A_0 = A[0..n/2-1, 0..n/2-1]$, $A_1 = A[0..n/2-1, n/2..n-1]$, $A_2 = A[n/2..n-1, 0..n/2-1]$, $A_3 = A[n/2..n-1, n/2..n-1]$. Any element in A_i proceeds any element in A_j if $i < j$. The indexing of the elements in the same subfile is recursively defined in the same way. The indexing of function $f_{i,j}$ is the number at the i-th row and j-th column of the array. After the bits at level 0 are fixed by our algorithm, functions indexed $4k$, $4k + 1$, $4k + 2$, $4k + 3$, $0 \leq k < n^2/4$ will be combined. After the combination of these functions they will be reindexed. The new index k will be assigned to the function combined from the original functions indexed $4k$, $4k + 1$, $4k + 2$, $4k + 3$. This allows the recursion in our algorithm to proceed.

Obviously we want the input to be arranged by the file-major indexing. When the input has been arranged by the file-major indexing, we are able to build a tree which reflects the way input functions $f_{i,j}$s are combined as the derandomisation process proceeds. We shall call this tree the derandomisation tree. This tree is built as follows.

We use one processor for each function $f_{i,j}$. These BPC functions are stored in an array. Let f_{i_1,j_1} be the function stored immediately before $f_{i,j}$ and f_{i_2,j_2} be the function stored immediately after $f_{i,j}$. By looking at (i_1, j_1) and (i_2, j_2) the processor could easily figure out at which step of the derandomisation $f_{i,j}$ should be combined with f_{i_1,j_1} or f_{i_2,j_2}. This information allows the tree to be built for the derandomisation process. This tree has $\log n + 1$ levels. The functions at the 0-th level (leaves) are those to be combined into a new function which will be associated with the parent of these leaves. The combination happens immediately after the random variables at level 0 in the random variable tree are fixed. In general, functions at level i will be combined immediately after the random variables at level i in the random variable tree are fixed. Note that the term level in the derandomisation tree corresponds to the level of the random variable tree. Thus, a node at level i of the derandomisation

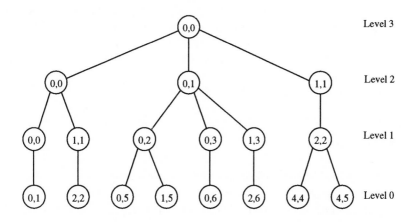

Figure 12.2: A derandomisation tree. Pairs in the circles are the subscripts of PROFIT/COST functions

tree could be at depth $\leq \log n - i$.

Since the derandomisation tree has height at most $\log n$, it can be built in $O(\log n)$ time using m processors. If the input is arranged by the file-major indexing, the tree can be built in $O(\log n)$ time using optimal $m/\log n$ processors by a careful processor scheduling.

A derandomisation tree is shown in Figure 12.2.

The derandomisation process can now be described in terms of the derandomisation tree. Combine the functions at level i of the derandomisation tree immediately after the random variables at level i of the random variable tree are fixed. The combination can be accomplished by the RAKE (Miller and Reif 1989) operation. The whole process of the derandomisation can now be viewed as a process of tree contraction which uses only the RAKE operation without using the COMPRESS operation (Miller and Reif 1989).

If the input is not arranged by the file-major indexing we could sort the input into the file-major indexing. Known sorting algorithms (Ajtai *et al.* 1983; Cole 1986) have time complexity $O(\log n)$ using a linear number of processors.

Theorem 12.3.2 (Han and Igarashi 1990) *A sample point* $\vec{r} = (r_0, r_1, ..., r_{n-1})$ *satisfying* $B(\vec{r}) \geq E[B]$ *can be found on the CREW PRAM in* $O(\log n)$ *time using m processors and $O(m)$ space.* □

We are able to obtain an optimal algorithm if the input is arranged by the file major indexing. Since no sorting is needed, the time complexity of the algorithm we described above takes $O(\frac{m}{p} + \frac{n \log n}{p} + \log n)$ time with p processors. The term $n \log n/p$ is incurred due to the fact that we have to keep updated information with which variable x_i, $0 \leq i < n$, has been combined into. This takes $O(n)$ operations in each step of the algorithm.

To obtain an optimal algorithm with time complexity $O(m/p + \log n)$ we first reduce the number of random variables to $n/\log n$. This is done by dividing n random variables into $n/\log n$ groups with random variable x_i, $k \log n \leq i < (k + 1) \log n$ in group k. We solve the BPC problem for each group, in parallel for all groups, using a modified version of Luby's algorithm (Luby 1988) which can be made to run in $O(\frac{m}{p} + \log \log n)$ time. Upon finishing there are $n/\log n$ random variables left and our original algorithm with time $O(\frac{m}{p} + \frac{n \log n}{p} + \log n)$ now takes $O(\frac{m+n}{p} + \log n)$ time when n is replaced with $n/\log n$.

Theorem 12.3.3 (Han and Igarashi 1990) *A sample point* $\vec{r} = (r_0, r_1, ..., r_{n-1})$ *satisfying* $B(\vec{r}) \geq E[B]$ *can be found on the CREW PRAM in* $O(\log n)$ *time using optimal* $m/\log n$ *processors and* $O(m)$ *space if the input is arranged by the file-major indexing.* $\square$

The General Pairs PROFIT/COST Problem

We shall present a scheme where the GPC problem is solved by pipelining the BPC algorithm to solve BPC problems in the GPC problem.

First we give a sketch of our approach. The incompleteness of the description in this paragraph will be fulfilled later. Let P be the GPC problem we are to solve. P can be decomposed into q BPC problems to be solved sequentially. Let P_u be the u-th BPC problem. Imagine that we are to solve P_u, $0 \leq u < k$, in one pass, *i.e.*, we are to fix $\vec{x}_0$, $\vec{x}_1$, ..., $\vec{x}_{k-1}$ in one pass, with the help of enough processors. For the moment we can have a random variable tree T_u and a derandomisation tree D_u for P_u, $0 \leq u < k$. In step j our algorithm will work on fixing the bits at level $j - u$ in T_u, $0 \leq u \leq \min\{k - 1, j\}$. The computation in each tree D_u proceeds as we have described in the last subsection. Note that BPC functions $f_{i_v, j_v}(x_{i_v}, x_{j_v})$ depends on the setting of bits x_{i_u}, x_{j_u}, $0 \leq u < v$. The main difficulty with our scheme is that when we are working on fixing $\vec{x}_v$, $\vec{x}_u$, $0 \leq u < v$, have not been fixed yet. The only information we can use when we are fixing the random variables at level l of T_u is that random variables at levels 0 to $l + c - 1$ are fixed in T_{u-c}, $0 \leq c \leq u$. This information can be accumulated in the pipeline of our algorithm and transmitted on the *bit pipeline trees*. Fortunately this information is sufficient for us to speed up the derandomisation process without resorting to too many processors. For the sake of a clear exposition we first describe a CREW derandomisation algorithm. We then show how to convert the CREW algorithm to an EREW algorithm.

Suppose we have $c \sum_{i=0}^{k} (m \times 4^i)$ processors available, where c is a constant. Assign $cm \times 4^u$ processors to work on P_u for $\vec{x}_u$. We shall work on $\vec{x}_u$, $0 \leq u \leq k$, simultaneously in a pipeline. The random variable tree for P_u (except that for P_0) is not constructed before the derandomisation process begins, rather it is constructed from a forest as the derandomisation process proceeds. A forest containing 2^u random variable trees corresponds to each variable x_{i_u} in P_u because there are 2^u bit patterns for $x_{i_j}, 0 \le j < u$. We use F_u to denote the random variable forest for P_u. We are to fix the random bits on the l-th level of F_v (for $\overrightarrow{x_v}$) under the condition that random bits from level 0 to level $l + c - 1, 0 \le c \le v$, in F_{v-c} have already been fixed. We are to perform this fixing in constant time. The 2^u random variable trees corresponding to each random variable x_{i_u} are built bottom up as the derandomisation process proceeds. Immediately before the step we are to fix the random bits on the l-th level of F_u, the 2^u random variable trees corresponding to x_{i_u} are constructed up to the l-th level. The details of the algorithm for constructing the random variable trees will be given later in this section.

Consider a GPC function $f_{i,j}(x_i, x_j)$ under the condition stated in the last paragraph. When we start working on $\overrightarrow{x_v}$ we should have the BPC functions $f_{i_v,j_v}(x_{i_v}, x_{j_v})$ evaluated and stored in a table. However, because $\overrightarrow{x_u}, 0 \le u < v$, have not been fixed yet, we have to try out all possible situations. There are a total of 4^v patterns for bits $x_{i_u}, x_{j_u}, 0 \le u < v$, we use 4^v BPC functions for each pair (i, j). By $f_{i_v,j_v}(x_{i_v}, x_{j_v})(y_{v-1}y_{v-2}\cdots y_0, z_{v-1}z_{v-2}\cdots z_0)$ we denote the function $f_{i_v,j_v}(x_{i_v}, x_{j_v})$ obtained under the condition that $(x_{i_{v-1}}x_{i_{v-2}}\cdots x_{i_0}, x_{j_{v-1}}x_{j_{v-2}}\cdots x_{j_0})$ is set to $(y_{v-1}y_{v-2}\cdots y_0, z_{v-1}z_{v-2}\cdots z_0)$.

For each pair $(w, w\#0)$ at each level l (this is the level in the random variable forest), $0 \le l \le \log n$, a *bit pipeline tree* is built (Figure 12.3) which is a complete binary tree of height $2k$. Nodes at even depth from the root in a bit pipeline tree are selectors, nodes at odd depth are fanout gates. A signal *true* is initially input into the root of the tree and propagates downward toward the leaves. The selectors at depth $2d$ select the output by the decision of the random bits which are the parents of random variables $x_{w_d}, x_{w\#0_d}$ in F_d. There is one random variable corresponds to each selector. Let random variable r corresponds to the selector s. If r is set to 0 then s selects the left child and propagates the true signal to its left child while no signal is sent to its right child. If r is set to 1 then the true signal will be sent to the right child and no signal will be sent to the left child. If s does not receive any signal from its parent then no signal will be propagated to s's children no matter how r is set. The gates at odd depth in the bit pipeline tree are fanout gates and pointers from them to their children are labelled with bits which are conditionally set. Refer to Figure 12.3 which shows a bit pipeline tree of height 4. If the selector at the root (node 0) selects 0 (which means that the random variable which is the parent of x_{w_0} and $x_{w\#0_0}$ in the random variable forest is set to 0), then $x_{w_0} = x_{w\#0_0}$, therefore the two random variables can only assume the patterns 00 or 11 which are labelled on the pointers from node 1. If, on the other hand, node 0 selects 1 then $x_{w_0} = 1 - x_{w\#0_0}$, the two random variables can only assume the patterns 01 or 10 which are labelled on the pointers of node 2. Let us take node 4 as another example. If node 4 selects 0 then $x_{w_1} = x_{w\#0_1}$, thus the pointers of node 9 are labelled with $\begin{smallmatrix} 1 & 1 \\ 0 & 0 \end{smallmatrix}$ and $\begin{smallmatrix} 1 & 1 \\ 1 & 1 \end{smallmatrix}$. This indicates that

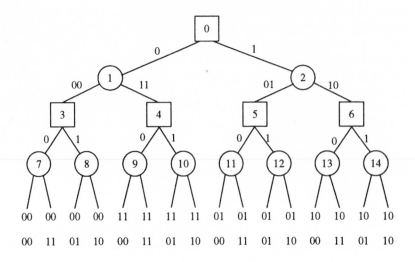

Figure 12.3: A bit pipeline tree of height 4

the bits for $(w_1w_0, w\#0_1w\#0_0)$ can have two patterns, $(01, 01)$ or $(11, 11)$.

The bit pipeline tree built for level $\log n$ has height k. No fanout gates will be used. This is a special and simpler case compared to the bit pipeline trees for other levels. In the following discussion we only consider bit pipeline tree for levels other than $\log n$.

Lemma 12.3.2 *In a bit pipeline tree there are exactly 2^d nodes at depth $2d$ which will receive the true signal from the root.*

Proof: Each selector selects only one path. Each fanout gate sends the true signal to both children. Therefore exactly 2^d nodes at depth $2d$ will receive the true signal from the root. □

For each node i at even depth we shall also say that it has the *conditional bit pattern* (or *conditional bits, bit pattern*) which is the pattern labelled on the pointer from $p(i)$. The root of the bit pipeline tree has empty string as its bit pattern.

Define step 0 as the step when the true signal is input to node 0. The function of a bit pipeline tree can be described as follows.

Step t: Selectors at depth $2t$ which have received true signal selects 0 or 1 for $(w_t, w\#0_t)$. Pass the true signal and the bit setting information to nodes at depth $2t + 2$.

Now consider the selectors at depth $2d$. By Lemma 12.3.5 a set of 2^d selectors at depth $2d$ receive the true signal. We call this set the *surviving set* $S_{w,d}^l$. We also denote by $S_{w,d}^l$ the set of bit patterns the 2^d surviving selectors have, where w in the subscript is for $(w, w\#0)$ and l is the level for which the pipeline tree is built. Let selector $s \in S^l_{w,d}$ have bit pattern $(y_{d-1}y_{d-2} \cdots y_0, z_{d-1}z_{d-2} \cdots z_0)$. s compares

$$f^{(l)}_{w_d, w\#0_d}(0,0)(y_{d-1}y_{d-2} \cdots y_0, z_{d-1}z_{d-2} \cdots z_0) +$$
$$f^{(l)}_{w_d, w\#0_d}(1,1)(y_{d-1}y_{d-2} \cdots y_0, z_{d-1}z_{d-2} \cdots z_0) +$$
$$f^{(l)}_{w\#0_d, w_d}(0,0)(z_{d-1}z_{d-2} \cdots z_0, y_{d-1}y_{d-2} \cdots y_0) +$$
$$f^{(l)}_{w\#0_d, w_d}(1,1)(z_{d-1}z_{d-2} \cdots z_0, y_{d-1}y_{d-2} \cdots y_0)$$

with

$$f^{(l)}_{w_d, w\#0_d}(0,1)(y_{d-1}y_{d-2} \cdots y_0, z_{d-1}z_{d-2} \cdots z_0) +$$
$$f^{(l)}_{w_d, w\#0_d}(1,0)(y_{d-1}y_{d-2} \cdots y_0, z_{d-1}z_{d-2} \cdots z_0) +$$
$$f^{(l)}_{w\#0_d, w_d}(0,1)(z_{d-1}z_{d-2} \cdots z_0, y_{d-1}y_{d-2} \cdots y_0) +$$
$$f^{(l)}_{w\#0_d, w_d}(1,0)(z_{d-1}z_{d-2} \cdots z_0, y_{d-1}y_{d-2} \cdots y_0)$$

and selects 0 if the former is no less than the latter and selects 1 otherwise. Note that the selectors which do not receive the true signal (there are $4^d - 2^d$ of them) have bit patterns which are eliminated.

Let $LS^l_{w,d} = \{\alpha | (\alpha, \beta) \in S^l_{w,d}\}$ and $RS^l_{w,d} = \{\beta | (\alpha, \beta) \in S^l_{w,d}\}$.

Lemma 12.3.3 $LS^l_{w,d} = RS^l_{w,d} = \{0,1\}^d$.

Proof: By induction. Assuming that it is true for bit pipeline trees of height $2d - 2$. A bit pipeline tree of height $2d$ can be constructed by using a new selector as the root, two new fanout gates at depth 1, and four copies of the bit pipeline tree of height $2d - 2$ at depth 2. If the root selects 0 then patterns 00 and 11 are concatenated with patterns in $S^l_{w,d-1}$, therefore both $LS^l_{w,d-1}$ and $RS^l_{w,d-1}$ are concatenated with $\{0,1\}$. The situation when the root selects 1 is similar. $\square$

Now let us consider how functions $f^{(l)}_{i_d, j_d}(x_{i_d}, x_{j_d})(\alpha, \beta)$ are combined. Take the difficult case where both i and j are odd. By Lemma 12.3.6 there is only one pattern $p_1 = (\alpha', \alpha) \in S^l_{i\#0,d}$ and there is only one pattern $p_2 = (\beta', \beta) \in S^l_{j\#0,d}$. If the selector having bit pattern p_1 selects 0 then $x_{i_d} = x_{i\#0_d}$ else $x_{i_d} = 1 - x_{i\#0_d}$. If the selector having bit pattern p_2 selects 0 then $x_{j_d} = x_{j\#0_d}$ else $x_{j_d} = 1 - x_{j\#0_d}$. In any case the conditional bit pattern is changed to (α', β'), i.e., $f^{(l)}_{i_d, j_d}(x_{i_d}, x_{j_d})(\alpha, \beta)$ will be combined into $f^{(l+1)}_{\lfloor i/2 \rfloor_d, \lfloor j/2 \rfloor_d}(x_{\lfloor i/2 \rfloor_d}, x_{\lfloor j/2 \rfloor_d})(\alpha', \beta')$. Note that $x_{\lfloor i/2 \rfloor_d}$ and $x_{\lfloor j/2 \rfloor_d}$ are new random variables and here we are not using superscript to denote this fact. The following lemma ensures that at most four functions will be combined into $f^{(l+1)}_{\lfloor i/2 \rfloor_d, \lfloor j/2 \rfloor_d}(x_{\lfloor i/2 \rfloor_d}, x_{\lfloor j/2 \rfloor_d})(\alpha', \beta')$.

Let $S = \{(\alpha', \beta') | (\alpha', \alpha) \in S^l_{i,d}, (\beta', \beta) \in S^l_{j,d}, \alpha, \beta \in \{0,1\}^d\}$.

Lemma 12.3.4 $|S| = 4^d$.

Proof: The definition of S can be viewed as a linear transformation. Represent $x \in \{0,1\}^d$ by a vector of 2^d bits with x-th bit set to 1 and the rest of the bits set to 0. The transformation $\alpha \mapsto \alpha'$ can be represented by a permutation matrix of order 2^d. The transformation $(\alpha, \beta) \mapsto (\alpha', \beta')$ can be represented by a permutation matrix of order 2^{d+1}. $\square$

Lemma 12.3.7 tells us that the functions to be combined are permuted, therefore no more than four functions will be combined under any conditional bit pattern.

We call this scheme of combining, *combining functions with respect to the surviving set*.

We have completed a preliminary description of our derandomisation scheme for the GPC problem. The algorithm for processors working on $\vec{x}_d$, $0 \le d < k$, can be summarised as follows.

Step t ($0 \le t < d$): Wait for the pipeline to be filled.

Step $d + t$ ($0 \le t < \log n$): Fix random variables at level t for all conditional bit patterns in the surviving set. (There are 2^d such patterns in the surviving set.) Combine functions with respect to the surviving set. (At the same time the bit setting information is transmitted to the nodes at depth $2d + 2$ on the bit pipeline tree.)

Step $d + \log n$: Fix the only remaining random variable at level $\log n$ for the only bit pattern in the surviving set. Output the good point for $\vec{x}_d$. (At the same time the bit setting information is transmitted to the node at depth $d + 1$ on the bit pipeline tree.)

Theorem 12.3.4 *The GPC problem can be solved on the CREW PRAM in time $O((q/k + 1)(\log n + \tau))$ with $O(4^k m)$ processors, where τ is the time for computing the BPC functions $f_{i_d, j_d}(x_{i_d}, x_{j_d})(\alpha, \beta)$.*

Proof: The correctness of the scheme comes from the fact that as random bits are fixed a smaller space with higher expectation is obtained, and thus when all random bits are fixed a good point is found. Since k $\vec{x}_u$s are fixed in one pass which takes $O(\log n + \tau)$ time, the time complexity for solving the GPC problem is $O((q/k+1)(\log n+\tau))$. The processor complexity is obvious from the description of the scheme. $\square$

We have not yet discussed explicitly the way the random variable trees are constructed. The construction is implied in the surviving set we computed. We now give the algorithm for constructing the random variable trees. This algorithm will help understand better the whole scheme.

The i-th node under conditional bit pattern j at the l-th level of the random variable trees for P_u is stored in $T_u^{(l)}[i][j]$. The leaves are stored in $T_u^{(-1)}$. Initially bit pipeline trees for level -1 are built such that $T_u^{(-1)}[i][j]$ has two children $T_{u+1}^{(-1)}[i][j0]$, $T_{u+1}^{(-1)}[i][j1]$, where $j0$ and $j1$ are the concatenations of j with 0 and 1 respectively. Note that the bit pipeline tree constructed here is different from the one we built before, but in principle they are the same tree and perform the same function in our scheme. The algorithm for constructing the random variable trees for P_u is below.

Procedure RV-Tree

begin

Step t $(0 \leq t < u)$: Wait for the pipeline to be filled.

Step $u + t$ $(0 \leq t < \log n)$:

(In this step we are to build $T_u^{(t)}[i][j]$, $0 \leq i < n/2^{t+1}$, $0 \leq j < 2^u$. At the beginning of this step $T_{u-1}^{(t)}[i][j]$ has already been constructed. Let $T_{u-1}^{(t-1)}[i0][j]$ and $T_{u-1}^{(t-1)}[i1][j']$ be the two children of $T_{u-1}^{(t)}[i][j]$ in the random variable tree. $T_u^{(t-1)}[i0][j0]$ and $T_u^{(t-1)}[i0][j1]$ are the children of $T_{u-1}^{(t-1)}[i0][j]$, $T_u^{(t-1)}[i1][j'0]$ and $T_u^{(t-1)}[i1][j'1]$ are the children of $T_{u-1}^{(t-1)}[i1][j']$, in the bit pipeline tree for level $t - 1$. The setting of the random variable r for the pair $(i0, i1)$ at level t for P_{u-1}, *i.e.* the random variable in $T_{u-1}^{(t)}[i][j]$, is known.)

make $T_u^{(t-1)}[i0][j0]$ and
$T_u^{(t-1)}[i1][j'r]$ as the children of
$T_u^{(t)}[i][j0]$ in the random variable forest for P_u;
(jr is the concatenation of j and r.)

make $T_u^{(t-1)}[i0][j1]$ and
$T_u^{(t-1)}[i1][j'\bar{r}]$ as the children of
$T_u^{(t)}[i][j1]$ in the random variable forest for P_u;
($\bar{r}$ is the complement of r.)

make $T_u^{(t)}[i][j0]$ and
$T_u^{(t)}[i][j1]$ as the children of
$T_{u-1}^{(t)}[i][j]$ in bit pipeline tree for level t;

Fix the random variables in $T_u^{(t)}[i][j0]$ and $T_u^{(t)}[i][j1]$;

Step $u + \log n$:

(At the beginning of this step the random variable trees have been built for T_i, $0 \leq i < u$. Let $T_{u-1}^{(\log n)}[0][j]$ be the root of T_{u-1}. The random variable r in $T_{u-1}^{(\log n)}[0][j]$ has been fixed. In this step we are to choose one of the two children of $T_{u-1}^{(\log n)}[0][j]$ in the bit pipeline tree for level $\log n$ as the root of T_u.)

make $T_u^{(\log n - 1)}[0][jr]$ as the child of
$T_u^{(\log n)}[0][jr]$ in the random variable tree;

make $T_u^{(\log n)}[0][jr]$ as the child of
$T_{u-1}^{(\log n)}[0][j]$ in the bit pipeline tree for level $\log n$;

fix the random variable in $T_u^{(\log n)}[0][jr]$;

output $T_u^{(\log n)}[0][jr]$ as the root of T_u;

end

Procedure **RV-Tree** uses the pipelining technique as well as the dynamic programming technique. These are some of the essential elements of our scheme.

We now show how to remove the concurrent read feature from the scheme. The difficulty here is in the step of combining functions with respect to the surviving set. The size of the surviving set $S_{u,k}^l$ is 2^k while there are 4^k conditional bit patterns. There are 4^k functions $f_{u_k,v_k}^{(l)}(x_{u_k}, x_{v_k})$, one for each bit pattern (α, β). All 4^k functions will consult the surviving set in order for them to be combined into new functions. The problem is how to do it in constant time without resorting to concurrent read.

We show how to let $f_{u_k,u\#0_k}^{(l)}(x_{u_k}, x_{u\#0_k})(\alpha, \beta)$ to acquire the bit pattern α' which satisfies $(\alpha', \beta) \in S_{u,k}^l$. Function $f_{u_k,v_k}^{(l)}(x_{u_k}, x_{v_k})(\alpha, \beta)$ can then obtain the bit pattern α' from $f_{u_k,u\#0_k}^{(l)}(x_{u_k}, x_{u\#0_k})(\alpha, \beta)$ by the pipeline scheme described by Han (1991a).

Suppose we are to solve P_u, $0 \le u \le k$, in one pass. We solve 4^k copies of P_{k-1}, one copy corresponds to one conditional bit pattern in P_k. $f_{u_k,v_k}^{(l)}(x_{u_k}, x_{v_k})(\alpha, \beta)$ in P_k can obtain α' by following the computation in the copy of P_{k-1} which corresponds to (α, β). This can be done without concurrent read. Now for each of the 4^k copies of P_{k-1} we solved 4^{k-1} copies of P_{k-2}, one copy corresponds to one conditional bit pattern in P_{k-1}. And so on. Thus to remove concurrent read we need $c^{k^2}(m+n)$ processors for solving P_u, $0 \le u \le k$, in one pass, where c is a suitable constant. Note also that it takes $O(k^2)$ time to make needed copies.

Theorem 12.3.5 *The general pairs PROFIT/COST problem can be solved on the EREW PRAM in time $O((q/\sqrt{k} + 1)(\log n + k + \tau))$ with $O(c^k m)$ processors, where c is a suitable constant and τ is the time for computing the bit pairs PROFIT/COST functions $f_{u_d,v_d}(x_{u_d}, x_{v_d})(\alpha, \beta)$.* □

12.3.2 $O(\log n)$-wise Independent Random Variables

We consider the derandomisation of functions of the form $B(\vec{x}) = \sum_i^{n^a} f_i(x_{i,1}, x_{i,2}, ..., x_{i,b\log n})$, where a, b are constants and $x_{i,j}$s are 0/1-valued uniformly distributed mutually independent random variables. The problem is to find a good point $\vec{y}$ such that $B(\vec{y}) \geq E[B(\vec{x})]$. The derandomisation algorithm for this problem is given by Berger and Rompel (1989) which is a generalisation of Luby's algorithm (Luby 1988).

Since each function $f_i(x_{i,1}, x_{i,2}, ..., x_{i,b\log n})$ contains at most $b\log n$ random variables, if every $b\log n$ random variables among the $x_{i,j}$s are mutually independent then the value of $E[B(\vec{x})]$ would be the same as that when all $x_{i,j}$s are mutually independent. A sample space containing n mutually independent random variables has $\Omega((n/k)^{\lceil k/2 \rceil})$ sample points (Alon *et al.* 1986). A sample space containing n 0/1-valued uniformly distributed $(b\log n)$-wise independent random variables can be constructed with $O(n^{\log n})$ sample points.

The sample space is thus constructed by taking n $(b\log n)$-wise linearly independent binary vectors $a_1, a_2, ..., a_n$, each of length $l = O(\log^2 n)$. The value of $a_i = <a_{i,1}, a_{i,2}, ..., a_{i,l}>$ at sample point $r = <r_1, r_2, ..., r_l>$ is $(\sum_{k=1}^{l} a_{i,k} r_k)$ mod 2. Using elementary linear algebra it is easy to show that a_is are $(b\log n)$-wise independent random variables.

The problem now is to find a good point r' such that $B(\vec{x}(r')) \geq E[B(\vec{x}(r))]$.

To determine r, one bit of r will be fixed at a time. Fixing one bit r_i of r is to partition the sample space to two subspaces—one is the subspace in which $r_i = 0$ and the other is the subspace in which $r_i = 1$—and to determine which subspace will be discarded. In the remaining subspace the value of r_i is fixed. Thus the fixing of binary bits can be viewed as a binary search into the sample space.

Assuming that $r_1 = s_1, ..., r_{t-1} = s_{t-1}$ have already been set, we compute $E[B(\vec{x})|r_1 = s_1, r_2 = s_2, ..., r_{t-1} = s_{t-1}, r_t = 0]$ and $E[B(\vec{x})|r_1 = s_1, r_2 = s_2, ..., r_{t-1} = s_{t-1}, r_t = 1]$. r_t is set to $s_t \in \{0, 1\}$ which maximizes $E[B(\vec{x})|r_1 = s_1, r_2 = s_2, ..., r_{t-1} = s_{t-1}, r_t = s_t]$. Assume that $E[B(\vec{x})|r_1 = s_1, r_2 = s_2, ..., r_{t-1} = s_{t-1}] \geq E[B(\vec{x})]$. We have $E[B(\vec{x})|r_1 = s_1, r_2 = s_2, ..., r_{t-1} = s_{t-1}, r_t = s_t] = \max\{E[B(\vec{x})|r_1 = s_1, r_2 = s_2, ..., r_{t-1} = s_{t-1}, r_t = 0], E[B(\vec{x})|r_1 = s_1, r_2 = s_2, ..., r_{t-1} = s_{t-1}, r_t = 1]\} \geq (E[B(\vec{x})|r_1 = s_1, r_2 = s_2, ..., r_{t-1} = s_{t-1}, r_t = 0] + E[B(\vec{x})|r_1 = s_1, r_2 = s_2, ..., r_{t-1} = s_{t-1}, r_t = 1])/2 = E[B(\vec{x})|r_1 = s_1, r_2 = s_2, ..., r_{t-1} = s_{t-1}] \geq E[B(\vec{x})]$.

Berger and Rompel gave the following method for evaluating conditional expectations. Since function B is the sum of function f_is, the conditional expectation can be evaluated on each f_i in parallel and then the conditional expectation of B can be obtained from the sum of the conditional expectations of f_is. Consider the problem of evaluating $E[f_i(x_{i,1}, ..., x_{i,b\log n})|r_1 =$

$s_1, ..., r_t = s_t]$. Let x be the vector $< x_{i,1}, ..., x_{i,b\log n} >$ and A be the matrix whose j-th row is $a_{i,j}$, the binary vector of length l corresponding to $x_{i,j}$. Then $x = Ar$, and

$$E[f_i(x_{i,1}, ..., x_{i,b\log n})|r_1 = s_1, ..., r_t = s_t]$$
$$= \sum_x f_i(x)Pr[Ar = x|r_1 = s_1, ..., r_t = s_t].$$

Let $r' =< r_1, r_2, ..., r_t >$, $r'' =< r_{t+1}, ..., r_l >$, $s =< s_1, ..., s_t >$, A' and A'' be the first t and the last $l - t$ columns of A respectively, then

$$\sum_x f_i(x)Pr[Ar = x|r_1 = s_1, ..., r_t = s_t]$$
$$= \sum_x f_i(x)Pr[A'r' + A''r'' = x|r' = s]$$
$$= \sum_x f_i(x)Pr[A''r'' = x - A's].$$

If $A''r'' = x - A's$ is solvable, then $Pr[A''r'' = x - A's] = 2^{-\text{rank}(A'')}$, otherwise $Pr[A''r'' = x - A's] = 0$.

The above scheme allow us to compute the conditional expectations in polylogarithmic time using a polynomial number of processors.

12.4 Applications

The idea of derandomisation has been successfully applied to the design of several efficient sequential and parallel algorithms (Alon *et al.* 1986; Berger and Rompel 1989; Berger *et al.* 1989; Han 1991a, 1991b; Han and Igarashi 1990; Karp and Wigderson 1985; Luby 1986, 1988, 1989; Motwani *et al.* 1985; Pantziou *et al.* 1988; Raghavan 1988; Spencer 1987). Spencer (1987) and Raghavan (1988) first used a binary search technique to locate a good sample point. Their binary search technique enables an algorithm to search an exponential sized sample space in polynomial time. Efficient sequential algorithms have been obtained by using this technique (Raghavan 1988; Spencer 1987).

In the design of efficient parallel algorithms, Karp and Wigderson (1985), Luby (1986) and Alon *et al.* (1986) designed small sample space by using limited independence. Since the sample space they designed contains a polynomial number of points, exhaustive searches were used to locate a good point to obtain a DNC algorithm. This technique has been used successfully by Karp and Wigderson (1985) and Luby (1986) to obtain DNC algorithms for the maximal independent set problem. By applying this technique Alon *et al.* (1986) obtained DNC algorithms for several problems, among them finding an independent set of size k for a hypergraph, finding a large d-partite subhypergraph, Siden-subsets of graphs and Ramsey-type problems.

In order to obtain processor efficient parallel algorithms through derandomisation, Luby (1988, 1989) used the idea of binary search on a small sample space. His idea yields efficient parallel algorithms for the $\Delta + 1$ vertex colouring problem, the maximal independent set problem and the maximal matching problem. His DNC algorithms for the maximal independent set problem and the maximal matching problem have time complexity which are fairly close to the time complexity of the algorithms for the two problems obtained through ad hoc designs (Goldberg and Spencer 1989a, 1989b; Israeli and Shiloach 1986).

Luby's technique was carried further by Berger and Rompel (1989) and Motwani *et al.* (1989) where $(\log^c n)$-wise independence among random variables was used. Because the sample space contains $n^{\log^{O(1)} n}$ sample points in the design of Berger and Rompel (1989) and Motwani *et al.* (1989), they used a thoughtfully designed binary search technique to obtain DNC algorithms. They presented DNC algorithms for the set discrepancy problem and the hypergraph colouring problem.

Recently Han and Igarashi (Han 1991a; Han and Igarashi 1990) showed how to obtain an efficient parallel algorithm for the bit pairs PROFIT/COST problem through derandomisation using an exponential sized sample space. Their derandomisation scheme gives a case in which the sample space can be reduced at the rate of $\sqrt{n}$. This technique can be developed further to obtain fast and efficient parallel algorithms for the $\Delta + 1$ vertex colouring problem, the maximal independent set problem and the maximal matching problem (Han 1991b).

In the rest of this section we will show several applications of derandomisation schemes.

12.4.1 Large d-Partite Subhypergraph

A hypergraph $H = (V, E)$ contains a set of vertices V and a set of edges E. Each edge is a subset of V. H is d-uniform if every edge has d elements. Given a d-uniform hypergraph H with $d \geq 2$, we intend to find a partition $(V_1, V_2, ..., V_d)$ of V such that the number of edges of H having precisely one vertex in each class V_i is at least $\lfloor |E|d!/d^d \rfloor$. We show the design of a deterministic parallel algorithm for this problem through derandomisation. This design is from Alon *et al.* (1986).

Let c_i, $1 \leq i \leq |V|$, be d-wise independent random variables uniformly distributed on $\{1, 2, ..., d\}$. Colour vertex i with colour c_i. A given edge is good if it has all the colours. The probability that a given edge is good is $d!/d^d$. Thus the expected number of good edges is $|E|d!/d^d$.

To obtain $|V|$ random variables which are d-wise independent and distributed almost equally on $\{1, 2, ..., d\}$, we pick a prime $q > 2d^2|E|$ and construct a sample space as given in Section 12.2.1. The probability that a given edge is good is greater than

$$d!(\frac{1}{d} - \frac{1}{2d^2|E|})^d > \frac{d!}{d^d}(1 - \frac{1}{2|E|}).$$

The expected number of good edges is greater than $|E|d!/d^d - 1/2$. That is, the expected number of good edges is at least $\lfloor |E|d!/d^d \rfloor$.

The sample space contains q^d sample points. Thus when d is a constant a polynomial number of processors is sufficient to perform an exhaustive search on all sample points and a NC algorithm can be derived.

12.4.2 The Vertex Partitioning Problem

The vertex partitioning problem (Luby 1988) is to label vertices of a graph $G = (V, E)$ with 0s and 1s such that the number of edges incident with both 0 and 1 labelled vertices is at least half of the total number of edges. This is a special case of the d-partite subhypergraph problem we discussed in the last subsection. Here we may use n 0/1-valued uniformly distributed mutually independent random variables, one for each vertex of the graph. Let $f(x_i, x_j) = x_i \oplus x_j$ and $F = \sum_{(i,j) \in E} f(x_i, x_j)$. The value of F is the number of edges incident with both 0 and 1 labelled vertices. Moreover, $E[F] = \sum_{(i,j) \in E} (f(0,0) + f(0,1) + f(1,0) + f(1,1))/4 = |E|/2$. Therefore there exists a sample point p such that $F(p) \geq |E|/2$. Function F is the BENEFIT function of the bit pairs PROFIT/COST problem. Thus for the case $d = 2$ the d-partite hypersubgraph problem can be solved by our fast derandomisation algorithm for the bit pairs PROFIT/COST problem.

12.4.3 Independent Set on Graphs

An independent set of a graph is a set I of vertices such that no two vertices in I are adjacent. I is maximal if I is not a proper subset of any other independent set. For certain restricted graphs a maximal independent set can be found by a parallel algorithm run in less than logarithmic time (Han 1989a, 1989b). We shall give a deterministic algorithm (Han 1991b) for computing a maximal independent set for general graphs through the derandomisation of a randomised algorithm due to Luby (1989).

Currently the fastest randomised algorithm for computing a maximal independent set is due to Luby (1986) which runs in expected $O(\log n)$ time using a linear number of processors. This algorithm uses a large sample space and no efficient derandomisation scheme is known to derandomise it. We will present another randomised algorithm, also due to Luby (1986), which runs in $O(\log^2 n)$ time.

Let $G = (V, E)$ be a graph. Let $d(i)$ be the degree of vertex i in G. Let k_i be such that $2^{k_i - 1} < 2d(i) \leq 2^{k_i}$. Let $q = \max\{k_i | i \in V\}$. Let $\vec{x} = < x_i \in \{0, 1\}^q, i \in V\}$. Each x_i is a random variable uniformly distributed on $\{0, 1, ..., 2^q - 1\}$. x_i is associated with vertex i. Define

$$Y_i(x_i) = \begin{cases} 1 & \text{if } x_i(k_i - 1) \cdots x_i(0) = 1^{k_i} \\ 0 & \text{otherwise} \end{cases}$$

where $x_i(p)$ is the p-th bit of x_i.

Luby used the following randomised algorithm to find an independent set I such that an expected constant fraction of the edges will be incident with vertices in $I \cup N(I)$, where $N(I) = \{j \in V | \exists i \in I, (i, j) \in E\}$.

Procedure *Independent*

begin

$I := \phi$;
for all i such that $d(i) = 0$ **do in parallel**
$\quad I := I \cup \{i\}$;

for all i such that $d(i) \neq 0$ **do in parallel**
$\quad$ generate random number x_i;
$\quad$ **if** $Y_i(x_i) = 1$ **then** $I := I \cup \{i\}$;

for all $(i, j) \in E$ **do in parallel**
$\quad$ **if** $i \in I$ and $j \in I$ **then**
$\quad\quad$ **if** $d(i) \leq d(j)$ **then** $I := I - \{i\}$;
$\quad\quad$ **else** $I := I - \{j\}$;

end

Through rather complicated probabilistic analysis Luby (1986) showed that after an execution of the above algorithm the expected number of edges deleted is at least a constant fraction of the number of edges in the input graph. The probabilistic analysis uses only the pairwise independence of the random variables. Procedure *Independent* can be executed in $O(\log n)$ time on an EREW PRAM using a linear number of processors (Luby 1986). A maximal independent set can be computed by the following procedure.

Procedure *Max-Independent*

begin

$I := \phi$;
$V' := V$;
while $V' \neq \phi$ **do**
$\quad$ **begin**
$\quad\quad$ Find an independent set $I' \subseteq V'$ using procedure *Independent*;
$\quad\quad$ $I := I \cup I'$;
$\quad\quad$ $V' := V' - (I' \cup N(I'))$;
$\quad$ **end**

end

We will use a fast derandomisation scheme to derandomise procedure *Independent* to obtain a deterministic parallel algorithm.

A vertex j will be in I after an execution of Independent if $Y_j(x_j) = 1$ and $Y_k(x_k) = 0$ for any neighbour k of j such that $d(k) \geq d(j)$. Define $Y_{j,k}(x_j, x_k) = -Y_j(x_j)Y_k(x_k)$. If $Y_j(x_j) + \sum_{k \in adj(j), d(k) \geq d(j)} Y_{j,k}(x_j, x_k) = 1$ then $j \in I$ and if it is ≤ 0 then $j \notin I$. A vertex i will be in $N(I)$ after an execution of Independent if any one of the neighbours of i is in I.

Therefore if

$$\sum_{j \in adj(i)} \left(Y_j(x_j) + \sum_{k \in adj(j), d(k) \geq d(j)} Y_{j,k}(x_j, x_k) \right.$$

$$\left. + \sum_{k \in adj(i)-\{j\}} Y_{j,k}(x_j, x_k) \right)$$

is 1 then $i \in N(I)$, if it is ≤ 0 then i may or may not be in $N(I)$. Now the following BENEFIT function gives a lower bound on the number of edges deleted after an execution of Independent (Luby 1986).

$$B(\vec{x}) = \sum_{i \in V} \frac{d(i)}{2} \sum_{j \in adj(i)} \left(Y_j(x_j) + \sum_{k \in adj(j), d(k) \geq d(j)} Y_{j,k}(x_j, x_k) \right.$$

$$\left. + \sum_{k \in adj(i)-\{j\}} Y_{j,k}(x_j, x_k) \right)$$

$\frac{d(i)}{2}$ is used because each edge has been counted twice at the two vertices it is incident with. The above formulation is due to Luby (1986, 1989) and he also showed that the expected value of B is $\geq |E|/c$ for a constant $c > 1$.

Function B can be written as,

$$B(\vec{x}) = \sum_{j \in V} \left(\sum_{i \in adj(j)} \frac{d(i)}{2} \right) Y(x_j) + \sum_{(j,k) \in E, d(k) \geq d(j)}$$

$$\left(\sum_{i \in adj(j)} \frac{d(i)}{2} \right) Y_{j,k}(x_j, x_k) + \sum_{i \in V} \frac{d(i)}{2} \sum_{j,k \in adj(i), j \neq k} Y_{j,k}(x_j, x_k)$$

It is now not difficult to see that we can obtain the following standard form of the BENEFIT function B by applying a matrix multiplication and then combining functions of the same subscript,

$$B(\vec{x}) = \sum_i f_i(x_i) + \sum_{(i,j)} f_{i,j}(x_i, x_j).$$

Such a matrix multiplication can be done in $O(\log n)$ time using $n^{2.376}$ processors (Coppersmith and Winograd 1987). By Theorem 12.3.8 the GPC problem can be solved in $O(\log n)$ time if we have $n^{2.376}$ processors. This eliminates a constant fraction of the edges from the graph. Therefore a maximal independent set can be found with $O(\log n)$ iterations of such processing.

Theorem 12.4.1 *A maximal independent set of a graph can be computed in* $O(\log^2 n)$ *time with* $n^{2.376}$ *processors on the CREW PRAM.* □

More sophisticated application of the derandomisation scheme yields fast and efficient parallel algorithms for the maximal independent set problem, the maximal matching problem and the $\Delta + 1$ vertex colouring problem. These results are given by Han (1991b).

Applications of derandomisation techniques to the design of several important parallel algorithms can be found in Alon *et al.* (1986), Berger and Rompel (1989), Berger *et al.* (1989), Han (1991a, 1991b), Han and Igarashi (1990), Karp and Wigderson (1985), Luby (1986, 1988, 1989), Motwani *et al.* (1989) and Pantziou *et al.* (1988).

References

Ajtai M., Komlós J. and Szemerédi E. (1983) An $O(N \log N)$ sorting network. *Proc. 15th ACM Symp. on Theory of Computing*, pp. 1–9.

Alon N., Babai L. and Itai A. (1986) A fast and simple randomized parallel algorithm for the maximal independent set problem. *J. of Algorithms* **7**, pp. 567–583.

Berger B. and Rompel J. (1989) Simulating $(\log^c n)$-wise independence in *NC*. *Proc. 30th Symp. on Foundations of Computer Science*, IEEE, pp. 2–7.

Berger B., Rompel J. and Shor P. (1989) Efficient NC algorithms for set cover with applications to learning and geometry. *Proc. 30th Symp. on Foundations of Computer Science*, IEEE, pp. 54–59.

Cole R. (1986) Parallel merge sort. *Proc. 27th Symp. on Foundations of Computer Science*, IEEE, pp. 511–516.

Cook S. A. (1979) Deterministic CFLs are accepted simultaneously in polynomial time and log square space. *Proc. of ACM STOC*, pp. 338–345.

Cook S. A. (1985) A taxonomy of problems with fast parallel algorithms. *Information and Control*, Vol. **64**, Nos. 1–3.

Coppersmith D. and Winograd S. (1987) Matrix multiplication via arithmetic progressions. *Proc. 19th Ann. ACM Symp. on Theory of Computing*, pp. 1–6.

Fortune S. and Wyllie J. (1978) Parallelism in random access machines. *Proc. 10th Annual ACM Symp. on Theory of Computing*, San Diego, California, pp. 114–118.

Goldberg M. and Spencer T. (1989a) A new parallel algorithm for the maximal independent set problem. *SIAM J. Comput.*, Vol. **18**, No. 2, pp. 419–427.

Goldberg M. and Spencer T. (1989b) Constructing a maximal independent set in parallel. *SIAM J. Dis. Math.*, Vol **2**, No. 3, pp. 322–328.

Han Y. (1989a) Matching partition a linked list and its optimization. *Proc. ACM Symposium on Parallel Algorithms and Architectures*, Santa Fe, New Mexico, pp. 246–253 (June, 1989).

Han Y. (1989b) Parallel algorithms for computing linked list prefix. *J. of Parallel and Distributed Computing* **6**, pp. 537–557.

Han Y. (1991a) A parallel algorithm for the PROFIT/COST problem. *Proc. of Int. Conf. on Parallel Processing*, Vol. III, pp. 107–114, St. Charles, Illinois.

Han Y. (1991b) A fast derandomization scheme and its applications. *Proc. Workshop on Algorithms and Data Structures*, Ottawa, Canada, Lecture Notes in Computer Science **519**, pp. 177–188. Full version in TR No. 180–90, Dept. Computer Sci., Univ. of Kentucky.

Han Y. and Igarashi Y. (1990) Derandomization by exploiting redundancy and mutual independence. *Proc. Int. Symp. SIGAL'90*, Tokyo, Japan, LNCS 450, pp. 328–337.

Israeli A. and Shiloach Y. (1986) An improved parallel algorithm for maximal matching. *Information Processing Letters* **22**, pp. 57–60.

Joffe A. (1974) On a set of almost deterministic k-independent random variables. *Ann. Probability* **2**, pp. 161–162.

Karp R. and Wigderson A. (1985) A fast parallel algorithm for the maximal independent set problem. *JACM* **32**:4, Oct., pp. 762–773.

Luby M. (1986) A simple parallel algorithm for the maximal independent set problem. *SIAM J. Comput.* **15**:4, Nov., pp. 1036–1053.

Luby M. (1988) Removing randomness in parallel computation without a processor penalty. *Proc. 29th Symp. on Foundations of Computer Science*, IEEE, pp. 162–173.

Luby M. (1989) *Removing randomness in parallel computation without a processor penalty.* TR–89–044, Int. Comp. Sci. Institute, Berkeley, California.

Miller G. L. and Reif J. H. (1989) Parallel tree contraction and its application. *Proc. 26th Symp. on Foundations of Computer Science*, IEEE, pp. 478–489.

Motwani R., Naor J. and Naor M. (1985) The probabilistic method yields deterministic parallel algorithms. *Proc. 30th Symp. on Foundations of Computer Science*, IEEE, pp. 8–13.

Pantziou G., Spirakis P. and Zaroliagis C. (1988) Fast parallel approximations of the maximum weighted cut problem through derandomization. *FST&TCS* 9 Bangalore, India, LNCS 405, pp. 20–29.

Raghavan P. (1988) Probabilistic construction of deterministic algorithms: approximating packing integer programs. *JCSS* **37**:4, pp. 130–143.

Spencer J. (1987) *Ten Lectures on the Probabilistic Method.* SIAM, Philadelphia.

CHAPTER 13

Parallel Algorithms for Matrix Operations and their Performance on Multiprocessor Systems

Jürgen-Friedrich Hake

Forschungszentrum Jülich (KFA), FRG

13.1 Introduction

An increased demand for computational capacity can either be satisfied by faster processors or by distributing the computational workload over a large number of processors. The first approach is limited by physical principles, e.g. the speed of electronic switches and chip complexity. The second approach is dominated by communication principles because the exchange of information between the processes becomes a limiting factor. Nevertheless, the latter way seems to be more promising for a successful treatment of problems arising from the *grand challenges* of computational sciences. Therefore, research on algorithms squeezing the most out of the existing multiprocessor systems represents a very active discipline. In this context, the term *parallel algorithm* refers to algorithms which can at least be partially executed concurrently on multiple processors in order to speed up the entire computation. The subsequent discussion is restricted to *parallel algorithms operating on matrix data structures*. For this topic, in particular, a large number of reports already exist and the results have been condensed into survey articles and books (Bertsekas and Tsitsiklis 1989; van der Vorst and van Dooren 1990; Gallivan *et al.* 1990; Heller 1978; McBryan and van de Velde 1987). The reader may find further information in these references.

In the field of numerical linear algebra, parallel algorithms and their implementation on multiprocessor systems can be approached from two directions:

- *Vector supercomputers* had strong impact on the efficient execution of matrix operations; regarding software developments, know-how has already been integrated in libraries like IMSL, NAG, and LAPACK. Moderately increasing the number of powerful processors, the vendors of vector supercomputers pursue a careful strategy towards new applications suitable for execution on a small number of processors with increased effectiveness. This is the domain of classical supercomputers like CRAY systems operating on a *shared memory*.

- A different view is taken if starting directly with a larger number of processors. While shared memory systems are moving towards a larger number of processors and probably a more elaborate memory hierarchy, massively parallel systems with *distributed memory* tend towards a more powerful and sophisticated architecture of the individual processor. Here, *hypercube systems* have to be mentioned. At the moment, the two most prominent vendors of hypercubes are Intel Supercomputer Systems Division and NCUBE.

- *Data parallel computing* systems like the Connection Machine associate one processor with each data element. This computing style exploits the fine-grained parallelism inherent in many data-intensive problems. A typical configuration consists of up to 65,536 processors. Each processor has its own bit-addressable memory of 32 Kbytes.

From these systems the CRAY Y-MP and the Intel iPSC have been selected in order to demonstrate the potential of parallel algorithms for matrix operations. The decision is based on two facts: both concepts represent the oldest commercially available systems in their field and their companies can be regarded as market leaders. Thus, it is of general interest to review and analyze the results obtained on these machines.

This chapter starts with a brief description of the architecture of CRAY Y-MP and iPSC multiprocessors including the corresponding parallel processing strategies. Additionally, a glance is cast at the tools supporting the implementation of parallel algorithms on both systems. Then, matrix multiplication is used as a model problem. Three different algorithms are considered: matrix multiplication based on elementary matrix-vector operations, matrix multiplication with recursive doubling, and Strassen's fast matrix multiplication. The efficient implementation of the three algorithms is also in the focus and their performance is compared with the performance of corresponding library routines. The results for matrix multiplication are related to the numerical solution of linear equations. Particularly, iterative methods for sparse problems are discussed. Finally, the numerical solution of eigenvalue problems on a multiprocessor system is outlined.

13.2 Architecture of Parallel Computers

Besides the number of processors and their architecture, two directions in the organisation of main memory can be observed. All processors of a multiprocessor configuration share a *global memory*, or each processor of the configuration has its *local memory*. Research on parallel systems with physically distributed memory but global address space has already been started. Detailed information on the architecture of supercomputers can be found in Hossfeld (1989) and Hwang and Briggs (1985).

Table 13.1: Hardware characteristics of CRAY multiprocessor systems

System	Proc. Cycle Time	No. of Funct. Units	No. of Procs	Peak Perform. (MFLOPS)	Main Memory (MB)	No. of Banks
CRAY-1 S (1976)	12.5	13 (3)	1	160	32	16
CRAY X-MP (1982)	8.5	13 (3)	4	940	128	64
CRAY Y-MP (1988)	6.0	13 (3)	8	2,664	1,024	256
CRAY C-90 (1992)	4.0	15 (6)	16	16,000	2,000	1,024

13.2.1 CRAY Shared Memory Multiprocessor Systems

The computer systems of Cray Research have evolved from a single processor CRAY-1—first installation in 1976—to a multiprocessor system CRAY Y-MP with up to eight processors; properties of the CRAY-2 are not considered here. A strong reduction in cycle time can be observed leading to improved performance of the individual processor while the parallelism represented by the number of functional units for floating point arithmetic per processor remains constant. Thus, the potential gained from parallelism originates from the growing number of processors. Details are given in Table 13.1.

The new generation of CRAY supercomputers, the CRAY Y-MP C90 series, have enhanced individual processors by internally doubling all vector functional units—given in parentheses in the third column of Table 13.1—i.e. two results can be delivered per clock period instead of one. The number of processors is increased to 16.

In addition to improvements to the processor, the bandwidth of data transfer from/to main memory has been enhanced. For CRAY multiprocessor systems, this can be characterised by the number of ports and the number of parallel banks in the main memory. A more detailed overview of the architecture of a CRAY Y-MP/832 is given later.

A CRAY Y-MP/832 represents an advanced X-MP-type architecture with 32 MW (= 256 MB) of interleaved main memory and eight processors; each processor has a cycle time of 6 nsec leading to a peak performance of 330 MFLOPS. Like the CRAY X-MP, each processor has four parallel memory ports: two for vector and scalar fetches, one for result store and one for bi-directional I/O operations. But in contrast to the X-MP, the main memory consists of 256 banks which for each CPU are grouped into four sections; each section is divided into eight subsections. The bank busy time is five cycles. Once a CPU references a subsection, this path to memory is busy for five cycles and succeeding requests to the same subsection will be delayed.

These *memory access conflicts* can significantly degrade the performance of user-written programs (Hake and Homberg 1991).

The CPUs of a CRAY Y-MP system are tightly coupled via the shared main memory and 9 identical groups of registers, called clusters, containing 8 shared address registers, 8 shared scalar registers, and 32 binary semaphore registers each. These registers can be accessed by all processors; depending on program requirements, either one cluster or multiple clusters, or possibly no cluster will be attached to a CPU. These hardware features provide the basis for high-speed communication between parallel tasks when multitasking is used (Faanes and Schwarzmeier 1989).

13.2.2 CRAY Multitasking Strategy

The multitasking concept provided by Cray Research supports the exploitation of parallelism on different programming levels. It is realised by means of subprogram libraries which allow the execution of one program in parallel using *tasks*. Depending on the granularity of parallelism, three options of multitasking are available on a CRAY multiprocessor system: *Macrotasking* for coarse-grain and *Microtasking* for fine-grain parallelism; *Autotasking* represents the latest development. Macrotasking works on the subroutine level and the user has to insert the appropriate calls to the library routines. Microtasking and Autotasking use compiler directives indicating segments of the code which can be executed concurrently. While Microtasking requires the programmer to insert the directives by hand, Autotasking performs the analysis automatically. Nevertheless, tuning by hand leads to optimal results.

Fine-grain parallelisation with autotasking is done by the automatic parallelisation of DO loops. The features are provided by the compiling system *cf77* which consists of three parts: *fpp*, *fmp*, and *cft77*. The preprocessor *fpp* analyzes FORTRAN programs; it is able to detect DO loops executable in parallel, and it marks these loops with preprocessor directives. These directives are translated by the *fmp* preprocessor which is the second part of the *cf77* compiling system, and the *cft77* generates the machine code for the modified FORTRAN program.

Autotasking supports flexible definition of parallel regions at any place within a subroutine. The preprocessor *fmp* generates a set of subroutines for each of the parallel regions, and all available CPUs will enter at the beginning of these subroutines. Implicit synchronisation takes place at the bottom of the code marked by the DO PARALLEL, the END CASE, or the DO ALL directives. For each of the parallel regions the data scope of the variables (SHARED or PRIVATE) is analyzed and can be explicitly defined.

The following list gives a short description of the autotasking primitives.

- CMIC$ PARALLEL
  marks the beginning of a parallel region where multiple CPUs may enter. The parallel processes within these parallel regions have to be marked with DO PARALLEL or CASE directives.

- CMIC$ END PARALLEL

  marks the end of a parallel region.

- CMIC$ DO PARALLEL

  marks a DO loop to be executed in parallel if more than one CPU is available within a parallel region.

- CMIC$ DO ALL

  defines a separate parallel region for the following DO loop only. In addition, this DO loop is marked as being executable in parallel by more than one CPU.

- CMIC$ CASE

  marks the beginning of a program part which could be executed in parallel to other processes. The directive may be used several times.

- CMIC$ END CASE

  marks the end of a CASE part.

- CMIC$ SOFT EXIT

  provides a way to stop parallel execution before all computations are finished.

Several directive options like threshold tests, flexible definition of chunk sizes (including the guided self-scheduling approach, automatic partitioning of long vector loops etc.) have been introduced to optimise the execution of parallel regions.

Autotasking can coexist with macrotasking and microtasking within one program system, but autotasking and microtasking are not allowed to be used together within one subroutine. The reader may find further information about the CRAY multitasking concept in Nagel and Szelenyi (1989) and Nagel (1990).

13.2.3 Intel Distributed Memory Multiprocessor Systems

The idea of designing a hypercube computer was first proposed in 1963. In 1985, Intel announced the first commercial hypercube, the iPSC/1. As a result of the Touchstone Project, Intel will deliver the Paragon XP/S supercomputer late in 1992, which will comprise up to 4,000 processors with a peak performance of 300 GFLOPS. Whereas the nodes of the previous systems are connected as a *hypercube*, the nodes of the Paragon are *mesh interconnected*. This change in interconnection topology leads to a better scalability, e.g. upgrades to the next larger model can be done in smaller steps.

The iPSC distributed memory multiprocessor system represents a multiple-instruction / multiple-data (MIMD), loosely coupled, message passing computer (Shih and Fier 1989). The connection network for the processing elements (also called nodes) is called hypercube because of its structure where each of the $p = 2^d$ nodes is connected by fixed communication paths to d other nodes; the

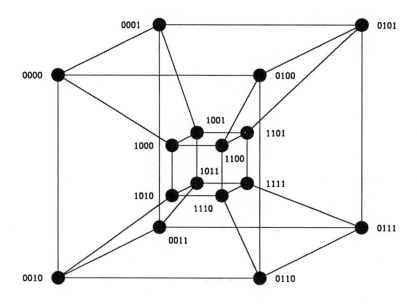

Figure 13.1: Interconnection topology of a 4-dimensional hypercube

value d is known as the dimension of the hypercube. Labelling the processing elements $0, 1, \ldots, 2^d - 1$, then processing element i has a communication link to processing element j if and only if $|i - j|$ is an integral power of two. If the labels of the nodes are viewed as d–bit strings, then two processing elements have a communication link between them if and only if their labels differ by a single bit (see Figure 13.1). The time to communicate k real numbers (4-byte words) to a neighbouring processor consists of a startup time t_{start} and the time t_{send} required to send one unit of data

$$t_{comm}(k) = t_{start} + k * t_{send}.$$

Since the iPSC systems distinguish between short ($\leq 100byte$) and long ($>100byte$) messages, there are two sets of parameters t_{start}, t_{send} given in the last column of Table 13.2.

The CPU of each node of the iPSC/860 is based on the Intel i860 microprocessor. While previous systems used CISC technology, the i860 processor contains a 64-bit RISC processor, a floating point adder, a floating point multiplier, and 8 and 16 Kbyte caches for instructions and data. The peak rate of a processor is 60 MFLOPS with 64-bit arithmetic resulting in a total of 7.6 GFLOPS for the overall system. The i860 microprocessor can have up to 64 Mbyte of on-board memory. I/O is performed with the help of specialised Intel 386 microprocessor-based I/O nodes. The iPSC/860's Direct-Connect internal network provides 5.6 Mbytes/sec high-level data pathways between the nodes

Table 13.2: Hardware characteristics of Intel multiprocessor systems

System	Max. No. of Nodes	Node Processor	Local Memory (MB)	Peak Performance (MFLOPS)	Communication (μs)
iPSC/1 (1985)	$2^7 = 128$	80286, 80287	0.5	3.2	1000 4.00 1000 7.50
iPSC/2 (1987)	$2^7 = 128$	80386, 80387	8	2,500	350 0.80 660 1.44
iPSC/860 (1989)	$2^7 = 128$	i860	64	7,600	70 1.6 175 1.44
Paragon (1992)	$\leq 4,000$	i860XP	128	300,000	

of the system. Automatic switching hardware frees the programmer from the details of message routing.

13.2.4 Intel Message Passing Strategy

For each run, it is possible to allocate an appropriate subcube of the entire system. This allocation has to take place at the start of the application and remains static for the whole run.

Information between the nodes of a hypercube is exchanged via messages. For this purpose, communication primitives are provided. They exist in a blocking (synchronous) and a non-blocking (asynchronous) version.

In the blocking version, the sending process is halted until the message has entered the communication network. From this moment on, the send buffer can be reused in the sending process. A blocking receive does not return until an appropriate message has arrived.

With the non-blocking primitives the program informs the operating system that a message could be sent or received. The processor is allowed to proceed with its computations while the communication processor handles the message request. The communication buffer should not be accessed or reused until the process has verified that the communication operation has been finished.

When a message is sent and the destination process is not expecting a message at the time of its arrival, the message is buffered by the node operating system. Messages are characterised by their *length, type,* and *id.* Messages between nodes can be of any length, but the length has to be exactly as specified; the type is a message identifier whose use is determined by the programmer; the id is used to check the completion of asynchronous messages.

The message passing primitives are summarised in the following:

- crecv()
  represents the name of a call for synchronous receive. A synchronous

receive means that the program executing the receive waits until the
message arrives in the specified buffer.

- csend()
  means a synchronous send. The program executing the send waits until
  it is complete.

- csendrecv()
  denotes a synchronous send followed by a receive.

- irecv()
  is the asynchronous receive.

- isend()
  represents the asynchronous send.

- isendrecv()
  is like an isend() followed by an irecv(); it returns one message id.

- msgdone()
  can be used to check if an asynchronous operation has been completed.

- msgwait()
  denotes the call to block the completion of an asynchronous call.

A detailed analysis concerning the costs of message passing on an Intel hypercube system has been carried out by Bomans and Roose (1989).

13.2.5 Performance of Parallel Algorithms

The performance of a program on a multiprocessor with p processors compared
to a program on a uniprocessor can be described by its *speedup* $S_p = t_1/t_p$
which is the number of times a given problem runs faster on a multiprocessor
than on a single node. The ratio $\varepsilon_p = S_p/p$ is termed *efficiency*.

A more detailed analysis has led to the formulation of *Amdahl's law* reflect-
ing the fact that in most cases the time consumed by a program consists of a
serial fraction which cannot be executed in parallel. This observation results in
modified speedup

$$
\begin{aligned}
S_A &= (\tilde{t}_s + \tilde{t}_p)/(\tilde{t}_s + \tilde{t}_p/p) \\
&= 1/(\tilde{t}_s + \tilde{t}_p/p)
\end{aligned}
$$

where p denotes the number of processors and $\tilde{t}_s, \tilde{t}_p$ represent the fractions
of time spent on serial and parallel work with respect to the normalization
$\tilde{t}_p + \tilde{t}_s = 1$.

Due to considerable scepticism regarding the viability of massive parallelism, Amdahl's law has been revised and an alternative has been formulated.
Analyzing the reverse question of how long a given parallel program would

have taken to run on a serial processor leads to the term of *scaled speedup* discussed by Gustafson *et al.* (1988).

$$
\begin{aligned}
S_{scal} &= (\hat{t_s} + \hat{t_p}p)/(\hat{t_s} + \hat{t_p}) \\
&= p + (1 - p)\hat{t_s}
\end{aligned}
$$

where $\hat{t_s}$ and $\hat{t_p}$ represent the serial and parallel time spent on the parallel system; with the normalization $\hat{t_s} + \hat{t_p} = 1$, a uniprocessor requires $\hat{t_s} + \hat{t_p}p$. This formulation is based on the observation that the serial fraction typically decreases as the problem size increases.

The differences of the two paradigms can be seen from Figure 13.2. On shared-memory systems with a few processors, Amdahl's law gives a reasonable interpretation. But on massively parallel computers, the fixed-size model does not represent an adequate approach because the same problem is treated with a more powerful tool increasing the cost.

An algorithm on concurrent processors exhibits a *load balance* if all the nodes perform approximately equal amounts of work, so that no node is idle for a significant amount of time.

In message passing systems, processes communicate by exchanging messages via communication channels. In this context, *communication overhead* is a measure of the workload incurred in a concurrent algorithm due to communication between the nodes of a multiprocessor. On shared memory systems, one can define *synchronisation overhead* analogously.

13.2.6 Tools for the Analysis and Implementation of Parallel Algorithms

Users of parallel computers have been hoping that parallelising compilers can convert sequential programs to their parallel version in a user transparent manner. Since compiler technology is still lagging behind the advances of computer architecture, tools have been developed to exploit the potential of parallelism in a sequential program and to support the user in the conversion of his software. There are four groups of tools under development:

- *Design and simulation tools* may help to evaluate the understanding of existing architectures as well as to support the development of new architectural concepts.

- After the completion of a parallel program, *static analysis tools* can be applied to check properties of this program without its execution. The analysis of data dependencies is of particular interest.

- During the execution of a parallel program, *monitoring tools* can be used to collect information, which is then analyzed for debugging and improving the performance.

- Even if most of today's supercomputer applications are written in Fortran or C, one should mention the field of *parallel languages* where constructs are developed to enhance existing languages or design new languages.

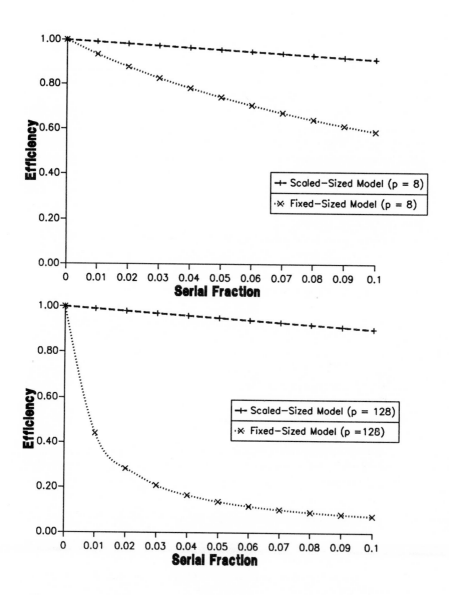

Figure 13.2: Efficiency given by Amdahl's law and by problem scaling

For a CRAY Y-MP, a family of tools is available. They help to analyze the performance of application programs. The performance data may reflect CPU execution time, I/O use or other types of data. Besides the measurement and display of data with FLOWTRACE and the performance monitor, there exists a tool ATEXPERT predicting the speedup of an autotasked program. The prediction is for a dedicated system and from data collected from a single run on a nondedicated system. Using the tools together, a detailed and systematic analysis can be rapidly carried out resulting in a better understanding and very often in an improved efficiency of the user's application.

The iPSC/860 software includes a development toolset, called Concurrent Workbench. The Concurrent Workbench toolset is based on C and Fortran compilers. In addition to the standard UNIX System V software development tools, the workbench provides an interactive parallel debugger (IPD) plus a node profiler for performance tuning. The parallel performance analysis tool produces CPU usage histograms, event tracing of multiple processors, and internode communications tracing. Further tools for efficient porting of existing codes to the iPSC systems are under development.

Moreover, there exists a small number of projects working independently from vendors on the development on tools.

13.3   Matrix Multiplication

Matrix multiplication is considered as a model problem; for simplicity, only square matrices are considered. The product $C = (\gamma_{ij}) \epsilon R^{n \times n}$ of two square matrices $A = (\alpha_{ik}) \epsilon R^{n \times n}$ and $B = (\beta_{kj}) \epsilon R^{n \times n}$ is defined by the relation $\gamma_{ij} = \sum_{k=1}^{n} \alpha_{ik} \beta_{kj}$. Although the structure of the algorithm is very simple, it can be implemented in six different ways depending on the order of the Fortran DO loops for the indices i, j, k. The different versions of the algorithm are denoted by the corresponding triple, i.e. the jki version means that the outer loop is referenced by index j and the innermost loop by index i (Figure 13.3). Additional options arise from the consideration of block matrix substructures leading to further improved performance particularly for systems with a hierarchical memory.

13.3.1   Matrix Multiplication Based on BLAS

The Basic Linear Algebra Subroutines (BLAS) represent a set of frequently occurring elementary vector and matrix operations which are used as building blocks for more complex numerical algorithms. At present, there are three levels of BLAS; the complexity of the operations increases from level one to level three. At level one, vector operations like SAXPY ($y \leftarrow \alpha x + y$) or SDOT ($x^T y$) are available; at level two, matrix-vector operations like the computation of a matrix-vector product SGEMV ($y \leftarrow \alpha A x + \beta y$) come into the focus. A discussion of Level 3 BLAS and its suitability for an efficient use of parallelism is given in Section 13.4.1.

```
SUBROUTINE MMJKI (N,A,B,C,)
INTEGER N, I, J, K
REAL     A(N,N), B(N,N), C(N,N)
DO 20 J=1,N
    DO 10 I=1,N
        C(I,J) = 0.0
10      CONTINUE
    DO 20 K=1,P
        DO 20 I=1,N
            C(I,J) = C(I,J) + A(I,K)*B(K,J)
20  CONTINUE
    RETURN
    END
```

Figure 13.3: jki version of matrix multiplication based on BLAS

Using the Level 1 BLAS notation, one can identify an SDOT (dot product) as a kernel of the ijk version and a SAXPY operation for the jki version of matrix multiplication, respectively. According to the Level 2 BLAS, the two innermost loops of subroutine MMJKI represent a matrix-vector multiplication (SGEMV) with columnwise access to the arrays. For one processor of a CRAY, the jki version has been recommended to achieve supervector speed because the coefficient matrices are referenced by columns and the vector length of the vector operations is n. This approach was chosen by the library vendors IMSL and NAG for the CRAY version of their routines MRRRR and F01CKF[1].

On a CRAY, the parallel implementation starts from the sequential jki version and is straightforward; the columnwise initialisation and the SAXPY operations can each be executed in parallel. The corresponding code is given in Figure 13.4; the performance data is included in Table 13.3.

On the iPSC, the product $C = AB$ of two $n \times n$ matrices A and B can be computed by distributing block columns to the processors. For simplicity let $n = k2^d$, d denoting the dimension of the hypercube. Initially, processor j, $0 \le j \le 2^d - 1$, contains the columns $jk + 1, \ldots, (j + 1)k$ of A and B, and upon completion the corresponding columns of the result C. During the formation of C, the columns of B remain at their places while the columns of A are passed along a ring from processor to processor, overwriting the previously held columns of A. Within the ring, processor j communicates with its neighbouring processors, $j - 1$ and $j + 1$, where the processor identifiers in a d−cube are taken modulo 2^d. The identifiers are generated according to a binary reflected Gray code sequence (see Figure 13.1). A pseudocode of this procedure is given in Figure 13.5.

---

[1]IMSL Edition 10 and NAG Mark 12

```
      SUBROUTINE MMJKI(N,A,B,C)
      INTEGER N,I,J,K
      REAL*8   A(N,N),B(N,N),C(N,N)
CMIC$ DO ALL SHARED (N,A,B,C) PRIVATE (J,K,I)
      DO 30, J = 1,N
         DO 10, I = 1,N
            C(I,J) = 0.0
 10      CONTINUE
         DO 20 K=1,N
            DO 20 I=1,N
               C(I,J) = C(I,J) + A(I,K)*B(K,J)
 20      CONTINUE
 30   CONTINUE
      RETURN
      END
```

Figure 13.4: Parallel jki version of matrix multiplication on CRAY Y-MP

Do on all processors j
  $C_j = 0$
  For $i = 0, \ldots, 2^{d-1}$
    $C_j = C_j + A_{j-i}B_{j-i,j}$
    send A_{j-i} to processor $j + 1$
    receive A_{j-i-1} from processor $j - 1$
  End
End

Figure 13.5: Parallel jki version of matrix multiplication for iPSC/860

13.3.2 Matrix Multiplication Based on a Divide-and-Conquer Strategy

Divide-and-conquer strategies represent a general technique for the efficient treatment of problems on a multiprocessor system. The problem, e.g. matrix multiplication, is split into a certain number of tasks which can be executed concurrently on the processors. After completion, the results of the tasks are combined in order to provide a solution for the original problem. This approach can be applied to a variety of problems; its simplicity permits great flexibility with regard to the implementation.

For matrix multiplication this means the introduction of block matrix sub-structures. For simplicity, it is assumed that all matrices have an order $n = 2^k$. Let $A, B, C \epsilon R^{n \times n}$ be partitioned into blocks, such that

Table 13.3: Performance (wall clock time in sec) of matrix multiplication on a CRAY Y-MP

$p\backslash n$	SMMJKI				SMMPW			
	128	256	512	1024	128	256	512	1024
1	.02	.15	1.18	9.23	.01	.11	.87	6.90
2	.01	.08	0.59	4.60	.01	.06	.43	3.45
4	.01	.05	0.30	2.30	.01	.03	.22	1.76
8	.01	.02	0.16	1.22	.01	.01	.11	0.88

$p\backslash n$	SSTRASS			
	128	256	512	1024
1	.01	.11	.77	5.49
2	.01	.05	.38	2.70
4	.01	.03	.19	1.35
8	.01	.02	.10	0.72

$$\begin{pmatrix} C_{11} & C_{12} \\ C_{21} & C_{22} \end{pmatrix} = \begin{pmatrix} A_{11} & A_{12} \\ A_{21} & A_{22} \end{pmatrix} * \begin{pmatrix} B_{11} & B_{12} \\ B_{21} & B_{22} \end{pmatrix}$$

where $A_{ij}, B_{ij}, C_{ij} \epsilon R^{n/2 \times n/2}$. Then, $C_{ij} = A_{i1}B_{1j} + A_{i2}B_{2j}$, where the eight submatrices $A_{ik}B_{kj}$ can be computed in parallel. This procedure can be recursively repeated for each product $A_{ik}B_{kj}$ until the block size is one (see Figure 13.6). The algorithm still has the complexity $O(n^3)$ but is well suited for an implementation on multiprocessors.

On a CRAY Y-MP with eight processors, $n_{min} = n/2$ was chosen to guarantee a maximum workload for the individual processor. Each processor can use SGEMM or MXMA from SCILIB for his subtask. The corresponding Fortran code is given in the Appendix. Performance data for this algorithm is given in Table 13.3.

13.3.3 Fast Matrix Multiplication

The divide-and-conquer strategy in conjunction with Strassen's fast multiplication of (2×2) matrices leads to an improved algorithm for matrix multiplication. In 1969, Strassen pointed out that two (2×2) matrices can be multiplied with seven instead of eight scalar multiplications resulting in an overall complexity of $O(n^{ld7})$ instead of $O(n^3)$ for conventional matrix multiplication (Kronsjö 1985). In this context one should also mention that reduced computational complexity is gained by a potential loss of accuracy which is due to a weaker stability condition (Higham 1990).

function: $C = \mathrm{mmpw}\,(A, B, n_{min})$
$n = rows(A)$
if $n \leq n_{min}$ **then** $C = AB$
else
$\quad m = n/2; u = 1 : m; v = m + 1 : n;$
$\quad P_1 = \mathrm{mmpw}\,(A(u, u), B(u, u), n_{min})$
$\quad P_2 = \mathrm{mmpw}\,(A(u, v), B(v, u), n_{min})$
$\quad P_3 = \mathrm{mmpw}\,(A(v, u), B(u, u), n_{min})$
$\quad P_4 = \mathrm{mmpw}\,(A(v, v), B(v, u), n_{min})$
$\quad P_5 = \mathrm{mmpw}\,(A(u, u), B(u, v), n_{min})$
$\quad P_6 = \mathrm{mmpw}\,(A(u, v), B(v, v), n_{min})$
$\quad P_7 = \mathrm{mmpw}\,(A(v, u), B(u, v), n_{min})$
$\quad P_8 = \mathrm{mmpw}\,(A(v, v), B(v, v), n_{min})$
$\quad C(u, u) = P_1 + P_2$
$\quad C(v, u) = P_3 + P_4$
$\quad C(u, v) = P_5 + P_6$
$\quad C(v, v) = P_7 + P_8$
end mmpw

Figure 13.6: Matrix multiplication based on a divide-and-conquer strategy

On one processor of a CRAY Y-MP, subroutine SGEMM is one of the fastest routines for matrix multiplication. Hence, approaches claiming to be more efficient have to compete with this routine. Bailey has pointed out that a fast sequential implementation of Strassen's algorithm can outperform corresponding library routines on one processor of a CRAY-2 (Bailey 1988). Thus, a combination of the divide-and-conquer strategy and fast matrix multiplication should result in a very efficient implementation on a parallel computer.

The subsequent results are based on a Fortran implementation SSTRASS of Strassen's algorithm as given in Figure 13.7. The size of the problem is recursively reduced until the actual order of the reduced problem is less than $n_{min} = 256$. Then, SGEMM is used as a kernel. Results on the performance of STRASS in comparison to the other methods considered so far are given in Table 13.3.

The subroutines SMMJKI, SMMPW, and SSTRASS operate at a speed of 230, 310, and 300 MFLOPS per processor on a CRAY Y-MP. In contrast to algorithm MMPW, algorithm STRASS does not automatically lead to a load balancing because in the first part only seven independent matrix operations are generated.

function: $C = \text{strass}\,(A, B, n_{min})$
$n = \text{rows}\,(A)$
if $n \leq n_{min}$ **then** $C = AB$
else
$\quad m = n/2; u = 1 : m; v = m + 1 : n;$
$\quad P_1 = \text{strass}\,(A(u, u) + A(v, v), B(u, u) + B(v, v), n_{min})$
$\quad P_2 = \text{strass}\,(A(v, u) + A(v, v), B(u, u), n_{min})$
$\quad P_3 = \text{strass}\,(A(u, u), B(u, v) - B(v, v), n_{min})$
$\quad P_4 = \text{strass}\,(A(v, v), B(v, u) - B(u, u), n_{min})$
$\quad P_5 = \text{strass}\,(A(u, u) + A(u, v), B(v, v), n_{min})$
$\quad P_6 = \text{strass}\,(A(v, u) - A(u, u), B(u, u) + B(u, v), n_{min})$
$\quad P_7 = \text{strass}\,(A(u, v) - A(v, v), B(v, u) + B(v, v), n_{min})$
$\quad C(u, u) = P_1 + P_4 - P_5 + P_7$
$\quad C(u, v) = P_3 + P_5$
$\quad C(v, u) = P_2 + P_4$
$\quad C(v, v) = P_1 + P_3 - P_2 + P_6$
end strass

Figure 13.7: Strassen's algorithm for fast matrix multiplication

13.4 Elementary Matrix Operations

Elementary matrix operations are used as building blocks for more complex algorithms; e.g. rank-one updates (Householder transformations) represent the constitutional part of the QR algorithm for the numerical solution of eigenvalue problems. This observation has led to the development of professional libraries. For problems with dense coefficient matrices, three levels of Basic Linear Algebra Subroutines (BLAS) document the progress in this field. Today, BLAS may be regarded rather as an interface than a library. Of course, the acceptance of this interface is strongly based on ease-of-use, reliability, and efficiency of its implementation.

13.4.1 Level 3 BLAS

The first two levels of BLAS have already been introduced in Section 13.3.1 dealing with matrix multiplication. Level 3 BLAS represents the matrix-matrix operations within BLAS (Dongarra *et al.* 1990; Gallivan *et al.* 1987).

The BLAS operations obey the following naming convention: each name consists of a root name describing the operation under consideration, e.g. MM for a generalised form of matrix multiplication $C \leftarrow \alpha AB + \beta C$, and two prefixes specifying the operation in more detail. The innermost prefix to the root name characterises the mathematical structure of suitable matrices, e.g. GE for a general quadratic matrix or SY for a real symmetric matrix. The

Table 13.4: Survey of level 3 BLAS

Operation	Data Type of Matrix	Prop. of Matrix	Name	Example
$C \leftarrow \alpha AB + \beta C$	S, D, C (, Z)	GE, SY, TR	MM	SGEMM
$C \leftarrow \alpha AA^T + \beta C$	S, D, C (, Z)	SY	RK	SSYRK
$C \leftarrow \alpha AB^T + \alpha BA^T$ $+\beta C$	S, D, C (, Z)	SY	R2K	SSYR2K
$B \leftarrow \alpha T^{-1} B$	S, D, C (, Z)	TR	SM	DTRSM

second prefix specifies the data type of the matrix on the computer system, e.g. S for single precision meaning 32-bit arithmetic on an iPSC hypercube and 64-bit arithmetic on a CRAY computer system. Hence on CRAY and iPSC, SGEMM and DGEMM represent corresponding versions. (See Table 13.4.)

On a CRAY, an optimised sequential version of all three levels of BLAS soon became available in the library SCILIB. High performance on a single processor can be achieved if the processor can operate on long contiguously stored vectors. A more detailed analysis shows that besides the storage locality of data the ratio of the number of references to memory and the number of operations represents a crucial point. Whereas a SAXPY operation (Level 1 BLAS) of length n requires $3n$ memory references and $2n$ arithmetic operations, a matrix multiplication (Level 3 BLAS) can be computed with $3n^2$ references and $2n^3$ operations. Hence, memory bottlenecks can be avoided and do not lead to a degradation of performance, if a higher level of BLAS is preferred.

With the advent of multitasking, work on parallel BLAS was started. First results show that only a parallel implementation of Level 3 BLAS leads to a sustained performance which is as close as possible to the peak performance because at this level the number of operations per processor, the number of references to memory, and synchronisation overhead can be balanced. Table 13.5 demonstrates the performance of different routines from Level 3 BLAS; a discussion of the performance of routine STRSM is postponed to Section 13.5.2.

On one processor of a CRAY Y-MP, SGEMM, SSYRK with $k = 32$, and SSYR2K ($k = 32$) operate at a speed of 310 MFLOPS, 290 MFLOPS, and 300 MFLOPS. Timings for these library routines are included in Table 13.5. Comparing the performance of subroutines SMMJKI and SGEMM yields a speedup of 1.3 for the benefit of SGEMM from SCILIB. The reader should note that the last two operations in the table have complexity $O(n^2)$ in contrast to the matrix multiplication.

Table 13.5: Performance (wall clock time in sec) of level 3 BLAS on a CRAY Y-MP

$p\backslash n$	SGEMM				SSYRK			
	128	256	512	1024	256	512	1024	2048
1	.01	.11	.87	6.90	.01	.03	.12	.46
2	.01	.06	.43	3.43	.01	.16	.06	.24
4	.01	.03	.22	1.73	.01	.01	.03	.12
8	.01	.01	.11	0.87	.01	.01	.02	.06

$p\backslash n$	SSYR2K			
	256	512	1024	2048
1	.01	.06	.22	.89
2	.01	.03	.12	.48
4	.01	.02	.06	.24
8	.01	.01	.03	.12

13.4.2 Beyond the BLAS

The three levels of the BLAS represent important milestones in the development of mathematical software. But from the table of contents of software libraries in this field one can also deduce that the libraries cover a much broader spectrum with more complex and advanced methods. The topics which can be treated with library software range from matrix multiplication, solution of linear equations and eigenvalue problems to the numerical integration of functions and differential equations. In most cases, the libraries offer more than one method for a particular problem. They also supply the user with information on how to select an appropriate algorithm. Many scientific application codes require access to subroutines from the field of numerical linear algebra which then are used as building blocks. Therefore, the linear algebra chapter represents a dominant part in most of the mathematical software libraries.

Mathematical software libraries originate from different sources. Vendors of a computer system provide a library on their machines (e.g. SCILIB on CRAY systems) in order to support efficient implementations of customer applications. Since these libraries are based on a machine-specific approach, portability is not a major concern but efficiency of the implemented algorithms is in the focus. Libraries like IMSL and NAG intend to complement the situation. In general, their scope is much broader. Today these libraries offer more than 500 driver routines for very diverse problems. Very often these subroutines can be regarded as an interface to a family of algorithms. The vendors deal with the efficiency aspect by providing for each major computer system a machine-specific version of the library (Hake and Homberg 1989). Thus, portability of user-written applications can be maintained by accessing these libraries.

On multiprocessor systems, the libraries have not only to deal with the general criteria of quality of mathematical software but also to support the parallel execution of more complex applications. On a CRAY multiprocessor system, the SCILIB provides library routines with multitasking capabilities. With the parallel implementation of Level 3 BLAS, a first step has been successfully taken. On the iPSC/860, the programming environment comprises similar tools. Obviously, the development of library software incorporating these features is still an ongoing process.

Work on BLAS has led to the design of a new software package LAPACK which will comprise a collection of black-box routines for a variety of problems from numerical linear algebra. This project will have a strong impact on the development of library software for parallel computers (Anderson *et al.* 1990).

13.5   Direct Solution of Linear Systems

The solution of linear systems $Au = b$ with a regular coefficient matrix $A \epsilon R^{n \times n}$ can be achieved by two subtasks: LU decomposition of the coefficient matrix and forward/backward substitution of two resulting systems with triangular matrices. For systems with special properties, e.g. sparse coefficient matrix, tridiagonal matrix, other algorithms may also be appropriate. Software for the direct solution of linear systems may be found in LINPACK. This package has represented the state of the art in the solution of systems of linear equations for several years. With the advent of LAPACK, new routines are under development for this important task.

The numerical solution of linear systems with a sparse matrix by means of direct methods has been extensively discussed. A comprehensive survey of this topic has been published by Heath *et al.* (1990).

13.5.1   LU Decomposition

Due to the dominant role LU decomposition plays in the solution of linear systems much attention has been paid to implement it well on supercomputers. The LU decomposition of a matrix $A \epsilon R^{n \times n}$ can be computed by means of stabilized elementary operations such that $A = PLR$, where L is unit lower triangular, R is upper triangular, and P represents a permutation matrix. Again, the ijk notation introduced for matrix multiplication helps to denote alternative implementations.

Detailed studies show that the kji version performs well on one processor of a vector computer; a pseudocode of the algorithm which is the starting point for a parallel implementation is given in Figure 13.8. In Figure 13.9, the implementation on a CRAY multiprocessor is indicated; Table 13.6 contains the corresponding performance data.

From the point of view of numerical analysis, a discussion of the LU decomposition has also to consider the implementation of a pivoting strategy in

```
for k = 1 : n - 1
    A(k + 1 : n, k) = A(k + 1 : n, k)/A(k, k)
    for j = k + 1 : n
        for i = k + 1 : n
            A(i, j) = A(i, j) - A(i, k)A(k, j)
        end
    end
end
```

Figure 13.8: kji version of LU decomposition

```
      DO 30 K=1,N-1

          C= 1.0 / A(K,K)
          DO 101 K1 = K+1,N
             A(K1,K) = C * A(K1,K)
101       CONTINUE

CMIC$ DO ALL NUMCHUNKS(NCPUS)
              SHARED(N,K,A) PRIVATE(J,K1)
          DO 102 J = K+1,N
             DO 102 K1 = K+1,N
                A(K1,J) = A(K1,J)-A(K1,K)*A(K,J)
102       CONTINUE
 30   CONTINUE
```

Figure 13.9: Parallel kji version of LU decomposition on CRAY

order to assure the existence of the decomposition and to guarantee numerical stability. A systematic analysis shows that for the kji variant, row and column pivoting is possible. For this variant, column pivoting vectorizes but implicit pivoting is not possible; for row pivoting, the situation is reversed.

The performance of LU decomposition can be further improved. Applying two successive rank-one updates $(I - m_{k+1}e_{k+1}^T) * (I - m_k e_k^T)$ instead of one increases the number of arithmetic operations per process; here, $m_k^T = (0, \ldots, 0, \mu_{k+1}, \ldots, \mu_n)$ and e_k denotes the k-th unit vector.

Whereas the LU decomposition based on rank-one updates operates with 150 MFLOPS per CRAY processor, the version with rank-two updates (see Figure 13.10) achieves approximately 200 MFLOPS per processor for the larger problems. Timings are included in Table 13.6; the first line of a row is for the rank-one updates, the second for the rank-two updates. A speedup of

Table 13.6: Performance (wall clock time in sec) of parallel LU decomposition

CRAY Y-MP	$n = 128$	$n = 256$	$n = 512$	$n = 1024$
1 proc	0.01	0.07	0.52	4.05
	0.01	0.06	0.45	3.47
4 procs	0.01	0.02	0.13	1.03
	0.01	0.02	0.12	0.89
8 procs	0.01	0.01	0.08	0.58
	0.01	0.01	0.06	0.45

1.2 is gained from rank-two updates in the LU decomposition. The various options for a parallel implementation of LU decomposition have been discussed in Frommer (1990) and Ortega and Romine (1988).

On the iPSC, a matrix of order n is wrapped onto the p nodes of the hypercube, i.e. the matrix is partitioned into quadratic blocks A_{ij} and the blocks are distributed to the nodes. If the nodes are indexed by (I, J), where $1 \leq I \leq M$ and $1 \leq J \leq N$ with $(NM = p)$, then *block-torus-wrapped* mapping means that block A_{ij} is assigned to node $((i-1)modM, (j-1)modN)$. Using this structure of communication, a right-looking LU factorization routine has been developed by van de Geijn.

Further experiments on a hypercube show that explicit pivoting is faster than implicit pivoting and should be preferred. With dynamic pivoting, the situation can be further improved. Here, pivoting is done implicitly if all the required information is local to the processor; otherwise explicit pivoting is chosen selecting a processor nearby such that the communication costs can be kept as low as possible.

13.5.2   Solution of Triangular Systems

The solution of triangular systems represents the second step in the solution of linear systems by means of Gaussian elimination. This step can be accomplished with Level 3 BLAS routine STRSM. Timings for STRSM on a CRAY Y-MP are included in Table 13.7.

On the iPSC, routines for the efficient solution of upper and lower triangular systems are based on the Li-Coleman algorithm (Li and Coleman 1989). The matrix is column-wrapped onto an embedded ring of nodes. Recent results by van de Geijn show that a block-torus-wrapped mapping combines the advantages of block matrix structures and an efficient communication scheme.

13.5.3   Solution of Linear Systems

A combination of the two subtasks—LU decomposition and solution of triangular systems—shows that a system of linear equations can be solved efficiently

```
      DO 100 K = 1,N,2
         DO 101 KP1 = K+1,N
            A(KP1,K) = (-1./A(K,K))*A(KP1,K)
101      CONTINUE
         DO 102 KP1 = K+1,N
            A(KP1,K+1) = A(KP1,K+1)+A(K,K+1)*A(KP1,K)
102      CONTINUE
         DO 103 KP2 = K+2,N
            A(KP2,K+1) = (-1./A(K+1,K+1))*A(KP2,K+1)
103      CONTINUE
         DO 104 KP2 = K+2,N
            A(K+1,KP2) = A(K+1,KP2)
                 + A(K,KP2) * A(K+1,K)
104      CONTINUE

CMIC$ DO ALL SHARED(N,K,A,NCPUS)
CMIC$1          PRIVATE(IPROC,J,KP2,J1,J2)
         DO 110 IPROC = 1 , NCPUS
            J1 = (IPROC-1)*(N-K-1)/NCPUS+K+2
            J2 = IPROC*(N-K-1)/NCPUS+K+1
            DO 111 J = J1, J2
               DO 111 KP2= K+2,N
                  A(KP2,J) = A(KP2,J)+A(K,J)*A(KP2,K)
                       +A(K+1,J)*A(KP2,K+1)
111         CONTINUE
110      CONTINUE
100   CONTINUE
```

Figure 13.10: Parallel LU decomposition with rank-2 updates on CRAY

on both CRAY and iPSC multiprocessor systems; see Table 13.8 (Dongarra 1991).

13.5.4 QR Decomposition

A numerically stable algorithm for the solution of overdetermined linear systems can be based on the QR decomposition. In analogy to the LU decomposition, a matrix $A \epsilon R^{n \times m}$ with $n > m$ is decomposed into its orthogonal factor $Q \epsilon R^{n \times m}$ and its upper triangular factor $R \epsilon R^{m \times m}$ such that $A = QR$. Householder transformations $H = I - 2ww^T$ with $w^T w = 1$ represent the basic matrix operation to construct the decomposition. For a CRAY Y-MP, performance data is given in Table 13.9.

Computation of the QR decomposition with Householder matrices involves

Table 13.7: Performance (wall clock time in sec) of parallel solution of triangular linear systems

CRAY Y-MP	$n = 128$	$n = 256$	$n = 512$	$n = 1024$
1 proc	0.01	0.08	0.50	3.69
2 procs	0.01	0.03	0.24	1.80
4 procs	0.01	0.02	0.12	0.91
8 procs	0.01	0.01	0.06	0.47

Table 13.8: Performance (MFLOP rate) on CRAY Y-MP and iPSC/860

CRAY Y-MP/832				iPSC/860			
procs	peak	$n = 100$	$n = 1000$	procs	peak	$n = 100$	$n = 1000$
1	333	90	324	1	40	4.5	10
2	667	144	604	8	320		63
4	1333	226	1159	16	640		99
8	2667	275	2144	32	1280		126

Table 13.9: Performance (MFLOP rate) of parallel QR decomposition

CRAY Y-MP	$n = 32$	$n = 64$	$n = 128$
1 proc	54	139	225
2 procs	50	134	256
4 procs	50	136	292
8 procs	50	133	328

CRAY Y-MP	$n = 256$	$n = 512$	$n = 1024$
1 proc	275	294	301
2 procs	391	505	562
4 procs	612	891	1060
8 procs	807	1476	1937

a matrix-vector multiplication and a rank-one update. Each of the operations requires the same order $O(n^2)$ of floating point operations and data movements. Accumulating a sequence of successive Householder transformations $\tilde{H} = H_1 H_2 \cdots H_r$ leads to the WY representation

$$\tilde{H} = I + WY^T,$$

where W and Y are $n \times r$ matrices. Let $W = YT$, where T is an $r \times r$ upper triangular matrix; then, one arrives at the more compact notation introduced by

Schreiber, van Loan, and DuCroz

$$\tilde{H} = I + YTY^T$$

In the typical case of $n \gg r$, the compact WY representation requires only about half as much storage as the original WY representation. Compared to the traditional Householder algorithm, the accumulation of T requires $O(nr^2)$ extra floating point operations and r^2 extra words for storage which is negligible in the case of $n \gg r$. The advantage of the compact WY representation is that the computation of the QR decomposition can now be regarded as a sequence of matrix-matrix multiplications and rank$-r$ updates. These block Householder transformations and their performance on parallel computers are extensively discussed by Bischof (1989).

13.6 Iterative Methods for the Solution of Linear Systems

Iterative methods for the solution of linear systems $Au = b$ represent an alternative in case of a large-order and/or sparse coefficient matrix. Starting from the classical methods of Jacobi and Gauss-Seidel, a large variety of algorithms have been developed and analyzed. The results have led to software packages like ITPACK and SLAP providing access to the most promising developments in this field. In contrast to packages like LINPACK, iterative methods have not yet been included in commercial software libraries because their use still requires some background. Both, ITPACK and SLAP focus on conjugate gradient techniques.

In order to benefit from the sparsity of the coefficient matrix, iterative methods are often implemented in conjunction with a special *storage format*. This approach leads to less storage requirements but introduces indirect addressing of the coefficients of the matrix because only the nonzero elements and their positions are stored.

13.6.1 Iterative Solution with Polynomial Acceleration

ITPACK is a research-oriented numerical software package of iteration methods for solving large sparse systems of linear equations. The ITPACKV 2D package combines three basic methods (Jacobi, Reduced System, Symmetric Successive Overrelaxation) with two acceleration procedures (Chebyshev and Conjugate Gradient). Users of the package can select from seven iterative routines:

- Jacobi Conjugate Gradient (JCG)

- Jacobi Semi-Iteration (JSI)

- Successive Overrelaxation (SOR)

- Symmetric SOR Conjugate Gradient (SSORCG)

- Symmetric SOR Semi-Iteration (SSORSI)

- Reduced System Conjugate Gradient (RSCG)

- Reduced System Semi-Iteration (RSSI).

The iterative methods of ITPACK start from a linear system of the form $Au = b$, where A is symmetric positive definite and sparse. From this equation, the standard form

$$u^{(\nu+1)} = Gu^{(\nu)} + k$$

of an iterative method with $G = I - Q^{-1}A$ and $k = Q^{-1}b$ is derived. For the Jacobi method, $Q = diag(A)$.

In many applications, the unknowns in the linear system correspond to grid points over a region. Natural ordering numbers the grid points in ascending order row by row from left to right starting at the left end of the bottom row. If the odd numbered grid points are labelled black and the even numbered points are labelled red, this is denoted as red-black ordering. The Reduced System method requires a red-black ordering of the variables

$$\left(\begin{array}{cc} D_r & C \\ C^T & D_b \end{array} \right) * \left(\begin{array}{c} u_r \\ u_b \end{array} \right) = \left(\begin{array}{c} b_r \\ b_b \end{array} \right),$$

where D_r and D_b denote diagonal matrices and C contains the interaction between the red and black variables. This system can be rewritten in the form

$$\left(\begin{array}{cc} I & -D_r^{-1}C \\ -D_b^{-1}C^T & I \end{array} \right) * \left(\begin{array}{c} u_r \\ u_b \end{array} \right) = \left(\begin{array}{c} D_r^{-1}b_r \\ D_b^{-1}b_b \end{array} \right).$$

Eliminating u_r will result in a reduced system involving only the black variables

$$(I - D_b^{-1}C^T D_r^{-1}C)u_b = (-D_b^{-1}C^T D_r^{-1}b_r + D_b^{-1}b_b).$$

Conjugate Gradient acceleration is applied to the least two iterates $u^{(\nu+1)}$ and $u^{(\nu)}$ of the basic iterative method in order to produce a better approximation

$$\tilde{u}^{(\nu+1)} = \rho_{\nu+1}(\gamma_{\nu+1}u^{(\nu+1)} + (1 - \gamma_{\nu+1})u^{(\nu)}) + (1 - \rho_{nu+1})u^{(\nu-1)}$$

to the solution of the linear system where $\rho_{\nu+1}$ and $\gamma_{\nu+1}$ denote scalars computed by dot products.

This brief presentation indicates that there are two principal operations: matrix-vector products for the basic iterative methods and scalar products for the coefficients $\gamma_{\nu+1}$ and $\rho_{\nu+1}$ in the acceleration procedure. Both steps have to take into consideration the *storage format*. ITPACKV 2D uses the ELLPACK sparse storage format storing the nonzero elements of the coefficient matrix A rowwise in a rectangular array $COEFF_{ITP}$. Another rectangular array $ICOL$ keeps the column numbers of the nonzero elements. This approach has the advantage of longer vector lengths and a smaller number of GATHER

Table 13.10: SLAP and ITPACK storage formats for sparse matrices

$$A = \begin{pmatrix} \alpha_{11} & \alpha_{12} & 0 & 0 & \alpha_{15} \\ \alpha_{21} & \alpha_{22} & 0 & 0 & 0 \\ 0 & 0 & \alpha_{33} & 0 & \alpha_{35} \\ 0 & 0 & 0 & \alpha_{44} & 0 \\ \alpha_{51} & 0 & \alpha_{53} & 0 & \alpha_{55} \end{pmatrix}$$

$$COEFF_{SLAP} = \begin{pmatrix} \alpha_{11} \\ \alpha_{21} \\ \alpha_{51} \\ \alpha_{22} \\ \alpha_{12} \\ \alpha_{33} \\ \alpha_{53} \\ \alpha_{44} \\ \alpha_{55} \\ \alpha_{15} \\ \alpha_{35} \end{pmatrix} \qquad COEFF_{ITP} = \begin{pmatrix} \alpha_{11} & \alpha_{12} & \alpha_{15} \\ \alpha_{21} & \alpha_{22} & 0 \\ \alpha_{33} & \alpha_{35} & 0 \\ \alpha_{44} & 0 & 0 \\ \alpha_{51} & \alpha_{53} & \alpha_{55} \end{pmatrix}$$

$$JROW = \begin{pmatrix} 1 \\ 2 \\ 5 \\ 2 \\ 1 \\ 3 \\ 5 \\ 4 \\ 5 \\ 1 \\ 3 \end{pmatrix} \qquad JCOL = \begin{pmatrix} 1 \\ 4 \\ 6 \\ 8 \\ 9 \\ 12 \end{pmatrix} \qquad ICOL = \begin{pmatrix} 1 & 2 & 5 \\ 1 & 2 & \times \\ 3 & 5 & \times \\ 4 & \times & \times \\ 1 & 3 & 5 \end{pmatrix}$$

operations in the routine computing the matrix-vector products; for details see Table 13.10.

A systematic analysis carried out by Ramdas and Kincaid on a CRAY Y-MP shows that the methods JCG and RSCG are at least slightly superior in their performance to the other methods (Ramdas and Kincaid 1991). The results are given in Table 13.11 and can be summarised as follows:

- with respect to wall clock time consumption and natural ordering, JCG represents the fastest method; the resulting speedup is significantly higher; on eight processors the speedup ranges from 4.49 to 5.74;

Table 13.11: Performance (wall clock time) of ITPACK on CRAY Y-MP

	JCG (natural ordering)				RSCG (red-black ordering)			
	prob1	prob2	prob3	prob4	prob1	prob2	prob3	prob4
n	99^2	139^2	20^3	30^3	99^2	139^2	20^3	30^3
$Itns$	343	480	107	161	172	242	54	81
1 proc	408.6	1087.4	138.0	660.4	197.5	500.9	100.8	406.3
2 procs	227.9	584.6	78.7	356.9	125.6	297.1	69.5	266.0
4 procs	129.0	318.6	46.5	195.0	85.5	189.1	52.7	193.0
8 procs	83.8	194.4	31.9	126.2	68.4	142.8	45.9	154.9

- with respect to wall clock time consumption and red-black ordering, RSCG is the fastest method for the test problems; the resulting speedup is moderately low; for RSCG on eight processors the highest speedup is 3.5;

- one should note that the methods differ in the number of iterations $Itns$ to accomplish the specified stopping criterion (first row of Table 13.11);

- the results also show that the overhead introduced with multitasking is negligible.

The model problems arise from a finite-difference discretization of partial differential equations in two (prob1 and prob2) and three dimensions (prob3 and prob4).

13.6.2 Preconditioned Conjugate Gradient Method

The conjugate-gradient method was developed by Hestenes and Stiefel to overcome difficulties related to other iterative methods for the solution of linear

$k = 0; x_0 = 0; r_0 = b;$
while $(r_k \neq 0)$
    Solve $M z_k = r_k$
    $k = k + 1$
    if $k = 1$
        $p_1 = z_0$
    else
        $\beta_k = r_{k-1}^T z_{k-1} / r_{k-2}^T z_{k-2}$
        $p_k = z_{k-1} + \beta_k p_{k-1}$
    end
    $\alpha_k = r_{k-1}^T z_{k-1} / p_k^T A p_k$
    $x_k = x_{k-1} + \alpha_k p_k$
    $r_k = r_{k-1} + \alpha_k A p_k$
end
$x = x_k$

Figure 13.11: Preconditioned conjugate gradient method

systems with A being symmetric and positive definite. An examination of its convergence properties shows that the method performs well when the coefficient matrix is near the identity either in the sense of a low-rank perturbation or in the sense of a norm. In order to enhance the range of matrices for which the conjugate gradient method performs well, preconditioners to a linear system have been considered. The choice of a good preconditioner can significantly improve the rate of convergence. In practice, different preconditioners are used; among them are incomplete Cholesky preconditioners and polynomial preconditioners.

From the pseudocode in Figure 13.11, one can conclude that for the preconditioned conjugate gradient method, time is consumed by four different types of operations: the preconditioning step (solve $M z_k = r_k$), an SGEMV operation for $A p_k$, three SAXPY operations for p_k, x_k, r_k, and two dot products for α_k, β_k. A parallel implementation of the algorithm can be based on the parallel execution of these tasks by using the standard techniques descibed in the previous chapters, i.e. a parallel execution of the SGEMV operation etc.

On a CRAY multiprocessor system, this approach was considered by A. Greenbaum and M. K. Seager using Macrotasking (Greenbaum *et al.* 1989; Seager 1986). With the advent of Autotasking, the results could be improved and the program was included in SLAP (Seager 1989).

In this context, the problem of choosing the appropriate *storage format* arises again. SLAP uses a column storage format where the nonzero coefficients are kept columnwise in a one-dimensional array $COEFF_{SLAP}$ and the information on the positions of the coefficients is stored in two separate vectors $JROW$ and $JCOL$. The SLAP column storage format and the ITPACK format are compared in Table 13.10. Both formats have been developed to benefit well from the vector processing capabilities of a single processor. However, the ITPACK format seems to be more efficient on one processor of a CRAY Y-MP.

For the iPSC hypercube architecture, Aykanat *et al.* discuss the implementation of the preconditioned conjugate gradient method; as a preconditioner, they use the diagonal of matrix A (Aykanat *et al.* 1988). The implementation is based on *strip mapping* of A and the vectors p, r, x, and b; additionally, one auxiliary vector is required. Strip mapping means that each processor gets the same number of successive rows of A and corresponding portions from the vectors. Hence, the DAXPY operations represent local operations because each processor can perform a local DAXPY operation on its part of the vectors. The computation of matrix-vector multiplication and scalar products consists of a local and a global part. In case of a dot product, each processor starts with computing its contribution to the global summation; then, the intermediate results are broadcasted to the other nodes to complete the operation. The performance can be further improved by introducing an exchange-add algorithm. The main idea in this algorithm is that each processor accumulates its own copy of the inner product instead of only one processor accumulating it and then broadcasting. With this modification, higher efficiencies can be reported because the computational workload per processor is increased. For small problems ($n \approx 600$), the efficiency decreases from 0.9 on a one-dimensional cube to 0.6 on a four-dimensional cube; for larger problems ($n \approx 4,500$), nearly maximum efficiency can be found on all configurations under consideration.

13.6.3 Conjugate Gradient Methods in Quantum Chromodynamics Computations

Conjugate gradient methods are used with great success in quantum chromodynamics (QCD) computations. QCD represents a theory of the strong interactions explaining the strong forces inside nuclei as interactions of quarks mediated through the exchange of gluons. In the regime where quark and gluon masses are small, solutions of the highly nonlinear equations of motions are found numerically by means of Monte Carlo methods, i.e. by simulating the theory on four-dimensional Euclidean space-time lattices. The computation of the effective fermion Hamiltonian $\phi^H (F^H F)^{-1} \phi$ represents the crucial step; here, F denotes the large sparse fermion or Dirac matrix. Projects concerned with this type of analysis consume large amounts of wall clock time and memory on today's supercomputers.

S. Knecht, E. Laermann, and W. E. Nagel ran such a project on a CRAY Y-MP of the HLRZ at KFA Jülich (Knecht *et al.* 1990). The solution of the system $F^H F x = \phi$ is computed with the conjugate gradient algorithm; preconditioning methods failed to improve convergence. Due to the large order of the fermion matrix ($n \approx 300,000$ on a $16^3 * 24$ lattice), it is not feasible to compute the product $F^H F$ first and then look for inversion. Hence,

Table 13.12: Performance of QCD computations

CRAY Y-MP	t_{tot} (sec)	t_{cg} (sec)	$No. of CG Steps$	$No. of Itns$	S_p
1 proc	1220	935	10	370	-
4 procs	324	246	10	370	3.7
8 procs	175	133	10	370	6.9

the matrix-vector product Ap_k is computed in two steps $F^H(Fp_k)$ using a storage format similar to ITPACK. Approximately 75% of the time for the entire simulation is spent on the conjugate gradient part of the code. Thus, an efficient parallel implementation of this step as outlined for the preconditioned conjugate gradient method could significantly speed up the entire application. Using Autotasking, the authors report the following results for a typical run of the QCD code on a CRAY Y-MP multiprocessor (Table 13.12); t_{tot} and t_{CG} denote the total wall clock time and the portion spent on the conjugate gradient method for this run, t_{CG} consists of 10 CG steps each performing 370 iterations. A speedup of $S_p = 6.9$ on an eight processor CRAY Y-MP can be observed.

A similar project has been carried out by the MIMD Lattice Collaboration on the iPSC/860 (Bernard *et al.* 1988). For the performance of the conjugate gradient part of their code, the authors report that the speed is between 11 and 15 MFLOPS per node for lattices of the size $6 * 12^2 * 6$ to $24^3 * 8$ ($n \approx 300,000$) on one to 32 nodes. The overall speed of the QCD code is in the same range which is due to the dominant contribution from the conjugate gradient part.

From Figure 13.12 one concludes that the speed of the calculation depends on the size of the problem and the number of nodes. For a fixed-size problem, the speed per node decreases slightly as the number of nodes devoted to the calculation increases since the communication demands increase. The authors compare the approach of distributing a $12^3 * 6$ lattice on various numbers of nodes (dotted line) with the case of putting 8^4 lattices on each node and changing the number of nodes (dashed line). The declining graph for the fixed-size QCD computation and the nearly horizontal curve for the scaled-sized case illustrate the advantage of a scaled-speedup calculation presented in the performance chapter.

13.7 Eigenvalue Problems

For the solution of eigenvalue problems on a uniprocessor, the state of the art is condensed in the EISPACK routines. This software package contains most of the methods which have become standard in this field. With the advent of vector computers, the basic operations have been recoded in order to benefit from the effort of the BLAS. Results will also be available through LAPACK.

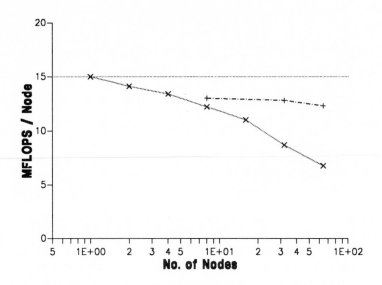

Figure 13.12: Performance of QCD code on the iPSC/860

From the point of view of their numerical solution, eigenvalue problems can be divided into two categories:

- the standard eigenvalue problem $Ax = \lambda x, x \neq 0$, where the case $A = A^H$ is treated separately, and

- the generalised eigenvalue problem $Ax = \lambda Bx, x \neq 0$.

Each category can be divided further depending on additional properties of the coefficient matrix.

The solution path for the standard eigenvalue problem starts with a reduction to simpler form which is tridiagonal in the hermitean case and otherwise of the Hessenberg form; then, a QR algorithm or LR algorithm with shift strategy is applied to the reduced matrix. Both steps are based on Householder or stabilized elementary similarity transformations. For the generalised eigenvalue problem, a QZ or LZ algorithm is recommended representing generalisations of the QR or LR algorithm, respectively.

In addition to these black-box solvers, several other methods for more specific questions are also on the workbench. In practice, one often encounters eigenvalue problems where the computation of a few dominant eigenvalues is required. Here, projection methods come into focus; subspace iteration methods or Lanczos methods computing a basis for the eigenspace are of particular interest.

13.7.1 Eigenvalue Problems with a Real Symmetric Tridiagonal Matrix

The numerical solution of eigenvalue problems with a real symmetric tridiagonal coefficient matrix T is based on two facts. If T is unreduced, the eigenvalues λ_i are distinct. The eigenvectors of T are orthogonal.

In the literature, the implementation of three different methods on multiprocessor systems is discussed:

- The QL method with Givens rotations represents the black-box approach from EISPACK. A sequence of orthogonal similarity transformations Q_ν is generated and applied to $T_0 = T$ in order to generate a sequence of similar tridiagonal matrices T_ν converging towards a diagonal matrix. Together with a shift strategy, the convergence of this method is of third order.

- Cuppen's method is based on a divide-and-conquer strategy. A given tridiagonal matrix T can be decomposed into two smaller submatrices by rank-one tearing

$$T = \begin{pmatrix} T_{11} & 0 \\ 0 & T_{22} \end{pmatrix} + \theta\alpha \begin{pmatrix} e_k \\ \theta^{-1}e_k \end{pmatrix} \begin{pmatrix} e_k^T, \theta^{-1}e_1^T \end{pmatrix}$$

where α is the off-diagonal element of T at position k and θ is a parameter used for improving numerical stability. Let $T_{11} = Q_1 D_1 Q_1^T$ and $T_{22} = Q_2 D_2 Q_2^T$ denote the eigendecompositions of the two smaller problems. Then,

$$T = \begin{pmatrix} Q_1 & 0 \\ 0 & Q_2 \end{pmatrix}$$

$$\times \left[\begin{pmatrix} D_1 & 0 \\ 0 & D_2 \end{pmatrix} + \theta\alpha \begin{pmatrix} q_1 \\ \theta^{-1}q_2 \end{pmatrix} \begin{pmatrix} q_1^T, \theta^{-1}q_2^T \end{pmatrix} \right]$$

$$\times \begin{pmatrix} Q_1 & 0 \\ 0 & Q_2 \end{pmatrix}$$

where $q_1 = Q_1^T e_k$ and $q_2 = Q_2^T e_1$. Hence, the remaining problem consists of computing the eigensystem of a matrix of the form

$$D + \rho z z^T,$$

where D is a diagonal matrix, ρ a nonzero scalar, and z is a real vector with $z^T z = 1$. The eigenvalues of $D + \rho z z^T$ can be found using a root-finding technique developed by Bunch *et al.*; the corresponding eigenvectors are computed from

$$u_i = \frac{(D - \lambda_i I)^{-1} z}{\| (D - \lambda_i I)^{-1} z \|_2}.$$

Let U denote the matrix of eigenvectors u_i and $\Lambda = diag(\lambda_1, \ldots, \lambda_n)$. The eigendecomposition of T can then be expressed as

$$T = (QU)\Lambda(QU)^T,$$

where Q is a block diagonal matrix with Q_1 and Q_2 as diagonal blocks.

This procedure can be recursively repeated until the created subtasks fit well on the processors of a parallel computer or the number of processors available is exhausted.

- Bisection and multisection with inverse iteration use properties of the characteristic polynomial to compute the spectrum of T; after the eigenvalues have been found, the corresponding eigenvectors are computed subsequently with inverse iteration.

The spectrum is computed in two steps:

1. The eigenvalues are isolated, i.e. starting from an interval containing the spectrum (e.g. Gerschgorin theorem), and intervals with a single isolated eigenvalue are computed by means of multisection (bisection). The number of eigenvalues in a given interval is determined by evaluating nonlinear recurrences derived from the Sturm sequences.

   Clusters of eigenvalues are treated as multiple eigenvalues if the prescribed accuracy is already reached during isolation.

2. The eigenvalues are extracted from intervals containing an isolated eigenvalue with multisection (bisection) or an iterative method with superlinear convergence to prescribed accuracy.

While the QL method offers only little potential for a successful parallel implementation, the performance of the other two algorithms can be significantly accelerated on multiprocessor systems.

On a CRAY, Cuppen's method has been evaluated in the early stage of multitasking by Dongarra *et al.* (1991). Bisection and multisection including iterative methods with superlinear convergence for extraction are discussed systematically by Basermann (1991). In Table 13.13, the performance of the bisection method is compared with bisection combined with the Pegasus method; computation of eigenvectors is excluded. Timings for two test problems of the order $n = 512$ are given in the table for runs on one and eight processors. The first matrix is two on the diagonal and one on the subdiagonal; the other tridiagonal matrix is generated with random numbers. Even though the speedup for bisection is higher, the Pegasus method should be preferred because of the smaller consumption of wall clock time.

Experiments on the solution of the symmetric eigenvalue problem on a hypercube are reported by Ipsen and Jessup (1990). In their paper they describe the implementation of Cuppen's method and multisection including bisection

Table 13.13: Comparison of the performance of bisection and pegasus method on CRAY Y-MP/832

	Bisection			Pegasus		
Problem	t_1 (ms)	t_8 (ms)	$S_8 = t_1/t_8$	t_1 (ms)	t_8 (ms)	$S_8 = t_1/t_8$
(1,2,1)	792	130	6.1	264	46	5.7
random	811	135	6.0	283	53	5.4

with inverse iteration for the computation of all eigenvalues and eigenvectors of a real symmetric tridiagonal matrix.

The results of Basermann and Ipsen/Jessup can be summarised as follows:

- Cuppen's method gives slightly more accurate results consistently yielding smaller residuals and deviations from orthogonality than both bisection and multisection.

- Cuppen's method is seen to be more efficient on shared-memory machines where the technique of a dynamic scheduler can be applied.

- Bisection with inverse iteration is the fastest method for finding all the eigenvalues and eigenvectors.

- Bisection combined with an iterative method with superlinear convergence (e.g. Pegasus method) for extraction is considerably faster than pure bisection.

- On a multiprocessor system with vector facility, bisection combined with an iterative method with superlinear convergence is superior to QR or LR algorithms, even on one processor.

- Cuppen's method requires slightly more storage because a matrix-matrix multiplication is involved.

13.7.2 Projection Methods

For large eigenvalue problems and in the case that only a few eigenvalues have to be computed, projection methods represent a competitive alternative to the QR algorithm. In this context, subspace iteration methods and Lanczos algorithms are discussed.

Lanczos methods can be used to solve certain large, sparse, symmetric eigenvalue problems $Ax = \lambda x, x \neq 0$. In contrast to the QL algorithm, the Lanczos methods compute partial tridiagonalizations of the given matrix A without generating intermediate full submatrices. The method is particularly useful in situations where a few of A's largest or smallest eigenvalues are desired, because it can be viewed as applying the Rayleigh-Ritz procedure

$r_0 = q_1; \beta_0 = 1; q_0 = 0; j = 0;$
while $(\beta_j \neq 0)$
$\qquad q_j + 1 = r_j/\beta_j$
$\qquad u_j = Aq_j - \beta_{j-1}q_{j-1}$
$\qquad \alpha_j = u_j^T q_j$
$\qquad r_j = u_j - \alpha_j q_j$
$\qquad \beta_j = \| r_j \|_2$
end

Figure 13.13: Lanczos iteration without reorthogonalization

to a Krylov subspace which is related to the eigenspace associated with the dominant eigenvalues.

Let $Q = (q_1, \ldots, q_n)$ with

$$
T = Q^T A Q = \begin{pmatrix} \alpha_1 & \beta_1 & & & \\ \beta_1 & \alpha_2 & & & \\ & & \ddots & \ddots & \ddots & \\ & & & & & \beta_{n-1} \\ & & & & \beta_{n-1} & \alpha_n \end{pmatrix}.
$$

Then, $AQ = QT$ implies

$$
Aq_j = \beta_{j-1}q_{j-1} + \alpha_j q_j + \beta_j q_{j+1}.
$$

From this formula together with orthogonality of the vectors q_j the Lanczos iteration as given in Figure 13.13 can be derived. The matrix-vector product Aq_j represents the dominant operation.

A parallel implementation of this step is straightforward. For the iPSC, the strategy is outlined by Scott (1989). Each processor gets a sequential part of elements of $q_1, \ldots, q_j$ in order to compute a corresponding set of the vector q_{j+1}. Similarly, the matrix A is sliced by rows and distributed to the nodes. On the other hand, the matrix T_j which must be accessed by all nodes should be replicated on each node. Then, the computation proceeds in five steps as indicated by Figure 13.13. The computation of q_{j+1} represents a local operation; each processor divides its part of r_{j-1} by its copy of β_{j-1}. The product Aq_j occurring in the formula for u_j is not local; each node has the part of A to compute the inner products but the information on q_j is not complete. A global collection of the missing parts in a temporary vector provides each processor with the required information to continue locally with the matrix-vector product. For α_j, each processor computes its contribution; the step is completed with a global summation. r_j is a local DAXPY operation on each processor; β_j is computed similarly to α_j. For the convergence test, it is less expensive to compute the eigensystem of the tridiagonal matrix T_j on each processor than to solve it once and distribute the result to all the other waiting processors.

The performance data from running experiments on an iPSC hypercube with up to 16 processors shows that eigenvalue problems can be solved with maximum efficiency. If the nodes can access optional vector coprocessors, the efficiency decreases from 0.9 for a configuration with two nodes to 0.3 on a four-dimensional hypercube. Again, the decrease in efficiency indicates that the problem is too small for the better equipped machine.

In practice, round-off errors lead to a loss of orthogonality but stability can be maintained by reorthogonalizing the vector q_{j+1} with respect to all its predecessors. Householder transformations can be used for this step. Since the Lanczos iteration is only performed in the first $j^* \ll n$ steps, the orthogonalization consumes a negligible part of the entire resources.

The Lanczos iteration can be generalised to a block Lanczos analog. In this case, a block tridiagonal matrix is computed. Block Lanczos methods are particularly useful in situations where multiple eigenvalues or clusters have to be detected. Moreover, they can benefit from Level 3 BLAS because matrix-matrix products replace the matrix-vector products.

A comparison of a block Lanczos method with subspace iterations for problems arising from reactive collision processes in chemistry is given by Kress *et al.* (1989). For problems of the order $n \approx 1,700$, up to 100 eigenvalues were computed showing that a block Lanczos method runs faster than a subspace iteration on a single processor of a CRAY X-MP. Hence, a sequential block Lanczos method represents a good starting point for a parallel implementation of the algorithm.

13.7.3 Eigenvalue Problems with a Real Nonsymmetric Matrix

A parallel QR algorithm for the nonsymmetric eigenvalue problem is discussed by Boley *et al.* (1991). The authors describe the parallel implementation of the explicitly shifted QR algorithm on a hypercube. The algorithm uses Givens rotations to generate a series of unitary similarity transformations. The rotations are passed between neighbouring processors of a processor array and applied in pipeline fashion to columns of the matrix. The rotations are accumulated in a unitary transformation matrix, enabling the computation of eigenvectors via back-substitution and back-transformation. The authors conclude that the high cost of messages can substantially reduce efficiency; in order to achieve practically useful efficiencies on currently available hypercube architectures, it will be necessary to reduce the message traffic relative to computation. Here, block matrix algorithms clustering together several columns of the matrix onto each processor provide a perspective with substantially reduced message passing and improved load balance.

13.8   Summary

With their products, Cray and Intel aim at the market segment of high-performance computing. The architecture of both systems is heavily based on the principle of parallelism. While today's shared memory systems can be characterised by a small number of powerful processors, distributed memory systems are delivered with a larger number of computational nodes. These massively parallel computers are of particular interest for the computational sciences because for these systems, the price/performance ratio is less than for the classical vector supercomputers.

The concurrency concepts provided by CRAY and Intel have already been extensively evaluated by users. Software for both computers has been developed; particularly, highly sophisticated software for problems in the field of numerical linear algebra is available. The three levels of BLAS together with LAPACK provide an optimistic perspective for the users of supercomputers. Iterative methods for sparse matrix problems (ITPACK) have also been implemented successfully on parallel computers. For the solution of eigenvalue problems, existing methods have been redesigned and efficiently implemented.

Comparing the performance of algorithms on these supercomputers, one can conclude that most of today's problems can be solved with a higher efficiency on a CRAY than on a hypercube. The reason for this is twofold: 1.) The i860 processor has a peak rate of 60 MFLOPS but for dense linear algebra computations, a speed of 30 to 35 MFLOPS can only be reported. 2.) Using the iPSC distributed memory MIMD architecture, communication overhead may further decrease the performance. However, the progress in approaching the optimum on hypercube systems is impressive. Facing the grand challenges, the massively parallel supercomputers offer an attractive and at present the only reasonable option for the computational sciences. Moreover, the new architectures have a strong impact on research in the field of parallel algorithms.

Acknowledgements

The author is indebted to A. Basermann, S. Knecht, W. E. Nagel, and Dr P. Weidner from the Zentralinstitut für Angewandte Mathematik (ZAM) for their substantial contributions. The author would also like to thank Dr E. Laermann from HLRZ for the information about QCD computations.

Several discussions with R. Chamberlain from Intel Supercomputer Systems Division and Dr W. Oed from Cray Research are gratefully acknowledged.

References

Anderson E., Bai Z., Bischof C., Demmel J., Dongarra J. J., DuCroz J., Greenbaum A., Hammarling S., McKenney A. and Sorensen D. (1990) *LAPACK: A Portable Linear Algebra Library for High-Performance*

Computers. Computer Science Department, University of Tennessee, pp. 90–105.

Aykanat C., Özgüner F., Ercal F. and Sadayappan P. (1988) Iterative algorithms for solutions of large sparse systems of linear equations on hypercubes. *IEEE Trans. Comput.* **37** pp. 1554–1568.

Bailey D. (1988) Extra High Speed Matrix Multiplication on the CRAY-2. *SIAM J. Sci. and Stat. Comp.* **9** pp. 603–607.

Basermann A. (1991) *Multisektionsverfahren zur Bestimmung der Eigenwerte von Tridiagonalmatrizen auf CRAY Multiprozessorrechnern.* Forschungszentrum Jülich, Jül-2427.

Bernard C. *et al.* (1988) QCD on the iPSC/860. *Intern. J. Mod. Phys.*

Bertsekas D. P. and Tsitsiklis J. N. (1989) *Parallel and distributed computation.* Prentice-Hall, Englewood Cliffs.

Bischof C. (1989) *A parallel QR factorization algorithm with controlled local pivoting.* Argonne National Laboratory, MCS–P21–1088.

Boley D., Maier R. and Kim J. (1991) Parallel QR algorithm for the nonsymmetric eigenvalue problem. *Comp. Phys. Comm.* **53** pp. 61–70.

Bomans L. and Roose D. (1989) Benchmarking the iPSC/2 hypercube multiprocessor. *Concurrency*, **1** pp. 3–18.

Dongarra J. J. (1991) Performance of various computers using standard linear equations software. *Supercomputing Review*, February, pp. 45–54.

Dongarra J. J., DuCroz J., Hammarling S. and Duff I. (1990) A Set of Level 3 BLAS. *ACM TOMS* **16**, pp. 1–17.

Dongarra J. J., Duff I. S., Sorensen D. C. and van der Vorst H. A. (1991) *Solving linear systems on vector and shared memory computers.* SIAM.

Faanes G. J. and Schwarzmeier J. L. (1989) Comparing the Performance of CRAY Y-MP and CRAY X-MP Computer Systems. *CRAY Channels* Winter 1989 pp. 26–30.

Frommer A. (1990) *Lösung linearer Gleichungssysteme auf Parallelrechnern.* Vieweg, Braunschweig.

Gallivan K., Jalby W. and Meier U. (1987) The Use of BLAS3 in Linear Algebra on a Parallel Processor with a Hierarchical Memory. *SIAM J. Sci. Stat. Comput.* **8** pp. 1079–1084.

Gallivan K. A., Plemmons R. J. and Sameh A. H. (1990) Parallel algorithms for dense linear algebra computations. *SIAM Rev.* **32** pp. 54–135.

Greenbaum A., Li C. and Chao H. Z. (1989) Parallelizing preconditioned conjugate gradient algorithms. *Comp. Phys. Comm.* **53** pp. 295–309.

Gustafson J. L., Montry J. L. and Benner R. E. (1988) Development of parallel methods for a 1024-processor hypercube. *SIAM J. Sci. Stat. Computing* **9** pp. 609–638.

Hake J.-Fr. and Homberg W. (1989) Linear Algebra Software on IBM and CRAY Computers. *J. Comp. Appl. Math.* **26** pp. 311–326.

Hake J.-Fr. and Homberg W. (1991) The Impact of Memory Organization on the Performance of Matrix Calculations. *Par. Comput.* **17** pp. 311–327.

Heath M. T., Ng E. and Peyton B. W. (1990) Parallel algorithms for sparse linear systems. In *Parallel algorithms for matrix computations*, SIAM, Philadelphia, pp. 83–124.

Heller D. (1978) A survey of parallel algorithms for numerical linear algebra. *SIAM Rev.* **20** pp. 740–777.

Higham N. J. (1990) Exploiting fast matrix multiplication within Level 3 BLAS. *ACM TOMS* **16**, pp. 352–368.

Hossfeld F. (1989) Vector Supercomputers. *Par. Comput.* **7** pp. 373–385.

Hwang K. and Briggs F. A. (1985) *Computer Architecture and Parallel Processing*. McGraw-Hill, New York.

Ipsen I. C. F. and Jessup E. R. (1990) Solving the symmetric eigenvalue problem on the hypercube. *SIAM J. Sci. Stat. Comput.* **11**, pp. 203–229.

Knecht S., Laermann E. and Nagel W. E. (1990) Parallelizing QCD with dynamical fermions on a CRAY multiprocessor system. *Par. Comput.* **15** pp. 3–20.

Kress J. D., Parker G. A., Pack R. T., Archer B. J. and Cook W. A. (1989) Comparison of Lanczos and subspace iterations for hyperspherical reaction path calculations. *Comp. Phys. Comm.* **53** pp. 91–108.

Kronsjö L. (1985) *Computational complexity of sequential and parallel algorithms*. Wiley, New York.

Li G. and Coleman T. (1989) A new method for solving triangular systems on distributed-memory message-passing multiprocessors. *SIAM J. Sci. Stat. Comput.* **10** pp. 382–396.

McBryan O. A. and van de Velde E. F. (1987) Hypercube algorithms and implementations. *SIAM J. Sci. Stat. Comput.* **8** pp. 227–287.

Nagel W. E. (1990) Parallelism on CRAY Multiprocessor Systems: Concepts of Multitasking. In *Proc. Ninth Int. Conf. on Computing Methods in Appl. Sciences and Engineering.*

Nagel W. E. and Szelenyi F. (1989) *Multitasking on Supercomputers: Concepts and Experiences.* IBM Tech. Rep. ICE–VS05, IBM ECSEC.

Ortega J. (1988) *Introduction to parallel and vector solution of linear systems.* Plenum, New York.

Ortega J. M. and Romine C. H. (1988) The ijk Forms of Factorization Methods II. *Par. Comput.* **7** pp. 149–162.

Ramdas M. and Kincaid D. R. (1991) *Parallelizing ITPACK 2D for the CRAY Y-MP.* University of Texas at Austin, Center for Numerical Analysis CNA–249.

Scott D. S. (1989) Implementing Lanczos-like algorithms on hypercube architectures. *Comp. Phys. Comm.* **53** pp. 271–281.

Seager M. K. (1986) Parallelizing conjugate gradient for the CRAY X-MP. *Par. Comput.* **3** pp. 35–47.

Seager M. K. (1989) A SLAP for the masses. In *Parallel supercomputing methods, algorithms and applications,* (Ed. by Carey G. F.), Wiley pp. 135–155.

Shih Y. and Fier J. (1989) Hypercube systems and key applications. In *Parallel Processing for Supercomputers and Artificial Intelligence,* (Ed. by Hwang K. and DeGroot D.), McGraw-Hill, New York, pp. 203–244.

van der Vorst H. A. and van Dooren P. (Eds.) (1990) *Parallel algorithms for numerical linear algebra.* North-Holland, Amsterdam.

Appendix

Parallel matrix multiplication with divide-and-conquer strategy on CRAY Y-MP.

```
      SUBROUTINE MMPW(FELDA,FELDB,FELDC,N)
      REAL FELDA(N,2*N),FELDB(N,N),FELDC(N,N)
      INTEGER ID(6,8)
      INTEGER NPROC, IPROC, ANZPRC, REST, ANFDO
      COMMON/NRTASK/NPROC
C
      NHALF = N / 2
      IND = 0
      DO 1 I = 0,1
        DO 1 J = 0,1
```

```
              IND = IND + 1
              ID(1,IND) = I * NHALF + 1
              ID(2,IND) = J * NHALF + 1
              ID(3,IND) = ID(2,IND)
              ID(4,IND) = ID(1,IND)
              ID(5,IND) = ID(1,IND)
              ID(6,IND) = N * J + ID(1,IND)
              IND = IND + 1
              ID(1,IND) = I * NHALF + 1
              ID(2,IND) = J * NHALF + 1
              ID(3,IND) = ID(2,IND)
              ID(4,IND) = (1-I) * NHALF + 1
              ID(5,IND) = ID(1,IND)
              ID(6,IND) = N * J + ID(4,IND)
1     CONTINUE
CFPP$ CNCALL
CMIC$ PARALLEL SHARED(N,FELDA,NHALF,ID,FELDB,FELDC)
     >PRIVATE(I,K,IPROC)
CMIC$ DO PARALLEL
      DO 20002 IPROC = 1, 8
        CALL MXMA(FELDB(ID(1,IPROC),ID(2,IPROC)),1,N,
     >           FELDC(ID(3,IPROC),ID(4,IPROC)),1,N,
     >           FELDA(ID(5,IPROC),ID(6,IPROC)),1,N,
     >           NHALF,NHALF,NHALF)
20002 CONTINUE
CFPP$ SELECT(CONCUR)
CMIC$ DO PARALLEL NUMCHUNKS(4)
CDIR$ IVDEP
      DO 77001 I = 1, N*N
         FELDA(I,1) = FELDA(I,1) + FELDA(I,1+N)
77001 CONTINUE
CMIC$ END PARALLEL
      END
```

Parallel matrix multiplication with Strassen's algorithm on CRAY Y-MP.

```
      SUBROUTINE SSTRASS(A,IA,LDA,B,IB,LDB,C,IC,
               LDC,NA,NB,NC,P,N)

      INTEGER N,I,J,K
      REAL*8  A(LDA,*),B(LDB,*),C(LDC,*)
      REAL*8  P(N,N,*)
      COMMON/INFO/ISWITCH
      IF (NA.LE.ISWITCH) THEN
%         CALL MXMAP (A,1,LDA,B,1,LDB,C,1,LDC,NA,NB,NC)
          CALL SGEMM (A,1,LDA,B,1,LDB,C,1,LDC,NA,NB,NC)
      ELSE
        N2=NA/2
        N2P=N2+1
```

```
      DO 1000, J = 1,N2
         DO 20, I = 1,N2
            P(I,J,2)  = A(I,J)  + A(N2+I,N2+J)
            P(I,J,3)  = B(I,J)  + B(N2+I,N2+J)
            P(I,J,4)  = A(N2+I,J)  + A(N2+I,N2+J)
            P(I,J,5)  = B(I,N2+J)  - B(N2+I,N2+J)
            P(I,J,6)  = B(N2+I,J)  - B(I,J)
            P(I,J,7)  = A(I,J)  + A(I,N2+J)
            P(I,J,8)  = A(N2+I,J)  - A(I,J)
            P(I,J,9)  = B(I,J)  + B(I,N2+J)
            P(I,J,10) = A(I,N2+J)  - A(N2+I,N2+J)
            P(I,J,11) = B(N2+I,J)  + B(N2+I,N2+J)
 20      CONTINUE
1000  CONTINUE
      CALL SSTRASS (P(1,1,2),1,N,P(1,1,3),1,N,P(1,1,1),
     >    1,N,N2,N2,N2,P(1,1,12),N/2)
      CALL SSTRASS (P(1,1,4),1,N,  B(1,1),1,LDB,
     >    P(1,1,2),1,N,N2,N2,N2,P(1,1,12),N/2)
      CALL SSTRASS (A(1,1),1,LDA,  P(1,1,5),1,N,
     >    P(1,1,3),1,N,N2,N2,N2,P(1,1,12),N/2)
      CALL SSTRASS (A(N2P,N2P),1,LDA,  P(1,1,6),1,N,
     >    P(1,1,4),1,N,N2,N2,N2,P(1,1,12),N/2)
      CALL SSTRASS (P(1,1,7),1,N,  B(N2P,N2P),1,LDB,
     >    P(1,1,5),1,N,N2,N2,N2,P(1,1,12),N/2)
      CALL SSTRASS (P(1,1,8),1,N,  P(1,1,9),1,N,
     >    P(1,1,6),1,N,N2,N2,N2,P(1,1,12),N/2)
      CALL SSTRASS (P(1,1,10),1,N,  P(1,1,11),1,N,
     >    P(1,1,7),1,N,N2,N2,N2,P(1,1,12),N/2)
      DO 3000, J = 1,N2
         DO 3000, I = 1,N2
            C(I,J)       = P(I,J,1) + P(I,J,4)
     >                   - P(I,J,5) + P(I,J,7)
            C(I,N2+J)    = P(I,J,3) + P(I,J,5)
            C(N2+I,J)    = P(I,J,2) + P(I,J,4)
            C(N2+I,N2+J) = P(I,J,1) + P(I,J,3)
     >                   - P(I,J,2) + P(I,J,6)
3000  CONTINUE
      ENDIF
      RETURN
      END
```

CHAPTER 14

Neural Networks for Robot Control

Dean Shumsheruddin

University of Birmingham, UK

14.1 Introduction

This chapter examines neural networks for robot control. It provides an introduction to neural network control and discusses a number of important neural network based robot arm controllers. It then describes the author's neural network based system for controlling robot arm tracking movements.

Robots are becoming increasingly important in industry. They are being used in a wide range of manufacturing processes, particularly in the automobile and electronics industries. Robots can perform repetitive tasks accurately and consistently over long periods of time, without suffering from boredom or going on strike. They can also work in environments that would be dangerous to human beings, such as areas with high concentrations of toxic chemicals or high levels of radiation. They can also be employed in locations that are inaccessible to humans such as gas pipelines. Rapid progress is being made in the field of robotics at the present time and increasingly complex, powerful and adaptable robots are being developed.

Robots need sophisticated control systems. Until recently, all robots were controlled by complex programs written in conventional languages, such as C or C++, running on sequential computers. Using conventional programs to control complex robots presents a number of serious problems. Robot control programs often need to employ numerical methods for the solution of differential equations, representing the kinematics of the robot system, and tend to be rather slow. Such programs are expensive to design and implement, and tend to be relatively inflexible and difficult to modify. It is also extremely difficult to prove their correctness for safety critical applications.

Neural network based robot controllers offer a new approach to the problems of robot control. They can overcome some of the limitations of conventional control programs. Neural network controllers can learn to perform complex tasks without explicit programming. They can also continuously adapt to changes in the system they are controlling, such as changes due to mechanical wear in components or changes in materials being processed. When implemented in hardware, neural network based controllers can run very quickly, and control rapidly moving systems. Little progress has been made on proving the correctness of neural network based control systems. However, when they receive unexpected input signals, they normally exhibit graceful degradation

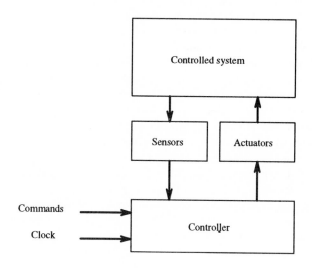

Figure 14.1: A control system

of performance rather than catastrophic failure, which is sometimes seen with conventional control programs.

Section 14.2 provides an introduction to neural network based controllers. It discusses the basic principles of the subject and explains how a backpropagation network can be used as a simple controller.

Section 14.3 examines neural networks for controlling multi-joint robot arm movements. It discusses the systems developed by Gaudiano and Grossberg (1991), Kuperstein (1991), Kawato *et al.* (1990), and Miller (1989) in some detail.

Section 14.4 describes the author's own neural network based system for the control of robot arm tracking movements. It describes the design of the system, its computer simulation, and the evaluation of its performance.

Finally, Section 14.5 draws some conclusions and attempts to predict likely future developments in the area of neural networks for robot control.

14.2   Neural Network Controllers

A controller controls the operation of a system or plant. The system may be as simple as a boiler or as complex as a nuclear reactor. Figure 14.1 shows the inputs and outputs of a typical controller. The inputs to the controller are signals from sensors monitoring the system under control, command signals and clock signals. The command signals may come from a human user or from a higher level controller. Some controllers rely on external clock signals to control the timing of system operation, others have their own built-in clock mechanisms.

The outputs from the controller are signals to actuators or effectors which influence the behaviour of the controlled system.

Control systems have a long history. The earliest controllers, such as James Watt's steam engine governor, were purely mechanical systems. These were followed by electro-mechanical controllers, and eventually by electronic controllers. With the advent of the microprocessor, computerised controllers superseded electro-mechanical and electronic controllers in many applications.

Computer based control systems are very flexible, generally reliable and relatively cheap to manufacture. Most of them employ a microprocessor with a control program in ROM and a small amount of RAM for working memory. They generally use analog to digital converters and parallel I/O interface chips to obtain information from sensors. They use parallel I/O interfaces, sometimes in conjunction with digital to analog converters, to send command signals to actuators and effectors. The early microprocessor based controllers were programmed in assembly language. Although this is sometimes necessary for high speed operation, the majority are now programmed in high-level languages such as C or C++.

A computerised control system can employ a very complicated control program, enabling it to produce very complex behaviours in the system under control. Over the last decade there has been a rapid expansion in the use of adaptive controllers. These controllers can adapt to long or medium term changes in the system under control.

Robots require complex control systems and are normally controlled by a computer based controller. A robot arm, for example, consists of a chain of mechanical links of various types. The links may undergo translational or rotational motion relative to each other. They may be driven by actuators controlling joint positions, joint velocities, or forces applied to joints. The most sophisticated robot arms allow control of joint torques, enabling rapid arm movements.

The traditional approach to controlling a complex dynamic mechanical system, such as a robot arm, is to derive a set of differential equations specifying the motion of the system under a given set of forces (the forward kinematics of the system). These can be used to derive equations specifying the forces required to produce any desired motion of the system (the reverse kinematics of the system). This is often a very complicated exercise for the control engineer.

Neural network based control systems employ a neural network to calculate the appropriate set of output signals for any given set of input signals. Figure 14.2 shows how a three-layer backpropagation network, of the type discussed in Chapter 3, can be used as a controller. The inputs to the network are signals from sensors, command signals and clock signals, if necessary. These inputs may be analog or digital. In order to obtain reasonable performance, it is usually necessary to convert an input signal into a pattern of activity over a number of input nodes using a coarse coding scheme. The outputs from the network may also be analog or digital signals, possibly encoded as patterns of activity over a number of units.

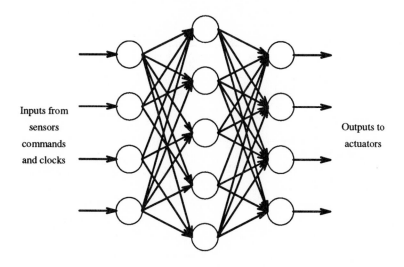

Inputs from
sensors
commands
and clocks

Outputs to
actuators

Figure 14.2: Using a backpropagation network as a controller

A neural network based controller is not explicitly programmed in the way that a conventional computer based controller is. Instead, it acquires the ability to generate an appropriate set of output signals for any given set of input signals during a learning or training process, which sets the strengths of the links between the processing units in the network. A backpropagation network controller needs to be trained with a large set of training cases. Each case contains a possible input pattern and the corresponding desired output pattern. The training cases may be collected during the operation of the system under manual control, or generated by a conventional computer program.

Simple backpropagation network controllers have a number of limitations. Even with external clock signals, it is difficult to train them to generate long sequences of outputs. Recurrent networks, such as those developed by Jordan and his colleagues (Jordan 1986; Jordan and Jacobs 1990) can overcome this problem. In a recurrent network, one or more of the output signals are fed back to extra input nodes. Werbos (1991) identifies five generic designs of neural network controller:

- Supervised control networks

- Direct inverse control networks

- Neural adaptive control networks

- Backpropagation networks

- Networks using adaptive critic methods

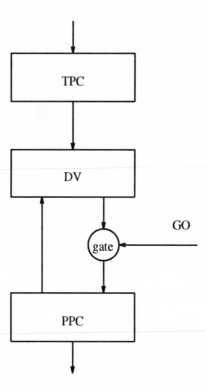

Figure 14.3: Structure of a VITE circuit, based on Bullock and Grossberg (1988)

Choosing the most appropriate type of controller for a specific application can be a difficult problem.

14.3   Neural Network Robot Arm Controllers

Most of the work on neural networks for robot control has focused on the problem of controlling robot arm movements. This work has proceeded in parallel with investigations of neural network models of motor control mechanisms in humans and animals. This section examines four of the most interesting neural network robot arm controllers.

14.3.1   VITE

Bullock and Grossberg (1988) developed a neural network model of human arm movement control called VITE (vector integration to endpoint). Gaudiano and Grossberg (1991) extended this model to produce a more general adaptive model of motor control called AVITE (adaptive VITE). This type of neural network can be used to control robot arm movements.

Figure 14.3 shows the basic structure of a VITE circuit. The rectangular boxes represent populations of neurons. The pattern of activity in the PPC neurons represents a command signal specifying the present position of the arm to a lower level servo mechanism controlling the position of the arm. The pattern of activity in the TPC neurons represents a target position for the arm to move to. The activity of the DV neurons represents a difference vector between the present position and the target position. The GO signal controls the timing and speed of the movement.

In order to make an arm movement, the target position is fed to the TPC neurons and the GO signal is applied. The DV neurons calculate the difference vector between the arm's present position and the target position. The gate multiplies this difference vector by the GO signal. The PPC neurons integrate the gated difference signal, gradually changing their activity until it matches the activity of the TPC neurons, and the arm is at the target position.

Bullock and Grossberg used computer simulations to show that the model accurately accounts for many of the characteristics of human arm movements, including their bell-shaped velocity profiles. The AVITE model learns to control the arm accurately by making random movements during a learning phase of operation.

14.3.2 INFANT

Kuperstein and his colleagues have developed an adaptive neural network based controller called INFANT (Kuperstein and Wang 1990; Kuperstein 1991). This system learns sensory-motor coordination from experience. Versions of the system have been developed for a number of different tasks.

One version of the INFANT system can learn to grasp an object, detected by a stereo pair of TV cameras, with a multi-joint robot arm. The system's neural network consists of a number of rectangular arrays of neurons termed neural maps. Images from the two TV cameras are fed into two neural maps. A third neural map calculates the disparity between the images in the first two maps. The outputs of these three maps project onto a target map and the outputs of the target map converge onto an array of output neurons. Each output neuron controls the angle of one joint in the robot arm.

During normal operation, an object is placed somewhere in the robot arm's three dimensional workspace. The images of the object detected by the TV cameras are fed into the network. Then the joint angles of the robot arm are slowly changed to the angles specified by the activities of the output neurons. As a result, the robot arm moves to and grasps the object.

In order to train the network, the object is grasped by the hand at the end of the robot arm. The arm is then moved through a long sequence of random positions in its workspace. At each position, the joint angles specified by the outputs of the network are compared with the actual joint angles in the arm.

The difference between the two angles provides an error signal and the weights in the network are adjusted according to the delta rule using this error signal. (The delta rule is described in Chapter 3.)

The system can adapt to unforeseen changes in the geometry of the physical motor system, the internal dynamics of the control circuits, and the shape, size and orientation of different objects.

14.3.3   Kawato *et al.*

Kawato and his colleagues have developed a neural network based controller for a two joint robot arm (Kawato *et al.* 1990). This system controls joint torques and can generate fast but smooth movements of the hand, avoiding obstacles or passing through via-points.

Investigations of human arm movements have shown that when the hand is moved from one point to another it approximately follows the trajectory with the minimum rate of change of torque, for all the joint actuators over the entire movement. Kawato's system is designed to generate trajectories obeying this minimum torque-change criterion.

The system uses a complex network consisting of an array of identical four-layer network modules. An individual module takes inputs representing the current position, and velocity, of the arm and the force being applied to it. It outputs the position and velocity the arm will have one time step later. The module can also work in reverse.

In order to generate a movement a sequence of modules, representing a sequence of time steps, are chained together. The position and velocity outputs of one module are linked to the inputs of the next. The inputs of the first module are set to the starting position of the arm, and the outputs of the last module are set to the target position of the arm. All the force inputs are then loosely linked together. A complex relaxation process then takes place. Each module represents one time step of the required movement, and its force input represents the joint torques to be applied for that time step of the movement. In order to produce a trajectory passing through a via-point, the position input of the appropriate module is clamped at the appropriate value during the relaxation process.

The four-layer network modules are trained by applying random torques to the arm at random positions and adjusting the output to match the correct position of the arm one time step later.

14.3.4   Miller

Miller (1989) has developed a robot arm controller based on two CMAC (cerebellar model arithmetic computer) neural networks. The system controls a robot arm with a TV camera attached to it. The arm moves in a horizontal plane over a conveyor belt and can follow objects moving along it.

During normal operation, signals from the TV camera are fed to a relatively slow image processing system. The image processing system outputs four parameters specifying the size and position of the object in the image. These parameters are fed into a CMAC network, along with signals specifying the angles of the four joints and command signals specifying desired changes in the image parameters over the next time step of operation. The network outputs four signals specifying control voltages for the four joint actuators. A second CMAC network is employed to correct for the delays in the image processing system.

The two networks are trained in parallel as the network makes trial movements, initially under the control of a fixed-gain linear error feedback controller. After training, the network is able to track objects moving along the conveyor belt with a high degree of accuracy.

14.4 Neural Network Control of Robot Arm Tracking Movements

This section describes the author's neural map based controller for tracking moving objects with a two-joint robot arm. Its single layer neural map contains pairs of cells with outputs connected to just two output neurons. The inputs to the network are the position of the hand, and the position and velocity of the object relative to the hand. The outputs from the network are the torques to be applied by the two joint actuators.

The network learns the mapping between joint torques and hand movements by making random movements of the arm. During learning, the delta rule is used to adjust the weights of the connections between the neural map and the output neurons.

The model has been tested by computer simulation of a system consisting of the control network, a two joint arm with inertia and damping, and a randomly moving object. In the simulation, the neural network controller tracks objects with position and velocity errors of the order of 0.5 per cent of the range of movement of the arm. This technique can be applied to a wide range of tracking control problems.

The controller performs the task of tracking a moving object with a two-joint robot arm as shown in Figure 14.4. The arm and the object are both confined to the two-dimensional plane, with coordinates x and y. The object moves around randomly within this plane. The shoulder of the arm is fixed at the origin of the plane and the hand is free to move within the rectangular workspace shown in the figure. The links of the arm are assumed to have masses and moments of inertia. θ and ϕ are the shoulder and elbow joint angles. The arm is moved by actuators which can apply variable torques to the two joints. The joints are assumed to have fixed damping factors.

The controller has to specify the torques to be applied to the joints in order for the hand to track the object within the workspace. The inputs to the controller during tracking are the position of the hand in the workspace, and the

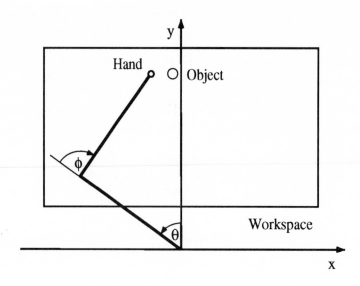

Figure 14.4: The robot arm under control

position and velocity of the object relative to the hand. These quantities are all vectors with real-valued x and y components, continuously changing in time. The inputs to the controller during learning are the position and acceleration of the hand in the workspace. It is assumed that these inputs are provided by the robot arm's sensory systems.

The controller assumes very little about the system. It needs to be programmed with the coordinates of the workspace and the range of activation of the joint actuators. However it does not need to be given the lengths, masses or moments of inertia of the arm segments, or the damping coefficients of the joints.

14.4.1 Design of the Controller

The structure of the neural network controller is shown in Figure 14.5. It consists of a two-dimensional neural map, of a similar type to that used in the INFANT system (Kuperstein 1991), and two torque output nodes, T_1 and T_2. The neural map consists of a rectangular array of pairs of neurons. Each pair of neurons represents the corresponding position in the workspace. During operation, inputs are gated into a circular region of the map centered on the position representing the current location of the hand. This is represented by the grey circle in Figure 14.5.

All the nodes in the map are connected to both torque output nodes. The strengths of these connections are set during the learning phase. Figure 14.6

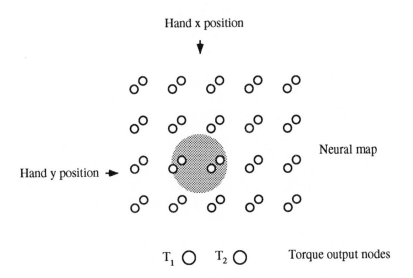

Figure 14.5: Structure of the controller's neural network

shows the connections of a single pair of nodes in the neural map. One neuron in the pair processes the x components of the input signals and the other processes the y components. Δx and Δy are input signals representing the x and y components of the position of the object relative to the hand. Δx_v and Δy_v are input signals representing the x and y components of the velocity of the object relative to the hand. The weights W_1 and W_2 are constant. The weights between the map nodes and the output nodes (W_a, W_b, W_c and W_d) are set during the learning phase. The torque output nodes, T1 and T2, are connected to the joint actuators at the shoulder and elbow joints respectively.

During tracking, the object position and velocity inputs are gated into the circular region of the map centered on the position representing the current location of the hand. The map nodes in this region calculate the x and y components of the hand acceleration required to track the object. The output nodes calculate the joint torques required to produce this acceleration.

The object position and velocity signals are gated into the neural map by being multiplied by a gating function. The gating function for the pair of nodes at location i,j in the map is given by the following formula:

$$G_{ij} = exp(-\lambda(\delta x^2 + \delta y^2))$$

where δx and δy are the x and y components of the difference between the position represented by node pair ij and the current hand position, and λ is a constant. This function has a value of one at the hand position and decays toward zero with increasing distance from the hand position. This gating

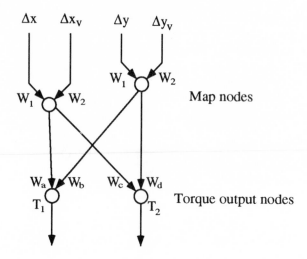

Figure 14.6: Connections of a single pair of neural map nodes

function is based on that used in the INFANT system (Kuperstein 1991).

The model assumes that the hand acceleration required to track the object is given by:

$$\mathbf{a} = k_1 \Delta \mathbf{p} - k_2 \Delta \mathbf{v}$$

where $\Delta \mathbf{p}$ is the position of the object relative to the hand and $\Delta \mathbf{v}$ is the velocity. k_1 and k_2 are constants, with values of 1.0 and 0.5 in the computer simulation. This simple function performs the task well and can easily be calculated by a pair of nodes. The gated input signals propagate, through the constant weights W_1 and W_2, into the map nodes. The activity of the x-component node of the pair ij is given by the following formula:

$$A_{ijx} = G_{ij} \Delta x W_1 + G_{ij} \Delta x_v W_2$$

where Δx is the x component of the position of the object relative to the hand and Δx_v is the x component of its velocity relative to the hand. The activity of the y node of the pair is calculated similarly.

Activity of the map nodes propagates into the torque output nodes according to the following formula:

$$T_i = \sum_{j=1}^{N} A_j w_{ij}$$

where T_i is the activity of the ith output node, j ranges over all the N nodes in the map, and w_{ij} is the weight of the link from node j to node i.

The controller learns the mapping between hand acceleration and joint torques over the workspace during a learning phase of operation. It does this by setting the hand to random positions in the workspace and making random movements of the arm at each position. During a learning trial, the hand is first set to a random position in the workspace. Then random torques, within the range of operation of the actuators, are applied to the two joints. The x and y components of the resulting hand acceleration are gated into the pairs of map nodes. Activity is propagated through the network to produce two joint torque signals. These are compared with the actual torques τ_i that were applied, to produce an error signal for each output node:

$$E_i = \tau_i - T_i.$$

Finally, the delta rule is used to correct the weights of the connections from all nodes in the map to the two output nodes:

$$\Delta w_{ij} = \eta E_i A_j$$

where η is a constant learning rate.

14.4.2 Computer Simulation of the System

A computer simulation was developed in order to test the neural network controller and investigate its performance on a number of tracking problems. The simulated system consisted of the neural network controller, a two-joint robot arm with inertia and damping, and a randomly moving object to be tracked by the arm.

The simulation was implemented in the C programming language on a Sun Microsystems SPARC workstation. The behaviour of the whole system was simulated with a time interval of one millisecond. All real-valued quantities in the simulation were represented by 64-bit floating point variables.

The neural map consisted of a 29 by 13 rectangular array of pairs of nodes, containing 754 nodes in total. The nodes represented an array of points in the workspace with a spacing of 5 cm. The constant λ was set to 500 to produce a gating function with a value of one half at a radius of 3.7 cm. The weights W_1 and W_2 were constant throughout the map with values of 10.0 and 5.0 respectively. The weights W_a to W_d were initialised with small random values and adjusted by the learning process described below.

The simulation of the arm was based on the mathematical model of the dynamics of a two-joint arm developed by Kalveram (1991a, 1991b). The model was simplified by assuming the arm and object moved in a horizontal plane and the effects of gravity could be ignored. The dynamics of the arm are described by two equations:

Joint 1: $A\ddot{\theta} + C\ddot{\phi} - D\dot{\phi}^2 - 2D\dot{\theta}\dot{\phi} + R_1\dot{\theta} = T_1$
Joint 2: $B\ddot{\phi} + C\ddot{\theta} + D\dot{\theta} + R_2\dot{\phi} = T_2$

where:
$$A = M_1 + M_2 + m_2 l_1^2 + l_1 a_2 m_2 cos\phi$$
$$B = M_2$$
$$C = M_2 + l_1 a_2 m_2 cos\phi$$
$$D = l_1 a_2 m_2 sin\phi$$

where θ, $\dot{\theta}$ and $\ddot{\theta}$ are the angle, angular velocity and acceleration at the shoulder joint. ϕ, $\dot{\phi}$ and $\ddot{\phi}$ are the angle, angular velocity and acceleration at the elbow joint, relative to the upper arm. M_1, m_1, l_1 and a_1 are the moment of inertia, mass, length, and distance from the joint to the centre of mass, of the upper arm. M_2, m_2, l_2 and a_2 are the corresponding quantities for the forearm. R_1 and R_2 are the moments of damping for the two joints and T_1 and T_2 are the torques exerted by the joint actuators. A and B represent the moments of inertial resistance against active acceleration torques about the joints. C, $-D$, D and $2D$ represent the dynamic coupling effects due to reactive acceleration, centrifugal and centripetal forces in the two-joint system.

In the simulation, the bottom left corner of the workspace was at (−0.7 m, 0.1 m) and the top right corner at (0.7 m, 0.7 m). Both arm segments were 0.5 m long, had masses of 0.5 kg and had centres of mass at their midpoints. The damping factors of both joints were 0.1 and the torque outputs of both actuators could vary between −1.0 Nm and +1.0 Nm.

The object was simulated as a fixed mass randomly changing its velocity every 500 ms. The maximum velocity of the object was about half that of the hand, with the arm in a typical position.

In order to simulate tracking movements, the states of the network, arm and object were calculated every millisecond. During each time interval, the inputs to the network were set from the simulated positions and velocities of the object and hand. Then the values of the gating function and the gated inputs were calculated for each pair of nodes in the map. Next, the activities of all the map nodes were calculated and propagated to the torque output nodes. To simulate the movement of the arm, angular accelerations were calculated from the torques output by the network, using the two formulae above, and angular velocities and positions were calculated by numerical integration. Then the object's velocity was determined and its position calculated by numerical integration.

14.4.3   Performance of the Controller

Figure 14.7 shows the distance between the hand and the object over the course of a typical tracking movement. The hand starts 0.1 m away from the object, moves towards it over the first 2 seconds, and then tracks it for 4 seconds. After 8 seconds the object moves out of the workspace and the error increases. Figure 14.8 shows the difference in the velocities of the hand and object during the same movement. The small spikes in the velocity error graph are due to the random changes in the object's velocity every 500 ms. The tracking errors were found to be minimal in the centre of the workspace with the elbow joint

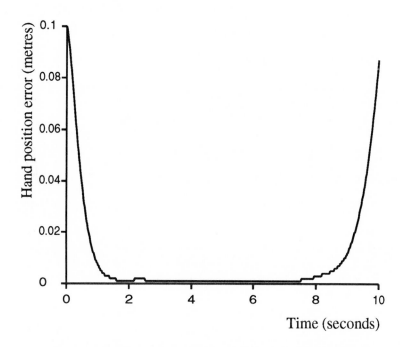

Figure 14.7: Position error during a typical tracking movement

angle close to $\pi/2$ radians. The errors were larger at the edges, and particularly in the corners of the workspace.

During learning, the hand was set to random positions in the workspace and random torques were applied to the two joints. The motion of the arm was simulated for three milliseconds and the hand acceleration measured. Then the gating function was calculated at each pair of nodes and the acceleration was multiplied by the gating function to set the activity of the nodes in the map. Activity was propagated to the output nodes and the errors were determined for each node. Finally the weights of all the connections from the neural map to output nodes were modified using the delta rule, as described above.

Figure 14.9 shows the mean error in the torques output by the network, for groups of 100 random movements, over the course of a typical learning phase. The optimum value for the learning rate η was found to be about 0.01. Larger values of η led to a failure of the weights to converge.

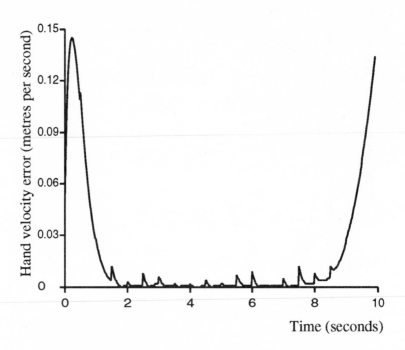

Figure 14.8: Velocity error during a typical tracking movement

14.5 Conclusions

There has been great progress in the field of robotics over the last ten years. Robot systems are being widely used in manufacturing industry. There is every reason to expect that more sophisticated, powerful and flexible robots will be developed over the next decade.

At present, the vast majority of robots are controlled by microcomputers running conventional control programs. Although these programs are satisfactory for many applications, they have a number of limitations which will become more obvious as time goes on. The main problems with conventional computer control systems are the high cost of developing software for them, and their lack of flexibility and adaptability to new and changing tasks.

Neural network based control systems offer the prospect of overcoming many of the limitations of current systems. Their ability to learn to perform a task makes them much more flexible, and greatly reduces the cost of developing software for a task. However, the field is still relatively new and many problems remain to be solved.

Many neural network based robot controllers use mechanisms discovered

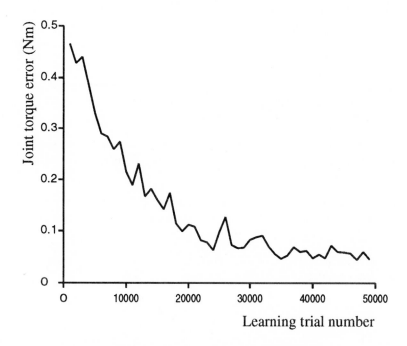

Figure 14.9: Mean torque output error during learning

in the motor control systems of humans and animals. Progress in understanding the biology of motor control should lead to advances in practical robot control.

One of the important characteristics of biological motor control systems is their hierarchical structure. Development of hierarchical neural network systems for controlling complex robot systems is likely to become an important area of research in the future.

References

Bullock D. and Grossberg S. (1988) Neural dynamics of planned arm movements: Emergent invariants and speed-accuracy properties during trajectory formation. *Psychological Review* **95**, pp. 49–90.

Gaudiano P. and Grossberg S. (1991) Vector associative maps: Unsupervised real-time error-based learning and control of movement trajectories. *Neural Networks* **4**, pp. 147–183.

Jordan M. I (1986) Attractor dynamics and parallelism in a connectionist sequential machine. In *Proc. of the 8th Annual Conference of the Cognitive*

Science Society.

Jordan M. I. and Jacobs R. A. (1990) Learning to control an unstable system with forward modeling. In *Advances in Neural Information Processing Systems 2*, edited by Touretzky D. S., Morgan Kaufmann, San Mateo, CA.

Kalveram K. T. (1991a) Pattern generating and reflex-like processes controlling aiming movements in the presence of inertia, damping and gravity. *Biological Cybernetics,* **64**, pp. 413–419.

Kalveram K. T. (1991b) Controlling the dynamics of a two-jointed arm by central patterning and reflex-like processing. *Biological Cybernetics,* **65**, pp. 65–71.

Kawato M., Maeda Y., Uno Y. and Suzuki R. (1990) Trajectory formation of arm movement by cascade neural network model based on minimum torque change criterion. *Biological Cybernetics,* **62**, pp. 275–288.

Kuperstein M. (1991) INFANT neural controller for adaptive sensory-motor coordination. *Neural Networks,* **4**, pp. 131–145.

Kuperstein M. and Wang J. (1990) Neural controller for adaptive movements with unforeseen payloads. *IEEE Transactions on Neural Networks,* **1**, pp. 137–142.

Miller W. (1989) Real-time application of neural networks for sensor-based control of robots with vision. *IEEE Transactions on Systems, Man and Cybernetics,* **19**, pp. 825–831.

Werbos P. J. (1991) An overview of neural networks for control. *IEEE Control Systems Magazine,* January 1991, pp. 40–41.

CHAPTER 15

Computational Complexity of Parallel Algorithms

Lydia Kronsjö
University of Birmingham, UK

15.1 Introduction

The overwhelming need for parallel algorithms arises from the fact that sequential algorithms are not efficient enough for many practically important problems. Parallel algorithms offer perhaps the greatest hope for major improvements. It is likely therefore that before the end of the century all developments in both hardware and software will be concentrating on parallel processing.

In general terms, parallel processing can be defined as an approach for increasing the computation speed for a problem, by dividing the algorithm into several sub-problems (grains) and allocating multiple processors to execute multiple grains simultaneously. Ideally, the best grain size is the one that results in the shortest overall execution time for an algorithm. In practice, communication time plays an important role in parallel processing, so that an optimal time of an algorithm can be achieved by carefully trading parallelism for communication time.

The difficulty is however that we still do not know how to recognize parallelism in a problem. For example, there is no unified set of characteristics that may be used to describe parallelism. We do not know what it is about a problem that lends itself to parallelism. We do not know how to decide on the granularity criteria: large, medium or fine grains. Which is the right level for a given problem?

Parallelism offers new degrees of freedom into algorithms not found in sequential algorithms. We would like to know whether parallelisation of sequential algorithms will lead to the greatest gains, or perhaps we need to recognise that only the completely new, parallel programming, mechanisms and strategies will achieve the aimed gains.

We would like to know under what circumstances a parallel algorithm can be designed that will tremendously reduce the amount of time required to solve a problem in a sequential environment. In the quest for answers to these problems theoretical computer scientists have begun developing a body of theory centered around parallel algorithms and parallel architectures. There is the study of various parallel architectures. There are questions posed about what the processors should look like and how they should be interconnected.

There are numerical analysis issues, complexity issues, parallel communication issues, and algorithm design issues.

While many of these areas correspond to serial algorithms' development, a parallel algorithm designer needs to know much more. Parallel algorithms, the languages they are written in, and the systems they run on are extremely more complex.

In analogy with the theory of sequential algorithms, the choice of a formal parallel computing model is the first step in the development of the theory of parallel algorithms.

15.2 Formal Computing Models

15.2.1 Fundamental Definitions

For every problem solved on a computer there exists a parameter that is a measure of the problem's size, such as the order of the matrix, number of digits, degree of the polynomial, number of graph vertices, and the like. The running time of an algorithm is measured with respect to problem size n, and so, say, for a sequential algorithm the (worst-case) time function is given by $T_{seq}(n)$.

For sequential algorithms an idealised computation model is a random-access machine (RAM), which is a model of a one-address computer, that consists of a memory, a read-only input tape, a write-only tape and a program. The success of the RAM model in the development of a powerful theory of sequential algorithms is attributed to the fact that the model encompasses successfully both software and hardware issues: high level languages are efficiently compileable on this model while the model can be efficiently implemented in hardware.

Parallel algorithms in addition to the problem size, have another parameter, the number of processors, p.

Definition 15.2.1 A given algorithm implemented on a p-processor computer is called p-parallel. The algorithm whose p-parallel implementation requires t parallel steps is described as p-computable in a time t.

A major goal in the design of parallel algorithms is to find optimal and efficient algorithms with t as small as possible. Kruskal *et al.* (1990) suggest that it is useful to distinguish between the problem size-dependent and the problem size-independent parallel algorithms.

Definition 15.2.2 A parallel algorithm is called size-dependent if its time function, $T_{par}(n)$, and the number of processors required, $p(n)$, both are functions of the problem size n.

A parallel algorithm is called size-independent if its time function, $T_{par}(n,p)$, is the function of both the problem size, n, and the number of processors required, p.

Kruskal considers that the definition of a size-independent algorithm is more natural, but the size-dependent algorithms are convenient in formal definitions. The two definitions, however, should not cause any confusion because in fact one type of the algorithm can be easily transformed to the other type. We shall use both definitions interchangeably as and when convenient.

A general purpose parallel processing model is needed for parallel computing. A number of theoretical parallel computational models have been proposed that assume an idealised environment of various degree.

A model that is as rich in flexibility as one can imagine is an ideal answer. However these aspirations are hindered by the existence of real technological constraints which need to be dealt with by parallel architecture engineers, such as memory contention, asynchronism and restricted communication geometry. This forces theoreticians to provide for these features in their computational models, hence a large variety of proposed models.

At the level nearest to the hardware there are VLSI models. They highlight the technological front line of today. At a higher level of the abstraction scale there are the models where a parallel machine is viewed as a set of processors in a fixed pattern, and each processor can access a small number of neighbours.

15.2.2 PRAM

At the top of abstraction scale we find the parallel random-access machine (PRAM), one of the most successful models developed yet (Fortune and Wyllie 1978; Savitch and Stimson 1979; Goldschlager 1982). The PRAM is a major class of models for general purpose parallel computers, and is a natural generalisation of the sequential RAM model.

Formally, PRAM consists of p general purpose processors, $P_1, P_2, ..., P_p$, all of which are connected to a large shared random access memory M. Ps have a private or local memory for their own computation, but all communication among them is done via the shared memory. (This model is sometimes called Shared-memory Single Instruction Stream, Multiple Data Stream Machine, SM SIMD machine.)

The PRAM has some unrealistic features and cannot be considered a physically realisable model; for example, for any reasonable number of processors in the model, it quickly becomes impossible to realistically retain a constant-length access time from any processor to any memory cell. Nevertheless the PRAM model is used widely in theoretical research on parallel computing. It is considered a reasonable model as the number of operations done by p processors per cycle is at most p. It has captured the concept of parallel computation on the level that allows study of the logical structure of parallel computation in a context divorced from the issues of parallel communication. Many other models can be seen as restricted versions of PRAM. Algorithms developed for other, more realistic, models are often based on algorithms originally designed for the PRAM.

PRAM processors are synchronised: all processors execute their instructions streams in lock-step, one instruction per time step. Each time step has three phases:

- the read phase—may read from a cell,
- the computation phase—does some calculations,
- the write phase—may write to a memory cell.

In the PRAM model there is the possibility of read- and write-conflicts, in which two or more processors try to read from or write into the same memory cell concurrently. These read and write conflicts are resolved by allowing one of the following mechanisms: Exclusive—Read, Exclusive—Write (EREW PRAM), Concurrent—Read, Exclusive—Write (CREW PRAM), Exclusive—Read, Concurrent—Write (ERCW PRAM), Concurrent—Read, Concurrent—Write (CRCW PRAM). Other versions for treating read/write conflicts are: COMMON, where all processors that simultaneously write to the same memory cell must write the same value. PRIORITY, where the write conflicts are resolved in favour of the processor that has the least index among those processors involved in the conflict. The distinct ways in which read/write conflicts are handled lead to a variety of PRAM models though these models do not differ very widely in their computational power. The EREW PRAM, in which concurrent reading and writing is forbidden, is the weakest of the models. The CRCW PRAM, which permits both kinds of concurrency is the strongest. Different methods in resolving write conflicts in CRCW PRAM model lead to several varieties of this model.

The first results of the hierarchy of computational power of various PRAM models have been obtained. So, it has been shown that the CREW PRAM is strictly more powerful than the EREW PRAM and is strictly less powerful than the CRCW PRAM (Valiant 1990a). It has also been shown that certain simple computations—for example, multiplying two n-bit numbers—inherently require time $\Omega(logn/loglogn)$ even on the strongest PRAM model provided the number of processors is polynomial-bounded in the size of the input.

It has also been shown that any algorithm for a CRCW PRAM in the PRIORITY model can be simulated by an EREW PRAM with the same number of processors and with the parallel time increased by only a factor of $O(log\ p)$, where p is the number of processors; and further any algorithm for a PRIORITY PRAM can be simulated by a COMMON PRAM with no loss in parallel time provided sufficient processors are available.

Thus, though it is easier to design optimal algorithms on a CRCW PRAM model than on a CREW or EREW PRAM, the simulations between the various PRAM models make the notion of an efficient algorithm invariant with respect to the particular PRAM model used.

It has also been observed that to some extent the PRAM allows practical realisation of it as large shared memory multiprocessors are actually being built.

Other models attempt to address both computational issues and interprocessor communication. We shall consider some of these models.

15.2.3 DRAM

The major drawback of the PRAM model is that it has no mechanism for representing communications between the processors. Its storage management and communication issues are hidden from the algorithm designer. A more restricted PRAM model, the so called distributed random-access machine (DRAM), is due to Leiserson and Maggs (1986). The model includes a communication network that conveys information between processors and memory banks.

The DRAM consists of a set of n processors. All memory in the DRAM is local to the processors, with each processor holding a small number of $O(log\ n)$-bit registers. A processor can read, write, and perform arithmetic and logical functions on values stored in its local memory. A processor can also read and write memory in other processors.

The cardinal difference between a DRAM and a PRAM is that the DRAM models communication costs. The DRAM model assumes an interconnection network, where the processors are interconnected as a graph, and routing of messages is performed by the processors.

The communication cost is measured in terms of the number of messages that must cross a cut of the network. A $cutS = (A, \overline{A})$ of a network is a partition of the network into two sets of processors A and $\overline{A}$. The capacity $cap(S)$ is the number of wires connecting processors in A with processors in $\overline{A}$, that is the bandwidth of communication between A and $\overline{A}$. For a set M of messages, the load of M on a $cutS = (A, \overline{A})$ is defined as the number of messages in M between a processor in A and a processor in $\overline{A}$.

A lower bound on the time required to deliver a set of messages is provided by the load factor of M on S, which is defined as the ratio of $load(M, S)$ and $cap(M)$, $L(M, S) = load(M, S)/cap(S)$. The load factor of M on the entire network is $L(M) = max\ L(M, S)$ for all S.

15.2.4 BSP Model

A more recent model for parallel computation has been put forward by Valiant (1990b). The so called bulk-synchronous parallel (BSP) model is suggested as candidate model of a unifying model for parallel computation. The argument for the necessity of such a model is based on an analogy with the von Neumann model of sequential computation.

The von Neumann machine serves equally well for hardware and software technologies. Hardware designers can share the common goal of realising efficient von Neumann machines using ever rapidly changing technology and architectural ideas, without having to be too concerned about the software that is going to be executed. Similarly, the software industry in all its diversity can aim to write programs that can be executed efficiently on this model, without explicit consideration of the hardware.

The BSP model is intended to act as a standard of what is meant by and expected from parallel processing, a standard on which people can agree.

The bulk-synchronous parallel computer (BSPC) includes three attributes:

- A number of components, each performing processing and/or memory functions,

- A router that delivers messages point to point between pairs of components,

- Facilities for synchronising all or a subset of the components at regular intervals of L time units where L is the periodicity parameter.

A computation consists of a sequence of supersteps. In each superstep, each component is allocated a task consisting of some combination of local computation steps, message transmissions and (implicitly) message arrivals from other components.

After each period of L time units, a global check is made to determine whether the superstep has been completed by all the components. If it has, the machine proceeds to the next superstep. Otherwise, the next period of L units is allocated to the unfinished superstep. Thus in the model the tasks of computation and communication can be separated.

The function of the router is to deliver messages point to point. No combining, duplicating, or broadcasting facilities are assumed.

A major feature of the BSP model is that it allows some parallel slackness, that is, it allows the algorithm designer not to worry about such details as memory management, communication assignment and realisation of low-level synchronisation which the model handles itself. A parallel algorithm can be written for a certain number p of processors and be efficiently executable on any universal parallel machine with fewer than $p/logp$ processes.

The compiler is assumed to be flexible enough to schedule and pipeline computation and communication efficiently.

Valiant stresses that so far in contrast to numerous other computational endevours no substantial impediments to the BSP's interpretation of the general-purpose parallel computation have been uncovered. The BSP model can be looked at as a pragmatic embodiment of these positive results much as the von Neumann model is a pragmatic embodiment of Turing's machine.

A formalization of perhaps the simplest instance of the BSP model is described in Valiant (1990a). An important aspect of this model is how to manage the memory. A mechanism is needed of allocating storage in such a way that the computation will not be slowed down by uneven memory accesses and overloading individual units. The most promising method known for distributing memory accesses about equally in arbitrary programs is hashing. Hash functions for this parallel context have been proposed and analyzed by Mehlhorn and Vishkin (1984), by Carter and Wegman (1979), and by Karlin and Upfal (1988).

Computation models related to BSP model are the phase PRAM of Gibbons (1989), which incorporates barrier synchronisation in a similar way to

BSP model, but uses a shared memory, the delay model of Papadimitriou and Yannakakis (1988), and the LPRAM of Aggarwal *et al*(1990).

The LPRAM of Aggarwal is a CREW PRAM in which each processor is provided with an unlimited amount of local random access memory. The input variables are initially available in global memory, and the final outputs must also be eventually stored in the global memory. At every time step the processors do one of the following:

- a communication step—a processor can write, and then read a word from global memory,

- a computation step—a processor can perform an operation on at most two words that are present in its local memory; the set of operations allowed depends on the application domain.

15.2.5   Example Problem

Example 15.2.1 In this example we illustrate how the run time of a parallel algorithm is affected by the number of processors available for a parallel computation (Gurari 1989). The assumed computational model is a minimal ERCW PRAM model.

Problem.   Consider the problem of finding the smallest value in a given set $S = S[1], S[2], ...S[n]$ of n elements.

To solve the problem at least three different parallel algorithms can be immediately thought of, depending on the number of processors that are assumed available for the computation.

Solution: Variant One This variant assumes that there are $p = n(n-1)/2$ processors available to solve the problem. This number of processors can be interpreted as unlimited, that is, that there are as many processors as required to accommodate the fastest possible computation. The parallel computation then obviously takes a small fixed number of steps.

Let each pair (i_1, i_2) of values of indices, such that $1 \leq i_1 < i_2 \leq n$, correspond to a different i, such that $1 \leq i \leq n(n-1)/2$. One such ordering i on the pairs (i_1, i_2) is given by the formula $i = 1 + 2 + \ldots + (i_2 - 2) + i_1 = (i_2 - 2)(i_2 - 1)/2 + i_1$. Here we have $i = 1$ for (1,2), $i = 8$ for (2,5), $i = 19$ for (4,7), $i = 27$ for (6,8), etc.

Algorithm 15.2.1
Begin
For i from 1 to $n(n-1)/2$ do in parallel
 P_i derives the pair (i_1, i_2) of indices that corresponds
 to a different i
 P_i reads values $S[i_1]$ and $S[i_2]$.
 If $S[i_1] \geq S[i_2]$
 then P_i sends NO-outcome to P_{i1}
 else P_i sends NO-outcome to P_{i2}

     EndIf
EndDo
// Comment. At this stage the only active processor is
$P_j, 1 \leq j \leq n$, which did not receive
a negative outcome. //
P_j reads the value $S[j]$ and writes it into the output cell
End

Solution: Variant Two This variant assumes that $p = |n/2|$ processors are available for the computation. The parallel computation then requires $O(logn)$ steps.

Algorithm 15.2.2
Begin
For i **from** 1 **to** |n/2| **do in parallel**
    P_i reads values $S[2i - 1]$ and $S[2i]$.
    If $S[2i - 1] < S[2i]$
    then P_i sends value $S[2i - 1]$ to $P_{|i/2|}$
    else P_i sends value $S[2i]$ to $P_{|i/2|}$
    EndIf
// Comment. At the end of the first step the processors
$P_1, \ldots, P_{|n/2|}$ hold the elements of S that
have not been eliminated. //
EndDo
For i **from** 1 **to** |n/2| **do in parallel**
    P_i determines itself *active* if and only if it has been
    sent some values of S in the previous stage. Two
    values of $S, S[i_1]$ and $S[i_2]$ should be present
    in an *active* P_i.
    If P_i is *active* **then**
      If $S[i_1] < S[i_2]$
      then P_i sends $S[i_1]$ to $P_{|i/2|}$
      else P_i sends $S[i_2]$ to $P_{|i/2|}$
      EndIf
    EndIf
EndDo
// Comment. After $O(logn)$ stages only P_1 is *active*, and
it holds a single value of S. //
P_1 returns the value to the output
End

Solution: Variant Three This variant assumes $p < |n/2|$ processors. The parallel computation then requires $O(n/g(p) + logp)$ steps.

Algorithm 15.2.3
Begin
For i **from** 1 **to** p **do in parallel**
    P_i computes (reads) p
    P_i finds independently in $O(n/p)$ steps the smallest value
    in $S[\lfloor n/p \rfloor \times (i-1) + 1], \ldots, S[\lfloor n/p \rfloor \times i]$
EndDo
Using $P_1, P_2, \ldots, P_p$ the smallest value among the p values is
determined as in Algorithm 15.2.2
End

15.3 Computational Complexity of Parallel Algorithms

For sequential algorithms computational complexity involves analysis of the amount of computing resources needed to solve a problem as a function of the problem's size. The most obvious and most important resource is time. Time function remains of fundamental importance for parallel algorithms. In addition, as seen in Example 15.2.1 the number of processors, p, is an important parameter, since the availability of many processors to carry out the computation simultaneously is a fundamental assumption in parallel processing. In an ideal model one can assume that the number of processors is infinite. Hence, an ideal optimal computation time. However in practice one will always reach a stage when the size of the problem exceeds the number of processors available. Thus realistic algorithms will inevitably run on an architecture with a very finite number of processors. We need to know the optimal values involved.

Example 15.3.1 *Example 15.2.1 Revisited* How quickly can the search for an item be performed on a parallel computer? Since the complexity of algorithms on the PRAM depends solely on the number of computations, the model provides a useful lower bound on the inherent computational complexity of the algorithm.

Theorem 15.3.1 *(Kruskal 1983).* Given positive integers k, n, and p, where $n = (p+1)^k - 1$, searching for an item in an n-item array while using the PRAM model requires $\leq \lfloor log(n+1)/log(p+1) \rfloor$ comparisons. This bound is tight.

    Proof. We shall give the proof on the lines of Quinn (1987).
    The first step is using induction on k to show that $\lfloor log(n+1)/log(p+1) \rfloor$ comparisons are sufficient.
    Basis. Let $k = 1$, then $n = (p+1)^1 - 1 = p$. It is obvious that one parallel comparison step is sufficient for p processors to determine whether the item is in the array, hence $\lfloor log(p+1)/log(p+1) \rfloor = 1$.
    Induction. Assume the statement is true for all arrays of size $(p+1)^j - 1$, where $1 \leq j < k$. Now consider an array of size $(p+1)^k - 1$. At the first

parallel comparison processor i, for $1 \leq i \leq p$, compares the search key with the array element indexed by $i \times (p+1)^{k-1}$.

After this step, either one of the array elements has matched the search key, or else the search key lies inside one of the unexamined subsections of the array. All these unexamined subsections have size $(p+1)^{k-1} - 1$. By the induction hypothesis, $k-1$ parallel comparison steps are sufficient to search any of these subarrays.

Next we show that the bound is tight. During the first parallel comparison step, only p elements of the array are compared with the search key. There must be one or more contiguous unexamined segments of the array with length at least

$$\lfloor (n-p)/(p+1) \rfloor \geq (n+1)/(p+1) - 1.$$

An inductive argument shows that after k parallel comparison steps there must be one or more contiguous unexamined seqments of the array with length at least

$$(n+1)/(p+1)^k - 1.$$

Thus the number of steps required by any parallel algorithm in the worst case is at least the minimum k satisfying

$$(n+1)/(p+1)^k - 1 \leq 0,$$

which gives $k \geq log(n+1)/log(p+1)$. $\square$

Two important rates of improvement in the run time of a parallel algorithm over that of a sequential algorithm that make the parallel algorithm worthwhile, are *polynomial* and *polylogarithmic* rates.

Definition 15.3.1 A parallel algorithm is said to be *polynomially fast* if its time, $T_{par}(n)$, is a polynomial function of the best sequential algorithm, $T_{seq}(n)$,

$$T_{par}(n) = (T_{seq}(n))^\epsilon, \epsilon < 1.$$

Definition 15.3.2 A parallel algorithm is said to be *polylogarithmically fast* if its time, $T_{par}(n)$, is a power $O(1)$ of the logarithmic function of the time of the best sequential algorithm, $T_{seq}(n)$,

$$T_{par}(n) = log^{O(1)} T_{seq}(n).$$

The fastest known algorithms determine the upper bounds on the time. The lower bound determines the minimum amount of time needed to solve the problem by an arbitrary parallel algorithm. This is the complexity of the problem studied; particularly important is the question of whether an algorithm with minimal execution time is known, that is, whether the lower bound of the execution time has been achieved.

Definition 15.3.3 The worst-case time, or the (time) complexity, of a parallel algorithm is a time function $T_{par}(n)$ that is maximum, over all inputs of size n, of the time elapsed from the moment when the first processor begins execution of the algorithm until the moment when the last processor terminates algorithm execution.

Definition 15.3.4 The computational complexity C_n of the parallel computation of a problem P_n of size n is the smallest number of parallel steps required to obtain a solution with arbitrary input data. Asymptotic complexity is the behaviour of C_n as n tends to ∞.

Example 15.3.2 *Broadcasting a Datum. (Akl 1989)*
In a typical parallel algorithm executed in the EREW SM PRAM model a situation often arises where all processors need to read a datum in a particular location of the common memory while the model of course does not allow access of more than one processor to the same memory location at any time. Thus a way needs to be found to simulate the operation, in the most efficient way. Here is one way of achieving this goal.

Let D be a location in memory holding a datum that all n processors, $P_i, 1 \leq i \leq n$, need at a given moment during the execution of an algorithm.

We shall assume that an array A of length n is present in memory. The array is initially empty and is used by the algorithm as a working space to distribute the contents of D to the processors. Its ith position is denoted by $A[i]$.

Algorithm 15.3.1 *Broadcast.*

Begin
P_1 reads the value in D and stores it in its own memory
P_1 writes the value in $A[1]$
For i **from** 0 **to** $log(n-1)$ **do**
    For j **from** $2^i + 1$ **to** 2^{i+1} **do in parallel**
        P_j reads the value in $A[j - 2^i]$ and stores it in its own memory
        P_j writes the value in $A[j]$
    EndDo
EndDo
End

When the algorithm terminates, all processors have stored the value of D in their local memories for later use. Since the number of processors having read D doubles in each iteration, the procedure terminates in $O(logn)$ time. The memory requirement of *Broadcast* is an array of length n. Strictly speaking, an array of half that length will do since in the last iteration of the algorithm all the processors have received the value in D and need not write it back in A. Hence use an array A of length $n/2$.

In the BSP model broadcasting the value can be done using a logical d-ary tree rather than the binary tree , so that sending one copy to each of the p processors (components) can be accomplished in $log(dp)$ supersteps. Appropriately chosen parameters for the computation model allow an optimal time, $O(n/p)$, for the implementation of the algorithm.

Example 15.3.3 *Computing All Sums* In this problem we assume the situation where each of n processors, P_i, holds in its local memory a number a_i, $1 \leq i \leq n$. It is often useful to compute, for each P_i, the sum $a_1 + a_2 + \ldots + a_i$; this problem is known as prefix sums.

The question is : Can the power of the shared-memory model be exploited to compute all sums of the form $a_1 + a_2 + \ldots + a_i$, $1 \leq i \leq n$, using n processors in $O(log n)$ time? The algorithm that follows shows that this is indeed possible.

Algorithm 15.3.2 *Allsums* $(a_1, a_2, \ldots, a_n)$

Begin
For j **from** 0 **to** $log n - 1$ **do**
    For i **from** $2^j + 1$ **to** n **do in parallel**
        P_i obtains the value $a[i - 2^j]$ from P_{i-2^j}
        through shared memory
        P_i overwrites $a[i]$ in its own memory with
        the value $a[i - 2^j] + a[i]$
    EndDo
EndDo
End

It is important to note that algorithm *Allsums* can be modified to solve any problem where the addition operation is replaced by any other (associative) binary operation. Examples of such operations on numbers are finding the larger or smaller of two numbers, multiplication and so on. Other operations that apply to a pair of logical quantities (or a pair of bits) are **and**, **or**, and **xor**.

15.4   Simulation of Different Parallel Computational Models

The theory of the PRAM model deals with how its various varieties can simulate each other.

Definition 15.4.1 A parallel model is self-simulating if a p-processor version can simulate one step of a q-processor version in time $O(q/p)$ for $p < q$ and in time $O(1)$ for $p \geq q$.

Suppose that a computation can be done in t parallel steps with dk primitive operations at step k. Its implementation direct on a PRAM to run in t steps will require d processors, where $d = max\{d_k\}$. With the number of processors $p < d$, the computation can still be simulated effectively by observing that the kth step can be simulated in time $|dk/p| \leq (dk/p) + 1$, and hence the total time to simulate the computation with p processors is no more than $|d/p| + t$ where $d = \Sigma k d k$. This observation is known as Brent's scheduling principle (Brent 1974). It is often used in the design of efficient parallel algorithms.

A serious flaw of this simulation model is that it does not take into account the processor allocation time. In other words the model assumes that it is possible for each of the p processors to determine directly, on line, the steps it needs to simulate.

Brent's scheduling principle implies that when the processor allocation is not a problem, a parallel algorithm requiring work $W(n)$ and time $T(n)$ can be simulated using p processors in time $W(n)/p + T(n)$. However if p is small compared to n, as it often is in practice, then $W(n)/p$ is much greater than $T(n)$. Therefore, as Kruskal *et al.* (1990) note, it may be too inflexible and unnecessary to focus on the design of parallel algorithms with the polylog time function. Instead one can usefully employ the concept of polynomial speedup.

15.5 Efficiency of Parallel Algorithms

One is interested in parallel algorithms as a means to improve the algorithm performance as compared to the performance of a sequential algorithm that solves the same problem. The running time of a sequential algorithm is used as a yardstick to evaluate a parallel algorithm. Ideally we want— when using p processors—to solve a problem p times faster.

Definition 15.5.1 The cost of a parallel algorithm is defined as its time, $T_{par}(n)$, times the number of processors, $p(n)$:

$$cost_{par}(n) = T_{par}(n) \times p(n).$$

Ultimately, we wish to have cost optimal parallel algorithms, that is the algorithms for which $cost_{par}(n)$ always matches a known lower bound on the number of sequential operations required in the worst case to solve the problem.

The two most important criteria for evaluation of parallel algorithms are speedup and efficiency of a parallel algorithm.

Definition 15.5.2 The speedup achieved by a parallel algorithm running on p processors is the ratio between the time $T_{seq}(n)$ taken by that parallel computer executing the fastest serial algorithm and the time $T_{par}(n)$ taken by the same parallel computer executing the parallel algorithm using p processors:

$$Speedup = T_{seq}(n)/T_{par}(n).$$

The speedup is said to be unbounded if $\lim_{n \to \infty}(T_{seq}(n)/T_{par}(n)) = \infty$

Definition 15.5.3 We say that a parallel algorithm with time $T_{par}(n)$ has a *polynomial speedup* over a sequential algorithm solving the same problem with time $T_{seq}(n)$ if $T_{par}(n) = O((T_{seq}(n))^\epsilon)$ for some $\epsilon < 1$.

The speedup measures an improvement in the running time of an algorithm due to parallelism.

Example 15.5.1   A parallel matrix multiplication.

Given matrices A of size $q \times r$ and B of size $r \times s$ one defines the product matrix C as the matrix of size $q \times s$ whose elements are given by:

$C_{ik} = \sum_{j=1}^{r} A_{ij} \times B_{jk}$

To appreciate the degree of parallelism inherent in this problem we note that the elements of the product matrix are completely independent of each other. Thus, given qr processors and an appropriate distribution of the input data, we could conceivably achieve an immediate speedup of qr. Furthermore if the inner products were performed in some sort of parallel divide-and-conquer manner we could achieve an additional speedup of $r/logr$.

Of course this is idealistic: in reality communications costs and other factors greatly affect the achievable speedup. These other factors account for the many variations of the parallel matrix multiplications algorithm. For example Aggarwal *et al.* (1990) show that on the LPRAM model two $n \times n$ matrices can be multiplied in $O(n^3/p)$ computation time and $O(n^2/p^{2/3})$ communication delay using p processors where $p \leq n^3/log^{3/2}n$.

Definition 15.5.4 The efficiency of a parallel algorithm running on p processors is the speedup divided by p, which is, of course, the ratio between the cost, $cost_{seq}(n)$, of a fastest serial algorithm executed on a parallel computer and the cost, $cost_{par}(n)$, of the parallel algorithm running on p processors executed on the same parallel computer:

$$Eff = Speedup/p = T_{seq}(n)/cost_{par}(n) = cost_{seq}(n)/cost_{par}(n).$$

Efficiency measures the work reduction achieved as a result of using several processors to run a parallel algorithm as opposed to a single processor to run a serial algorithm.

Kruskal defines another useful measure, inefficiency, as a ratio inverse to that of efficiency:

$$Ineff = cost_{par}(n)/cost_{seq}(n).$$

Three desirable bounds are then defined on evaluation of the performance of a parallel algorithm.

An algorithm is said to be of a *constant inefficiency*, of a *polylogarithmically bounded inefficiency* or of a *polynomially bounded inefficiency* if its cost

$$cost_{par}(n) = T_{par}(n) \times p(n) = \begin{cases} O(T_{seq}(n)), \\ T_{seq}(n) \times log^{O(1)}(T_{seq}(n)), \\ T_{seq}(n)^{O(1)}, \end{cases}$$

respectively. Similar polynomial bounds can be defined on the number of available processors.

Kruskal *et al.* suggest that algorithms with polynomial speedup and constant or polylog inefficiency may be extremely useful, even if they do not run in polylog time.

Definition 15.5.5 Let a parallel algorithm solve a problem of size n on $p(n)$ processors. If for a certain polynomial K and for all $n, p(n) \leq K(n)$, then the number of processors is considered polynomially bounded and in other cases polynomially unbounded.

The interest in algorithms for unbounded parallelism in terms of the number of processors is mainly of theoretical nature since these algorithms give the limits of parallel computations and allow a deeper understanding of the intrinsic structure of algorithms.

However of real importance are parallel algorithms of bounded parallelism, where the number of processors is finite and much smaller than the problem size. In practice even if one could afford to have as many processors as data for a particular problem instance size, it may not be desirable to design an algorithm based on that assumption: a larger problem instance would render the algorithm totally useless.

Generally, we would expect that an efficient parallel algorithm should exhibit the following properties:

1. The running time of the parallel algorithm is considerably smaller than that of the comparable sequential algorithm.

2. The running time is decreasing with the number of parallel processors increasing.

3. The number of parallel processors required by a parallel algorithm is considerably fewer than the input of the problem instance.

4. The number of parallel processors is decreasing with the size of the input data increasing.

15.6 Measuring the Performance of a Parallel Algorithm

A researcher involved in a development of a new application in his or her need for a desired level of performance turns to parallelism. He or she has a conceptual model of the problem's solution and a proposed algorithm, which includes a mental model of where parallelism can be helpfully used (Pancake 1991). An efficient implementation of a parallel algorithm on a given parallel architecture is inhibited by many factors.

A parallel algorithm running on a real-life parallel processor, in addition to its speedup and efficiency, is measured by other useful measurements such as its elapsed time and the price/performance ratio (Dongarra 1989 and Karp and Flatt 1990).

The elapsed time it takes to run a particular program on a given machine is the most convincing metric. It is simply the time between the start of the computation and the end of the computation. Karp and Flatt (1990) note that a CRAY Y-MP/1 solves the order 1000 linear system in 2.17 seconds compared to 445 seconds for a Sequent Balance 21000 with 30 processors. It is obvious thus that if one can afford a CRAY and if one's computing requirements are prevailing in terms of large matrices then one should buy a CRAY.

Price/performance of a parallel system is given by the ratio of the program's elapsed time and the cost of the machine that ran the job. The elapsed time and price/performance ratio are two measurements helpful to make a decision on which machine to buy. The next step is to apply the speedup and efficiency metrics to monitor how effectively the parallel machine is used.

The theoretical definition of the speedup is adapted to make use of real-life parameters available. The speedup estimate is obtained as the ratio of the elapsed time required to run a sequential algorithm on *one* processor and the time required to run the parallel algorithm on p processors,

$$Speedup = \frac{Elapsed\ time\ for\ 1\ processor,\ T(1)}{Elapsed\ time\ on\ p\ processors,\ T(p)}.$$

Strictly speaking, the correct time for the one-processor run would be the time for the best serial algorithm; in practice this can be achieved only rarely. When studying algorithms for parallel processors, the number of processors is of highest interest. An estimate of the speedup of the same program run on a varying number of processors, $p \geq 1$:

$$Speedup = \frac{T(1)}{T(2)},\ \frac{T(1)}{T(3)},\ \frac{T(1)}{T(4)},\ \cdots,$$

when used to determine the efficiency,

$$Eff = \frac{Speedup}{p} = \frac{T(1)}{pT(p)},$$

shows that the issue of efficiency is related to that of the price/performance ratio.

If efficiency for a given system is close to unity it means that the hardware is used effectively; low efficiency means that the resources are wasted.

It is obvious that adding processors should reduce the elapsed time, but by how much? We need to know how to tune the application to get the best speedup. If the speedup is close to linear, it is good, but how close to linear is good enough? If the efficiency is not particularly high, why?

These questions emphasise the complexity of both theoretical analysis and practical implementation issues relating to parallel algorithms, and hence the need for effective, properly defined, metrics to measure all aspects of parallel processing.

Karp and Flatt (1990) define a new metric:

$$f = \frac{1/Speedup - 1/p}{1 - 1/p},$$

which is experimentally determined for a given machine. They describe a case of a performance bug discovered in one of their applications thanks to their new metric.

Alan Karp noticed an irregularity in the behaviour of the metric f which led him to examine the way the IBM Parallel Fortran prototype compiler was splitting up the work in the loops. It transpired that the compiler erroneously truncated the result of dividing the number of loop iterations by the number of processors, p. As a result, one processor had to do one extra pass to finish the work in the loop. The remainder for some values of p was small, for others large. To resolve the inconsistency, the problem size was increased from the value that balances only for some ps to the value that gives perfect balance for all values of p.

For distributed programming much of the research on performance measurement of programs shows that the behaviour of a distributed program can be described at many levels of abstraction: system programming, operating system design, hardware architecture and hardware engineering.

The measurement results from most performance systems are presented in some form of performance metrics, such as the process execution time, message traffic statistics, and overall parallelism.

These metrics are good for the evaluation of the performance outcome of a program's execution. They tell much about how good or bad the program's execution is but little about why. The efforts are needed in the direction of the study to combine performance measurement tools with techniques for locating performance problems and improving the behaviour of parallel and distributed programs.

Another concept which tries to capture and measure the increase in performance of parallel processing in relation to communication patterns in a program and the communication infrastructure provided by the parallel machine is that of *scalability* of a parallel machine (Nussbaum and Agarwal 1991).

In many cases, researchers initially develop applications on essentially small machines. This means that the choice of an algorithm is influenced by the number of processors. Transferring this algorithm onto a super-machine with a huge number of processors is not at all an obviously efficiency gaining exercise. For example, in a linear algebra algorithm one may find that a small sequential portion that accounts for less than 1 % of execution when the algorithm is implemented on a four-processor machine, becomes a bottleneck when the algorithm is run on a machine with thousands of processors (Pancake 1991).

The scalability of a given architecture can be defined as the fraction of the parallelism inherent in a given algorithm which can be exploited by any machine of that architecture as a function of problem size. In turn, for a given algorithm and problem size, the inherent parallelism can be defined as the ratio of the serial execution time and the run time on an ideal realisation of a PRAM.

Furthermore, the asymptotic speedup is defined to be the ratio between the serial running time and the minimum parallel time. Then an architecture's scalability is the ratio of the algorithm's asymptotic speedup when run on the architecture in question to its corresponding asymptotic speedup when run on an EREW PRAM, as a function of problem size.

Scalability is a useful metric because once the architecture is shown to be scalable, then machines whose size varies over a wide range can use that same architecture.

On the other hand, it is believed that it is unlikely that practical machines differing in size by several orders of magnitude will have the same structure: there are size-related issues that are simply not relevant when building small machines.

Nevertheless, scalability can aid the study of large machines: if a system can be shown to scale well for a certain class of algorithms, then measurements from simulations of those algorithms on relatively few processors can be used to predict the behaviour of systems in which the number of processors is too large for simulation to be practical.

15.7   Optimisation of Parallel Processing

Optimisation of parallel processing is a fundamental consideration. It concerns the implementation of a parallel algorithm so as to achieve the shortest execution time. In this context, one or more concurrently executing program module came to be known as a grain. For example, in general, large-grain processing means that the amount of processing a lowest level program performs during an execution cycle is large compared to the overhead of scheduling its execution (Babb II 1984). There are also distinguished medium and fine grain sizes.

Theoretically for a given problem one strives to determine the best grain size which then will result in the shortest execution time. In practice parallelism can be carefully traded for communication time via scheduling which reduces delays caused by communication and execution.

A large grain parallel model based on the data flow approach has been discussed by Babb II (1984).

More recently Kruatrachue and Lewis (1988) proposed an optimisation model for parallel programs called grain packing. Grain packing starts from the smallest grain size, schedules those small grains and then defines larger grains. Thus there are four main steps:

- Constructing a small-grain task graph

- Scheduling small grains

- Grain packing

- Generating parallel modules.

The execution time is reduced by balancing execution time and communication time. The model is applicable to extended serial and concurrent programming languages. A grain begins executing as soon as all of its inputs are available, and terminates only after all of its outputs have been computed. The method is language-independent and is applicable to both extended serial and concurrent programming languages, including Occam, Fortran and Pascal.

Based on the DRAM model an interesting method for determining the grain size for parallel processes is proposed by McGreary and Gill (1989). The authors use an aggregation approach to determine the partitioning of an algorithm. Using the data flow graph of the algorithm, each node represents a sequential process, the node labels provide the execution time, the edges represent the data flow, and the edge labels represent the cost of communications.

The algorithm is partitioned using a graph grammar decomposition in such a way as to reduce total execution time. The method balances processing time with communication time to realise an effective grain size. It is general enough to accommodate very large or very small grain size.

Yet another important development in the area of optimisation of parallel processing is a realisation of the crucial significance of an efficient *cache memory management*.

A few definitions are in order:

Memory locality takes several forms:

Spatial locality arises because two contemporaneous memory references are likely to access nearby words.

Temporal locality occurs because a recently referenced memory word is likely to be accessed again.

A *cache* maintains a copy of data in a local memory that requires less time to access than the original data. Typically, a cache is a high-speed buffer adjacent to a processor that holds copies of memory locations likely to be used soon.

Parallel computing introduces a new type of locality, called *processor locality*, in which contemporaneous references to a memory word come from a single processor, rather than many different ones.

Typically, the cache-oriented features are considered to be too low-level details for the algorithm designer to be aware of. They are therefore hidden from the programmer. And yet at the end of the day efficiency of a parallel algorithm can be gained due to efficient use of the cache facilities as communications within a processor are much faster than between processors.

In the latest developments on parallel computing these features are emphasised to alert the programmers for more efficient ways of programming to fully benefit from the multiprocessor environment. In many applications scientific computations can be partitioned for parallel machines in such a way that almost all operations are done locally, and those that are not can access data from neighbouring processors. Accessing data from far processors is rarely necessary (Fox *et al.* 1988, Hoshino 1989, Hill and Larus 1990).

As an alternative to program structuring, *cache-coherence* techniques (that is techniques which allow shared data to be cached thereby significantly reducing the latency of memory accesses and yielding higher processor utilisation) are proving to be effective for providing a fast local memory to processors within a complex of closely coupled processors and for easing the burden of managing local memory. These techniques continue to evolve, with concentration on effective techniques for moderate and large-scale parallelism (Lenoski *et al.* 1992).

The following three rules-of-thumb are suggested by professionals to be useful when writing a parallel algorithm:

- Try to perform all operations on a datum in the same processor to avoid unnecessary communication.

- Align data to prevent locations used by different processors from occupying the same cache block.

- Cluster work and re-use parts of the data quickly, instead of making long passes over all the data.

15.8   Complexity Classification of Problems

In analogy with the theory of sequential algorithms, classification of problems into the parallel complexity classes is an important direction of study.

The theory of computational complexity of sequential algorithms classifies the problems as *tractable*, if solvable within reasonable time, or *intractable*, where the tractability is defined in terms of the class P of problems solvable in polynomial time. In analogy to the class P for sequential algorithms, one seeks a complexity class that would encapsulate the problems which can be solved *efficiently* by parallel algorithms.

One notion of a class of such problems is associated with the so called Nick's class (NC). These are the problems which can be solved on PRAM in *polylogarithmic time* using *polynomially many* processors. This means that for the *NC class problems* the parallel computation time is bounded by a fixed power of the logarithm of the size of the input, and the processor-time product exceeds the number of steps in an optimal sequential algorithm by at most a polylog factor.

P-completeness and Nick's class relationship is as follows: *a problem has an efficient parallel solution only if it is tractable:* $NC \subseteq P$.

A number of important problems lie in the class NC, such as the basic arithmetic operations, transitive closure and Boolean matrix multiplication, the computation of the determinant, the rank or the inverse of a matrix, the elevation of certain classes of straight-line programs, and the construction of a maximal independent set of vertices in a graph.

The criticism that this classification attracts is that an NC algorithm can be polynomially wasteful; it can perform polynomially more operations than a sequential algorithm that solves the same problem. Using a theory of reducibility analogous to the theory of NP-completeness, it has been possible to identify certain problems as P-complete; such problems are solvable sequentially in polynomial time, but do not lie in the class NC unless every problem solvable in sequential polynomial time lies in NC.

It has been shown that P-complete problems may be solved by efficient parallel algorithms, while on the other hand a problem like search on an ordered list, which runs in logarithmic sequential time, is in NC, irrespective of the existence of efficient parallel algorithms. In fact, searching does not admit efficient parallel algorithms (Kruskal 1983).

For tractable parallel algorithms a computational taxonomy needs to be developed, so that computer architects can then match classes of algorithms to various architectures for the most effective or optimal use of resources. One will aim at a good match between algorithms and architectures.

15.9 The Parallel Computation Thesis

An important breakthrough on the theoretical front is formulation of the Parallel Computation Thesis. The Parallel Computation Thesis highlights a connection between sequential and parallel computation.

Two functions are said to be polynomially equivalent if each is bounded above by a polynomial function of the other.

The size of circuits is a major resource for parallel computations, as is time for sequential computations. It has been shown that these two types of resources are polynomially related (Gurari 1989).

Sequential computations are considered feasible only if they are polynomially time-bounded. Similarly, families of circuits are considered feasible only if they are polynomially size-bounded. As a result, parallelism does not seem to have a major influence on problems that are not solvable in polynomial time. Yet for those problems that are solvable in polynomial time, parallelism is of central importance when it can significantly increase computing speed. One such class of problems is that which can be solved by uniform families of circuits, simultaneously having polynomial size complexity and polylog time complexity (Gurari 1989).

The Parallel Computation Thesis states that parallel time is polynomially equivalent to sequential space.

Computation on a PRAM can be simulated by a Turing machine with space polynomially bounded in the parallel time, and conversely, provided the number of processors is no more than an exponential in the parallel time (Fortune and Wyllie 1978, Goldschlager 1982).

15.10   Nondeterminism

A *deterministic algorithm* is an algorithm where for any given step in the sequence of the algorithmic steps the next step is uniquely determined. The *nondeterministic algorithms* are *more abstract* than corresponding deterministic versions. They do not include a provision (a constraint) for a uniquely determined next step. They simply assume that a *valid* next step from a range of possible steps will be carried out.

A nondeterministic algorithm can be seen as a more abstract tool to describe problem solution. Whenever it may be useful, the designer may forget about specifying a unique sequence of moves for the processor that lead to a problem solution. One just defines a general, nondeterministic strategy to look for the solution, leaving to a later stage its translation into a deterministic algorithm that can be run effectively on an actual computer.

Parallel processing research advances in several directions. With a given number of processors, how fast can you solve a problem? A distinct development pursues the issue: what happens if a program is running asynchronously and there is no way of telling whether A happens before B and B happens before A? Can one still have a determinate result for the whole computation even though one cannot control the order in which these various events are happening in parallel?

Thus there is an interesting dichotomy emerging in the search for a unified parallel formalism. On the one hand, it is agreed that the development of such formalism would prove invaluable. On the other hand it is felt that with the intricacies and inter-dependencies to be found in many parallel applications it is impossible to predict the full behaviour of an algorithm running on a parallel architecture. Hence a fully unified formalism seems impossible.

Nondeterminism can be an advantage since it gives an increase of expressive power, or even simplicity of analysis. It may however be of little help or worse when one looks at the problem from the point of view of computational complexity, say time complexity. It is not yet known whether a nondeterministic algorithm of time complexity $\Theta(T)$ can always be translated into a deterministic version of complexity $\Theta(g(T))$, $g(T)$ being a polynomial function.

If this were a case, we would have $P = NP$ and a lot of problems would become tractable, thanks to the existence of efficient nondeterministic algorithms for their solution. Unfortunately, the conjecture is $P \subseteq NP$. And so results on time complexity of nondeterministic solution do not give any insight into the complexity that can be achieved with its deterministic counterparts.

15.11   Conclusions

Computational complexity addresses itself to the quantitative aspects of the solution of computational problems.

In this chapter basic notions and concepts of computational complexity for solution of problems by parallel algorithms have been explored. The theory of parallel computational complexity is only beginning to emerge while at the same time its necessity cannot be overestimated for the development of parallel algorithms of truly superior performance compared with the sequential algorithms solving the same problems.

Theory-based metrics for measuring performance of parallel algorithms are continuously tested experimentally on real-life parallel architectures so that all aspects of efficient parallel processing—performance of the computers themselves, communications between parallel processors and compact storage for huge masses of data—can be fully mastered.

References

Aggarwal A, Chandra A, and Snir M. (1990) Communication complexity of PRAMs. *Theory Comp. Sci.*, 71, pp. 3–28.

Akl S.G. (1989) *The Design and Analysis of Parallel Algorithms.* Prentice-Hall.

Babb II R.G. (1984) Parallel Processing with Large-Grain Data Flow Techniques. *Computer,* 17, 7, pp. 55–61.

Brent R.P. (1974) The Parallel Evaluation of General Arithmetic Expressions. *J. ACM*, 21, pp. 201–206.

Carter J.L. and Wegman M.N. (1979) Universal Classes of Hash Functions. *J. Comput. Syst. Sci.* 18, pp. 143–154.

Dongarra J.J. (1989) Performance of various computers using standard linear equation software. *Report CS-89–85, Computer Science Department,* Univ. Tennessee, Knoxville, October, 12.

Fortune S. and Wyllie J. (1978) Parallelism in Random Access Machines. *Proc. 10th Ann. ACM Symp. on Theory of Computing,* pp. 114–118.

Fox G. C., Johnson M., Lyzenga G., Otto S., Salmon J. and Walker D. (1988) *Solving Problems on Concurrent Processors,* Vol. 1. Prentice-Hall, New Jersey.

Gibbons PB (1989) A more practical PRAM model. *Proc. of the 1989 ACM Symposium on Parallel Algorithms and Architectures.* pp. 158–168.

Goldschlager L.M. (1982) A Unified Approach to Models of Synchronous Parallel Machines. *J. ACM,* 29, pp. 1073–1086.

Gurari E.M. (1989) *An Introduction to Theory of Computation.* Comp Sci. Press.

Hill M.D. and Larus J. R. (1990) Cache Considerations for Multiprocessor Programmers. *CACM,* 33, 8, pp. 97–102.

Hoshino T. (1989) *PAX Computer*. Addison-Wesley, Reading, Mass.

Karlin A. and Upfal E. (1988) Parallel Hashing—An efficient implementation of shared memory. *J. ACM*, 35, 4, pp. 876–892.

Karp A. and Flatt H. P. (1990) Measuring Parallel Processor Performance. *CACM*, 33, 5, pp. 539–543.

Kruatrachue B. and Lewis T. (1988) Grain Size Determination for Parallel Processing. *IEEE Software*, 5, 1, pp. 23–32.

Kruskal C.P. (1983) Searching, Merging and Sorting in Parallel Computation. *IEEE Trans. Comput.*, 32, pp. 942–946.

Kruskal C. P., Rudolph L. and Snir M. (1990) A Complexity Theory of Efficient Parallel Algorithms. *Theor. Comp. Sci.*, 71, pp. 95–132.

Leiserson C. and Maggs B. (1986) Communication-efficient Parallel Graph Algorithms. *International Parallel Processing Conf., IEEE*

Lenoski D., Laudon J., Gharacorloo K., Weber W.-D., Gupta A., Henessy J., Horowitz M. and Lam M. S. (1992) The Stanford Dash Multiprocessor. *Computer* **25**(3) 63-79.

McGreary C. and Gill H. (1989) Automatic Determination of Grain Size for Efficient Parallel Processing. *CACM*, 32, 9, 1073-1078.

Mehlhorn K. and Vishkin U. (1984) Randomized and Deterministic Simulations of PRAMs by Parallel Machines with Restricted Granularity of Parallel Memories. *Acta Inf.*, 21, pp. 339–374.

Nussbaum D. and Agarwal A.(1991) Scalability of Parallel Machines. *CACM*, 34, 3, pp. 56–61.

Pancake C.M.(1991) Where Are We Headed? *CACM*, 34, 11, pp. 53–64.

Papadimitriou C. H. and Yannakakis M.(1988) Towards an Architecture-independent Analysis of Parallel Algorithms. *Proc. of the 20th ACM Symposium on Theory of Computing*, pp. 510–513.

Quinn M.J. (1987) Designing Efficient Algorithms for Parallel Computers. McGraw-Hill, New York.

Savitch W. J. and Stimson M. J. (1979) Time bounded random access machines with parallel processing. *J. ACM*, 26, pp. 103–118.

Valiant L. G. (1990a) General purpose parallel architectures. In *Handbook of Theoretical Computer Science* (Ed. by van Leeuwen, J.), North Holland, Amsterdam, 945–969.

Valiant L. G. (1990b) A Bridging Model for Parallel Computation. *CACM*, 33, 8, pp. 103–111.

Index